Lectures on the electrical properties of materials

L. SOLYMAR *and* D. WALSH

Lectures on the electrical properties of materials

SECOND EDITION

OXFORD UNIVERSITY PRESS

Oxford University Press, Walton Street, Oxford OX2 6DP

LONDON GLASGOW NEW YORK TORONTO
DELHI BOMBAY CALCUTTA MADRAS KARACHI
KUALA LUMPUR SINGAPORE HONG KONG TOKYO
NAIROBI DAR ES SALAAM CAPE TOWN SALISBURY
MELBOURNE AUCKLAND

and associates in
BEIRUT BERLIN IBADAN MEXICO CITY NICOSIA

First edition 1970
Reprinted (with revisions) 1975
Second edition 1979
Reprinted 1981, 1982

British Library Cataloguing in Publication Data

Solymar, Laszlo
Lectures on the electrical properties of materials. — 2nd ed.
1. Solids — Electric properties
I. Title II. Walsh, Donald
620.1'1297 QC176.8.E35 79—40479
ISBN 0—19—851144—2
ISBN 0—19—851145—0 Pbk

Printed in Hong Kong

Preface to the second edition

WE have tried to incorporate into this new edition many of the recent advances in this field. We have, therefore, expanded considerably the chapters on semiconductor devices, magnetism, and lasers, and added a few sections to other chapters.

We would like to thank Dr. J. P. Jakubovics of the Metallurgy Department and members of the Clarendon Laboratory for helpful comments on Chapters 11 and 12 respectively.

Contents

INTRODUCTION xiii

1. THE ELECTRON AS A PARTICLE 1

1.1. Introduction 1

1.2. The effect of an electric field—conductivity and Ohm's law 3

1.3. The hydrodynamic model of electron flow 6

1.4. The Hall effect 6

1.5. Electromagnetic waves in solids 8

1.6. Waves in the presence of an applied magnetic field 17

1.7. Cyclotron resonance 19

1.8. Plasma waves 22

1.9. Specific heat of metals 25

EXAMPLES 25

2. THE ELECTRON AS A WAVE 27

2.1. Introduction 27

2.2. The electron microscope 31

2.3. Some properties of waves 32

2.4. Applications to electrons 35

2.5. Two analogies 38

EXAMPLES 40

3. THE ELECTRON 41

3.1. Introduction 41

3.2. Schrödinger's equation 44

3.3. Solutions of Schrödinger's equation 45

3.4. The electron as a wave 47

3.5. The electron as a particle 48

3.6. The electron meeting a potential barrier 48

3.7. Two analogies 51

3.8. The electron in a potential well 53

3.9. The potential well with a rigid wall 55
3.10. The uncertainty relationship 56
3.11. Philosophical implications 57
EXAMPLES 59

4. THE HYDROGEN ATOM AND THE PERIODIC TABLE 62
4.1. The hydrogen atom 62
4.2. Quantum numbers 68
4.3. Electron spin and Pauli's exclusion principle 70
4.4. The periodic table 71
EXAMPLES 77

5. BONDS 78
5.1. Introduction 78
5.2. General mechanical properties of bonds 79
5.3. Bond types 82
5.4. Ionic bonds 82
5.5. Metallic bonds 84
5.6. The covalent bond 84
5.7. The Van der Waals bond 86
5.8. Feynman's coupled mode approach 87
5.9. Nuclear forces 94
5.10. The hydrogen molecule 94
5.11. An analogy 96
EXAMPLES 97

6. THE FREE ELECTRON THEORY OF METALS 98
6.1. Free electrons 98
6.2. The density of states and the Fermi–Dirac distribution 99
6.3. The specific heat of electrons 103
6.4. The work function 104
6.5. Thermionic emission 105
6.6. The Schottky effect 109
6.7. Field emission 112
6.8. The field-emission microscope 113
6.9. The photoelectric effect 114

6.10. The junction between two metals 116

EXAMPLES 117

7. THE BAND THEORY OF SOLIDS 118

7.1. Introduction 118

7.2. The Kronig–Penney model 119

7.3. The Ziman model 124

7.4. The Feynman model 128

7.5. The effective mass 131

7.6. The effective number of free electrons 135

7.7. The number of possible states per band 137

7.8. Metals and insulators 138

7.9. Holes 139

7.10. Divalent metals 140

7.11. Finite temperatures 142

7.12. Concluding remarks 144

EXAMPLES 145

8. SEMICONDUCTORS 146

8.1. Introduction 146

8.2. Intrinsic semiconductors 147

8.3. Extrinsic semiconductors 152

8.4. Scattering 159

8.5. A relationship between electron and hole densities 161

8.6. Non-equilibrium processes 163

8.7. Real semiconductors 164

8.8. Measurement of semiconductor properties 166

8.9. Preparation of pure and controlled-impurity single-crystal semiconductors 175

EXAMPLES 181

9. PRINCIPLES OF SEMICONDUCTOR DEVICES 183

9.1. Introduction 183

9.2. The p–n junction in equilibrium 183

9.3. Rectification 188

9.4. Injection 191

9.5. Junction capacity 193

9.6. The transistor 196
9.7. Metal–semiconductor junctions 203
9.8. The role of surface states; real metal–semiconductor junctions 207
9.9. Metal–insulator–semiconductor junctions 209
9.10. The tunnel diode 213
9.11. The backward diode 217
9.12. The Zener diode and the avalanche diode 218
9.13. Varactor diodes 220
9.14. Photo-diodes and lamps 223
9.15. Infrared detectors 225
9.16. Field-effect transistors 226
9.17. Charge-coupled devices 230
9.18. Silicon controlled rectifier 233
9.19. The Gunn effect 235
9.20. Strain gauges 240
9.21. Measurement of magnetic field by the Hall effect 241
9.22. Microelectronic circuits 242
EXAMPLES 246

10. DIELECTRIC MATERIALS 250
10.1. Introduction 250
10.2. Macroscopic approach 250
10.3. Microscopic approach 251
10.4. Types of polarization 253
10.5. The complex dielectric constant and the refractive index 254
10.6. Frequency response 256
10.7. Polar and non-polar materials 258
10.8. The Debye equation 259
10.9. The effective field 260
10.10. Dielectric breakdown 262
10.11. Piezoelectricity 264
10.12. Ferroelectrics 270
10.13. Optical fibres 272
10.14. The Xerox process 274
EXAMPLES 275

11. MAGNETIC MATERIALS 278
11.1. Introduction 278
11.2. Macroscopic approach 279
11.3. Microscopic theory (phenomenological) 279
11.4. Domains and the hysteresis curve 285
11.5. Microscopic theory (quantum-mechanical) 296
11.6. The Stern–Gerlach experiment 300
11.7. Magnetic resonance 307
11.8 Some applications 310
EXAMPLES 316

12. INTRODUCTION TO MASERS AND LASERS 318
12.1. Equilibrium 318
12.2. Two-state systems 319
12.3. Resonators 324
12.4. Some practical laser systems 325
12.5. Laser modes and mode control techniques 334
12.6. Masers 338
12.7. Noise 341
12.8. Applications 343
EXAMPLES 349

13. SUPERCONDUCTIVITY 351
13.1. Introduction 351
13.2. The effect of a magnetic field 355
13.3. Microscopic theory 357
13.4. Thermodynamic treatment 358
13.5. Surface energy 364
13.6. The Landau–Ginzburg theory 366
13.7. The energy gap 373
13.8. Some applications. 379
EXAMPLES 385

EPILOGUE 387
APPENDIX I: SYMBOLS, UNITS, AND CONSTANTS 388
APPENDIX II: VARIATIONAL CALCULUS 389

APPENDIX III: SUGGESTIONS FOR FURTHER READING 391

ANSWERS TO EXAMPLES 392

INDEX 395

Introduction

ENGINEERING used to be a down-to-earth profession. The Roman engineers who provided civilized Europe with bridges and roads did a job comprehensible to all. And this is still true in most branches of engineering today. Bridge-building has become a sophisticated science, the mathematics of optimum structures is formidable; nevertheless the basic relationships are not far removed from common sense. A heavier load is more likely to cause a bridge to collapse, and the use of steel instead of wood will improve the load-carrying capacity.

Solid-state electronic devices are in a different category. In order to understand their behaviour you need to delve into quantum mechanics. Is quantum mechanics far removed from common sense? Yes, for the time being, it is. We live in a classical world. The phenomena we meet every day are classical phenomena. The fine details represented by quantum mechanics are averaged out, we have no first-hand experience of the laws of quantum mechanics; we can only infer the existence of certain relationships from the final outcome. Will it be always this way? Not necessarily. There are quantum phenomena known to exist on a macroscopic scale as, for example, superconductivity, and it is quite likely that certain biological processes will be found to represent macroscopic quantum phenomena. So a ten-year-old might be able to give a summary of the laws of quantum mechanics—half a century hence. For the time being there is no easy way to quantum mechanics; no short cuts and no broad highways. We just have to struggle through. I believe it will be worth the effort. It will be your first opportunity to glance behind the scenes, to pierce the surface and find the grandiose logic of a hidden world.

Should engineers be interested at all in hidden mysteries? Isn't that the duty and privilege of the physicists? I don't think so. If you want to invent new electronic devices you must be able to understand the operation of the existing ones. And perhaps you need to more than merely understand the physical mechanism. You need to grow familiar with the world of atoms and electrons, to feel at home among them, to appreciate their habits and characters.

We shall not be able to go very deeply into the subject. Time is short and few of you will have the mathematical apparatus for the frontal assault. So we shall approach the subject in carefully planned steps. First we shall try to deduce as much information as possible on the basis of the classical picture. Then we shall talk about a number of phenomena that are clearly in contrast with classical ideas and introduce quantum mechanics, starting with Schrödinger's equation. You will become acquainted with the properties of individual atoms and what happens when they conglomerate and take the form of a solid. You will hear about conductors, insulators, semiconductors, p–n junctions, transistors, masers, lasers, and a number of related solid-state devices. Sometimes the statement will be purely qualitative but in most cases we shall try to give the essential quantitative relationships.

These lectures will not make you an expert in quantum mechanics nor will they enable you to design a computer the size of a matchbox. They will give you no more than a general idea.

If you elect to specialize in solid-state devices you will, no doubt, delve more deeply into the intricacies of the theory and into the details of the technology. If you should work in a related subject then, presumably, you will keep alive your interest, and you may occasionally find it useful to be able to think in quantum-mechanical terms. If your branch of engineering has nothing to do with quantum mechanics, would you be able to claim in ten years' time that you profited from this course? I hope the answer to this question is *yes*. I believe that once you have been exposed (however superficially) to quantum-mechanical reasoning, it will leave permanent marks on you. It will influence your ideas on the nature of physical laws, on the ultimate accuracy of measurements, and, in general, will sharpen your critical faculties. If you intend to move into the higher echelons of industrial management then the answer is a little ambiguous. Captains of British industry are usually bred on Plato and Aristotle; so if you entertain such high ambitions, you might be well advised not to display any knowledge of engineering. Familiarity with certain tenets of quantum mechanics might turn out to be an advantage even under those circumstances. If you can comment on the uncertainty principle as a modern version of one of Zeno's paradoxes, you might rise in your boss's esteem.

1. The electron as a particle

And I laugh to see them whirl and flee,
Like a swarm of golden bees.

SHELLEY *The Cloud*

1.1. Introduction

IN the popular mind the electron lives as something very small that has something to do with electricity. Studying electromagnetism doesn't change the picture appreciably. You learn that the electron can be regarded as a negative point charge and it duly obeys the laws of mechanics and electromagnetism. It is a particle that can be accelerated or decelerated but can't be taken to bits.

Is this picture likely to benefit an engineer? Yes, if it helps him to produce a device. Is it a *correct* picture? Well, an engineer is not concerned with the truth; that is left to philosophers and theologians: the prime concern of an engineer is the utility of the final product. If this physical picture makes possible the birth of the vacuum tube, we must deem it useful; but if it fails to account for the properties of the transistor then we must regard its appeal as less alluring. There is no doubt, however, that we can go quite far by regarding the electron as a particle even in a solid—the subject of our study.

What does a solid look like? It consists of atoms. This idea originated a few thousand years ago in Greece, and has had some ups and downs in history, but today its truth is universally accepted. Now if matter consists of atoms, they must be somehow piled upon each other. The science that is concerned with the spatial arrangement of atoms is called crystallography. It is a science greatly revered by crystallographers; engineers tend to look at it with awe and respect but with no more than a moderate amount of enthusiasm. This is because the existence of various crystal structures tremendously complicates the engineer's job. From the point of view of the electronic engineer the situation is not too bad. If all materials would suddenly decide to crystallize in the simple cubical structure shown in Fig. 1.1, the electronics industry would not suffer too great a loss. Some parameters would change but most of the devices would be able to continue their useful existence. So when we are trying to explain the fundamental

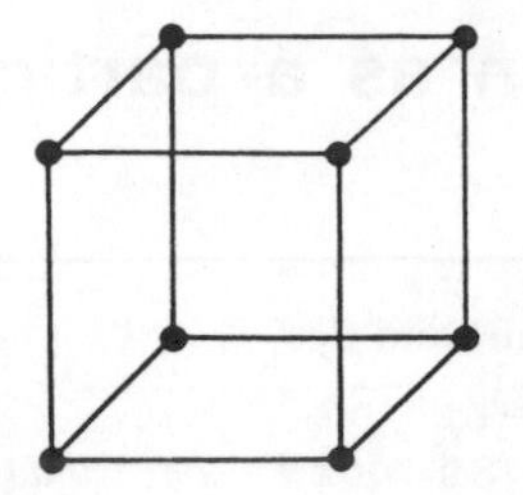

FIG. 1.1. Atoms crystallizing in a cubical lattice.

properties of materials and the working of the more illustrious solid-state electronic devices, we don't really need to go into the details of the positions of the atoms. A solid from now on, unless otherwise stated, will crystallize in a cubic crystal structure.

Let's specify our model a little more closely. If we postulate the existence of a certain number of electrons capable of conducting electricity, we must also say that a corresponding amount of positive charge exists in the solid. It must look electrically neutral to the outside world. Secondly, in analogy with our picture of gases, we may assume that the electrons bounce around in the interatomic spaces, colliding occasionally with lattice atoms. We may even go further with this analogy and claim that in equilibrium the electrons follow the same statistical distribution as gas molecules (that is the Maxwell–Boltzmann distribution), which depends strongly on the temperature of the system. The average kinetic energy of each degree of freedom is then $\frac{1}{2}kT$ where T is absolute temperature and k is Boltzmann's constant. So we may say that the mean thermal velocity of electrons is given by the formula†

$$\tfrac{1}{2}mv_{\mathrm{th}}^2 = \tfrac{3}{2}kT \qquad (1.1)$$

because particles moving in three dimensions have three degrees of freedom.

We shall now calculate some observable quantities on the basis of this simplest model and see how the results compare with experiment. The success of this simple model is somewhat surprising, but we shall see as we proceed that viewing a solid (or at least a metal) as a fixed lattice of positive ions held together by a jelly-like mass of electrons approximates well to the modern view of the electronic structure of solids. Some books discuss mechanical properties in terms of dislocations that can move and spread; the solid is then pictured as a fixed distribution

† We shall see later that this is not so for metals but it is nearly true for conduction electrons in semiconductors.

of negative charge in which the lattice ions can move. These views are almost identical; only the external stimuli are different.

1.2. The effect of an electric field—conductivity and Ohm's law

Suppose a potential difference U is applied between the two ends of a solid of length L. Then an electric field

$$\mathscr{E} = \frac{U}{L} \tag{1.2}$$

is present at every point in the solid, causing an acceleration

$$a = \frac{e}{m}\,\mathscr{E}. \tag{1.3}$$

Thus the electrons, in addition to their random velocities, will acquire a velocity in the direction of the electric field. We may assume that this directed velocity is completely lost after each collision, because an electron is much lighter than a lattice atom. Thus only the part of this velocity that is picked up in between collisions counts. If we write τ for the average time between two collisions, the final velocity of the electron will be $a\tau$ and the average velocity

$$v_{\text{average}} = \tfrac{1}{2}a\tau. \tag{1.4}$$

This is simple enough but not quite correct. We shouldn't use the *average* time between collisions to calculate the average velocity but the actual times, and then average. The correct derivation is fairly lengthy but all it gives is a factor of 2.† Numerical factors like 2 or 3 or π are generally not worth worrying about in simple models but just to agree with the formulae generally quoted in the literature we shall incorporate that factor 2, and use

$$v_{\text{average}} = a\tau. \tag{1.5}$$

The average time between collisions, τ, has many other names; for example, mean free time, relaxation time, and collision time. Similarly, the average velocity is often referred to as the mean velocity or drift velocity. We shall call them 'collision time' and 'drift velocity' (denoting the latter by v_{D}).

The relationship between drift velocity and electric field may be obtained from eqns. (1.3) and (1.5), yielding

$$v_{\mathrm{D}} = \left(\frac{e}{m}\,\tau\right)\mathscr{E}, \tag{1.6}$$

† See, for example, W. Shockley, *Electrons and holes in semiconductors*, D. van Nostrand, New York, 1950, pp. 191–5.

where the proportionality constant in parentheses is called the 'mobility'. This is the only name it has, and it's quite a logical one; the higher the mobility, the more mobile the electrons.

Assuming now that all electrons drift with their drift velocity, the total number of electrons crossing a plane of unit area per second may be obtained by multiplying the drift velocity by the density of electrons, N_e. Multiplying further by the charge on the electron we obtain the electric current density

$$J = N_e e v_D. \tag{1.7}$$

Notice that it is only the drift velocity, created by the electric field, that comes into the expression. The random velocities do not contribute to the electric current because they average out to zero.†

We can derive similarly the relationship between current density and electric field with the aid of eqns. (1.6) and (1.7) in the form

$$J = \frac{N_e e^2 \tau}{m} \mathscr{E}. \tag{1.8}$$

This is a linear relationship which you may recognize as Ohm's law

$$J = \sigma \mathscr{E}, \tag{1.9}$$

where σ is the electrical conductivity. When first learning about electricity you looked upon σ as a bulk constant; now you can see what it comprises. We can write it in the form

$$\begin{aligned} \sigma &= \left(\frac{e}{m}\tau\right)(N_e e) \\ &= \mu_e(N_e e). \end{aligned} \tag{1.10}$$

That is, we may regard conductivity as the product of two factors, charge density ($N_e e$) and mobility (μ_e). Thus we may have high conductivities because there are lots of electrons around or because they can acquire high drift velocities (by having high mobilities). In metals, incidentally, the mobilities are quite low (about two orders of magnitude below those of semiconductors); so their high conductivity is due to the high density of electrons.

Ohm's law further implies that σ is a constant, which means that τ must be independent of electric field.‡ From our model so far it is more

† They give rise however to *electrical noise* in a conductor. Its value is usually much smaller than the signals we are concerned with.

‡ It seems reasonable at this stage to assume that the charge and mass of the electron and the number of electrons present will be independent of the electric field.

reasonable to assume that l, the distance between collisions (usually called the mean free path) in the regularly spaced lattice, rather than τ, is independent of electric field. But l must be related to τ by the relationship

$$l = \tau(v_{\rm th}+v_{\rm D}). \tag{1.11}$$

Since $v_{\rm D}$ varies with electric field, τ must also vary with the field unless

$$v_{\rm th} \gg v_{\rm D}. \tag{1.12}$$

As Ohm's law is accurately true for most metals this inequality should hold. In a typical metal $\mu_{\rm e} = 5\times10^{-3}\ {\rm m^2\ V^{-1}\ s^{-1}}$, which gives a drift velocity $v_{\rm D}$ of 5×10^{-3} m/s for an electric field of 1 V/m. The thermal velocity at room temperature according to eqn. (1.1) (which actually gives too low a value for metals) is

$$v_{\rm th} = \left(\frac{3kT}{m}\right)^{\frac{1}{2}} \simeq 10^5\ {\rm m/s}. \tag{1.13}$$

Thus there will be a constant relationship between current and electric field accurate to about 1 part in 10^8.†

This important consideration can be emphasized in another way. Let us draw the graph (Fig. 1.2) of the distribution of particles in velocity space, i.e. with rectilinear axes representing velocities in three

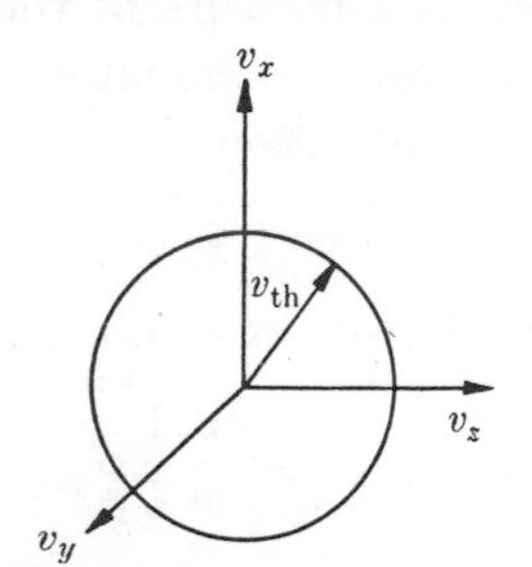

FIG. 1.2. Distribution of electrons in velocity space.

dimensions, v_x, v_y, v_z. With no electric field present the distribution is spherically symmetric about the origin. The surface of a sphere of radius $v_{\rm th}$ represents all electrons moving in all possible directions with that r.m.s. speed. When a field is applied along the x-axis (say), the distribution is minutely perturbed (the electrons acquire some additional velocity in the direction of the x-axis) so that its centre shifts from (0, 0, 0) to about ($v_{\rm th}/10^8$, 0, 0).

† This is less true for semiconductors as they violate Ohm's law at high electric fields.

Taking copper, a field of 1 V/m causes a current density of 10^8 A/m^2. It is quite remarkable that a current density of this magnitude can be achieved with an almost negligible perturbation of the electron velocity distribution.

1.3. The hydrodynamic model of electron flow

By considering the flow of a charged fluid a highly sophisticated model may be developed. We shall use it only in its crudest form, which does not give much of a physical picture but leads quickly to the desired result.

The equation of motion for an electron is

$$m\frac{\mathrm{d}v}{\mathrm{d}t} = e\mathscr{E}. \tag{1.14}$$

If we now assume that the electron moves in a viscous medium, then the forces trying to change the momentum will be resisted. We may account for this by adding a 'momentum-destroying' term, proportional to v. Taking the proportionality constant as ζ, eqn. (1.14) modifies to

$$m\left(\frac{\mathrm{d}v}{\mathrm{d}t}+\zeta v\right) = e\mathscr{E}. \tag{1.15}$$

ζ may be regarded here as a measure of the viscosity of the medium. In the limit, when viscosity dominates, the term $\mathrm{d}v/\mathrm{d}t$ becomes negligible, resulting in the equation

$$mv\zeta = e\mathscr{E}, \tag{1.16}$$

which gives for the velocity of the electron

$$v = \frac{e}{m}\frac{1}{\zeta}\mathscr{E}. \tag{1.17}$$

It may be clearly seen that by taking $\zeta = 1/\tau$ eqn. (1.17) agrees with eqn. (1.6); hence we may regard the two models as equivalent and in any given case use whichever is more convenient.

1.4. The Hall effect

Let us now investigate the current flow in a rectangular piece of material, as shown in Fig. 1.3. We apply a voltage so that the right-hand side is positive. Current, by convention, flows from the positive side to the negative side, that is in the direction of the negative z-axis. But electrons, remember, flow in a direction opposite to conventional

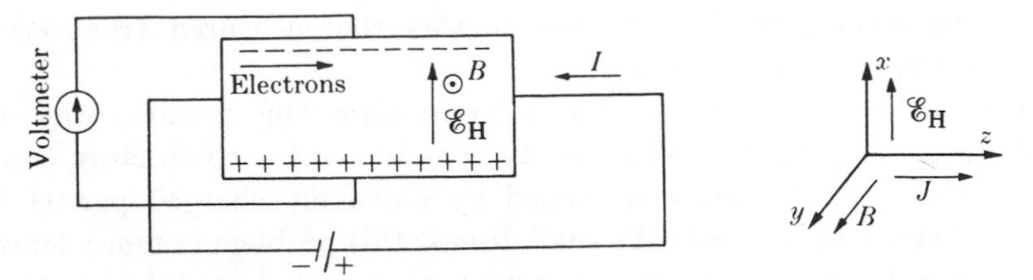

FIG. 1.3. Schematic representation of the measurement of the Hall effect.

current, that is from left to right. Having sorted this out let us now apply a magnetic field in the positive y-direction. The force on an electron due to this magnetic field is

$$e(\mathbf{v}\times\mathbf{B}). \tag{1.18}$$

To get the resultant vector we rotate vector $\mathbf{v}$ into vector $\mathbf{B}$. This is a clockwise rotation giving a vector in the negative x-direction. But the charge of the electron, e, is negative; so the force will point in the positive x-direction; the electrons are deflected upwards. They cannot move farther than the top end of the slab, and they will accumulate there. But if the material was electrically neutral before, and some electrons have moved upwards, then some positive ions at the bottom will be deprived of their compensating negative charge. Hence an electric field will develop between the positive bottom layer and the negative top layer. Thus, after a while, the upward motion of the electrons will be prevented by this internal electric field. In other words, equilibrium is established when the force due to the transverse electric field just cancels the force due to the magnetic field. This happens when

$$\mathscr{E}_{\mathrm{H}} = vB. \tag{1.19}$$

Expressed in terms of current density,

$$\mathscr{E}_{\mathrm{H}} = R_{\mathrm{H}}JB, \qquad R_{\mathrm{H}} = \frac{1}{N_{\mathrm{e}}e}, \tag{1.20}$$

where R_{H} is called the *Hall coefficient*. In this experiment $\mathscr{E}_{\mathrm{H}}$, J, and B are measurable; thus R_{H}, and with it the density of electrons, may be determined.

What can we say about the direction of $\mathscr{E}_{\mathrm{H}}$? Well, we have taken meticulous care to find the correct direction. Once the polarity of the applied voltage and the direction of the magnetic field are chosen, the electric field is well and truly defined. So if we put into our measuring apparatus one conductor after the other, the measured transverse voltage should always have the same polarity. Yes ... the logic seems unassailable. Unfortunately, the experimental facts do not conform.

For some conductors and semiconductors the measured transverse voltage is in the *other* direction.

How could we account for the different sign? One possible way of explaining the phenomenon is to say that in certain conductors (and semiconductors) electricity is carried by positively charged particles. Where do they come from? We shall discuss this problem in more detail some time later; for the moment just accept that mobile positive particles may exist in a solid. They bear the unpretentious name 'holes'.

To incorporate holes in our model is not at all difficult. There are now two species of charge carriers bouncing around, which you may imagine as a mixture of two gases. Take good care that the net charge density is zero, and the new model is ready. It is actually quite a good model. Whenever you come across a new phenomenon, try this model first. It might work.

Returning to the Hall effect, you may now appreciate that the experimental determination of $R_{\rm H}$ is of considerable importance. If only one type of carrier is present, the measurement will give us immediately the sign and the density of the carrier. If both carriers are simultaneously present it still gives useful information but a little more algebra needs to be done to extract it.

In our previous example we took a typical metal where conduction takes place by electrons only, and we got a drift velocity of 5×10^{-3}m s^{-1}. For a magnetic field of 1 weber m^{-2} (10^4 gauss if you still prefer cgs units) the transverse electric field is

$$\mathscr{E}_{\rm H} = Bv = 5\times10^{-3}\ \text{V m}^{-1}. \tag{1.21}$$

The corresponding electric field in a semiconductor is considerably higher because of the higher mobilities.

1.5. Electromagnetic waves in solids

So far as the propagation of electromagnetic waves is concerned our model works very well indeed. It can explain all the major important experimental observations I know of with the exception of such esoteric phenomena as the anomalous skin effect.

All we need to assume is that our holes and electrons obey the equations of motion and when they move they give rise to fields in accordance with Maxwell's theory of electrodynamics.

It is perfectly simple to take holes into account but the equations, with holes included, would be considerably longer so we shall confine our attention to electrons.

We could start immediately with the equation of motion for electrons but let us first review what you already know about wave propagation in a medium characterized by the constants permeability μ, dielectric constant ϵ, and conductivity σ (it will not be a waste of time).

First of all we shall need Maxwell's equations:

$$\frac{1}{\mu}\,\nabla\times\mathbf{B} = \mathbf{J}+\epsilon\,\frac{\partial\mathscr{E}}{\partial t}, \tag{1.22}$$

$$\nabla\times\mathscr{E} = -\frac{\partial\mathbf{B}}{\partial t}. \tag{1.23}$$

Secondly, we shall express the current density in terms of the electric field as

$$\mathbf{J} = \sigma\mathscr{E}. \tag{1.24}$$

It would now be a little more elegant to perform all the calculations in vector form but then you would need to know a few vector identities, and tensors (quite simple ones, actually) would also appear. If we use coordinates instead it will make the treatment a little lengthier but not too clumsy if we consider only the one-dimensional case, when

$$\frac{\partial}{\partial x} = 0, \qquad \frac{\partial}{\partial y} = 0. \tag{1.25}$$

Assuming that the electric field has only a component in the x-direction (see the coordinate system in Fig. 1.3), then

$$\nabla\times\mathscr{E} = \begin{vmatrix} \mathbf{e}_x, & \mathbf{e}_y, & \mathbf{e}_z \\ 0 & 0 & \dfrac{\partial}{\partial z} \\ \mathscr{E}_x & 0 & 0 \end{vmatrix} = \frac{\partial\mathscr{E}_x}{\partial z}\,\mathbf{e}_y, \tag{1.26}$$

where $\mathbf{e}_x$, $\mathbf{e}_y$, $\mathbf{e}_z$ are the unit vectors. It may be seen from this equation that the magnetic field can have only a y-component. Thus eqn. (1.23) takes the simple form

$$\frac{\partial\mathscr{E}_x}{\partial z} = -\frac{\partial B_y}{\partial t}. \tag{1.27}$$

We need further

$$\nabla\times\mathbf{B} = \begin{vmatrix} \mathbf{e}_x, & \mathbf{e}_y, & \mathbf{e}_z \\ 0 & 0 & \dfrac{\partial}{\partial z} \\ 0 & B_y & 0 \end{vmatrix} = -\frac{\partial B_y}{\partial z}\,\mathbf{e}_x, \tag{1.28}$$

which, combined with eqn. (1.24), brings eqn. (1.22) to the scalar form

$$-\frac{\partial B_y}{\partial z} = \mu\sigma\mathscr{E}_x + \mu\epsilon\frac{\partial \mathscr{E}_x}{\partial t}. \qquad (1.29)$$

Thus we have two fairly simple differential equations to solve. We shall attempt the solution in the form†

$$\mathscr{E}_x = \mathscr{E}_{x_0}\exp\{-\mathrm{i}(\omega t - kz)\} \qquad (1.30)$$

and

$$B_y = B_{y_0}\exp\{-\mathrm{i}(\omega t - kz)\}. \qquad (1.31)$$

Then

$$\frac{\partial}{\partial z} \equiv \mathrm{i}k, \qquad \frac{\partial}{\partial t} \equiv -\mathrm{i}\omega, \qquad (1.32)$$

which reduces our differential equations to the algebraic equations

$$\mathrm{i}k\mathscr{E}_x = \mathrm{i}\omega B_y \qquad (1.33)$$

and

$$-\mathrm{i}kB_y = (\mu\sigma - \mathrm{i}\omega\mu\epsilon)\mathscr{E}_x. \qquad (1.34)$$

This is a homogeneous equation system. By the rules of algebra there is a solution (apart from the trivial $\mathscr{E}_x = B_y = 0$) only if the determinant of the coefficients vanishes, i.e.

$$\begin{vmatrix} -\mathrm{i}k & \mathrm{i}\omega \\ \mu\sigma - \mathrm{i}\omega\mu\epsilon & \mathrm{i}k \end{vmatrix} = 0. \qquad (1.35)$$

Expanding the determinant we get

$$k^2 - \mathrm{i}\omega(\mu\sigma - \mathrm{i}\omega\mu\epsilon) = 0. \qquad (1.36)$$

Different people call this equation by different names. Characteristic, determinantal, and dispersion equation are among the names more frequently used. We shall call it the *dispersion equation* because that name describes best what is happening physically. Essentially, the equation gives a relationship between the frequency ω and the wave number k (which is related to phase velocity by $v_{\mathrm{p}} = \omega/k$). Thus,

† We have here come face to face with a dispute that has raged between physicists and engineers for ages. For some odd reason the physicists (aided and abetted by mathematicians) use the symbol i for $\sqrt{-1}$ and the exponent $-\mathrm{i}(\omega t - kz)$ to describe a wave travelling in the z-direction. The engineers' notation is j for $\sqrt{-1}$ and $\mathrm{j}(\omega t - kz)$ for the exponent. In this course we have, rather reluctantly, accepted the physicists' notations so as not to confuse you further when reading books on quantum mechanics. We have, however, refused to go as far as using gaussian units. Fortunately, common sense has made some advance among the physicists; an increasing number of them are being converted to the advantages of the MKS or SI systems.

unless ω and k are linearly related, the various frequencies propagate with different velocities and at the boundary of two media are refracted at different angles. Hence the name dispersion.

A medium for which $\sigma = 0$ and μ and ϵ are independent of frequency is nondispersive. The relationship between k and ω is simply

$$k = \omega\sqrt{\mu\epsilon} = \frac{\omega}{c_{\mathrm{m}}}, \tag{1.37}$$

where $c_{\mathrm{m}} \leqslant c$ is the velocity of the electromagnetic wave in the medium.

Solving eqn. (1.36) formally, we get

$$k = (\omega^2\mu\epsilon + \mathrm{i}\omega\mu\sigma)^{\frac{1}{2}}. \tag{1.38}$$

Thus whenever $\sigma \neq 0$, the wave number is complex. What is meant by a complex wave number? We can find this out easily by looking at the exponent of eqn. (1.30). The spatially varying part is

$$\begin{aligned}\exp(\mathrm{i}kz) &= \exp\{\mathrm{i}(k_{\mathrm{real}} + \mathrm{i}k_{\mathrm{imag}})z\}\\ &= \exp(\mathrm{i}k_{\mathrm{real}}z)\exp(-k_{\mathrm{imag}}z).\end{aligned} \tag{1.39}$$

Hence if the imaginary part of k is positive the amplitude of the electromagnetic wave declines exponentially.†

If the conductivity is large enough, the second term is the dominant one in eqn. (1.38) and we may write

$$k \cong (-\mathrm{i}\omega\mu\sigma)^{\frac{1}{2}} = \frac{\pm(\mathrm{i}-1)}{\sqrt{2}}(\omega\mu\sigma)^{\frac{1}{2}}. \tag{1.40}$$

So if we wish to know how rapidly an electromagnetic wave decays in a good conductor, we may find out from this expression. Since†

$$k_{\mathrm{imag}} = \left(\frac{\omega\mu\sigma}{2}\right)^{\frac{1}{2}} \tag{1.41}$$

the amplitude of the electric field varies as

$$|\mathscr{E}_x| = \mathscr{E}_{x_0}\exp\left\{-\left(\frac{\omega\mu\sigma}{2}\right)^{\frac{1}{2}}z\right\}. \tag{1.42}$$

† The negative sign is also permissible though it does not give rise to an exponentially increasing wave as would follow from eqn. (1.39). It would be very nice to make an amplifier by putting a piece of lossy material in the way of the electromagnetic wave. Unfortunately, it violates the principle of conservation of energy. Without some source of energy at its disposal no wave can grow. So the wave which seems to be exponentially growing is in effect a decaying wave which travels in the direction of the negative z-axis.

The distance δ at which the amplitude decays to $1/e$ of its value at the surface is called the *skin depth* and may be obtained from the equation

$$1 = \left(\frac{\omega\mu\sigma}{2}\right)^{\frac{1}{2}}\delta, \tag{1.43}$$

yielding

$$\delta = \left(\frac{2}{\omega\mu\sigma}\right)^{\frac{1}{2}}. \tag{1.44}$$

You have seen this formula before. You need it often to work out the resistance of wires at high frequencies. I derived it solely to emphasize the major steps that are common to all these calculations.

We can now go further, and instead of taking the constant σ, we shall look a little more critically at the mechanism of conduction. We express the current density in terms of velocity by the equation

$$\mathbf{J} = N_e e\mathbf{v}. \tag{1.45}$$

This is really the same thing as eqn. (1.7). The symbol $\mathbf{v}$ still means the average velocity of electrons but now it may be a function of space and time, whereas the notation v_D is generally restricted to d.c. phenomena. The velocity of the electron is related to the electric and magnetic fields by the equation of motion

$$m\left(\frac{d\mathbf{v}}{dt}+\frac{\mathbf{v}}{\tau}\right) = e(\mathscr{E}+\mathbf{v}\times\mathbf{B}), \tag{1.46}$$

where $1/\tau$ is introduced again as a 'viscous' or 'damping' term. You may notice that if the terms $d\mathbf{v}/dt$ and $\mathbf{v}\times\mathbf{B}$ are negligible, then the last two equations reduce again to the $\mathbf{J} = \sigma\mathscr{E}$ relationship.

Now we meet something new. In eqn. (1.46) $\mathbf{v}$ is expressed in the co-moving coordinate system. This is the sort of equation one plays with in electron optics when following the motion of a single electron. The electric and magnetic fields are given as functions of space, and we calculate the position of the electron as a function of time. In our case it will be more convenient to work in a stationary coordinate system. Instead of following one electron we would like to know the velocity of the electron that happens to be at the point z at the time t. This is generally called a Lagrangian to Eulerian transformation. Its essence is that $\mathbf{v}$ becomes a function of z and t. Mathematically it means that we change over to a function with two variables; the total differential needs to be expressed with the aid of partial differentials

$$\left(\frac{\mathrm{d}v}{\mathrm{d}t}\right)_{\text{co-moving}} = \frac{\partial v}{\partial t}+\frac{\partial v}{\partial z}\frac{\mathrm{d}z}{\mathrm{d}t}$$

$$= \frac{\partial v}{\partial t}+v\frac{\partial v}{\partial z}. \tag{1.47}$$

We now have two terms, $v(\partial v/\partial z)$ and $\mathbf{v}\times\mathbf{B}$, containing the product of two wave amplitudes. Since we are interested only in wave-like solutions that represent *small* perturbations of the existing order, we shall always neglect these products. The general rule is usually quoted in the form: cross-products of a.c. quantities are negligible. Thus eqn. (1.46) reduces to

$$m\left(\frac{\partial \mathbf{v}}{\partial t}+\frac{\mathbf{v}}{\tau}\right) = e\mathscr{E}. \tag{1.48}$$

Assuming again that the electric field is in the x-direction, eqn. (1.48) tells us that the electron velocity must be in the same direction. Using the rules set out in eqn. (1.32) we get the following algebraic equation

$$mv_x\left(-\mathrm{i}\omega+\frac{1}{\tau}\right) = e\mathscr{E}_x \tag{1.49}$$

The current density is then also in the x-direction:

$$J_x = N_e e v_x$$

$$= \frac{N_e e^2\tau}{m}\frac{1}{1-\mathrm{i}\omega\tau}\mathscr{E}_x$$

$$= \frac{\sigma}{1-\mathrm{i}\omega\tau}\mathscr{E}_x, \tag{1.50}$$

where σ is defined as before. You may notice now that the only difference from our previous (J–$\mathscr{E}$) relationship is a factor $(1-\mathrm{i}\omega\tau)$ in the denominator. Accordingly, the whole derivation leading to the expression of k in eqn. (1.38) remains valid if σ is replaced by $\sigma/(1-\mathrm{i}\omega\tau)$. We get

$$k = \left(\omega^2\mu\epsilon+\mathrm{i}\omega\mu\frac{\sigma}{1-\mathrm{i}\omega\tau}\right)^{\frac{1}{2}}$$

$$= \omega(\mu\epsilon)^{\frac{1}{2}}\left(1+\frac{\mathrm{i}\sigma}{\omega\epsilon(1-\mathrm{i}\omega\tau)}\right)^{\frac{1}{2}} \tag{1.51}$$

If $\omega\tau \ll 1$, we are back where we started from, but what happens when $\omega\tau \gg 1$? Could that happen at all? Yes, it can happen if the signal frequency is high enough or the collision time is long enough. Then unity is negligible in comparison with $i\omega\tau$ in eqn. (1.51), leading to

$$k = \omega(\mu\epsilon)^{\frac{1}{2}}\left(1-\frac{\sigma}{\omega^2\epsilon\tau}\right)^{\frac{1}{2}}. \tag{1.52}$$

Introducing the new notation

$$\omega_{\mathrm{p}}^2 \equiv \frac{N_{\mathrm{e}}e^2}{m\epsilon} = \frac{\dfrac{N_{\mathrm{e}}e^2}{m}\tau}{\epsilon\tau} = \frac{\sigma}{\epsilon\tau} \tag{1.53}$$

we get

$$k = \omega(\mu\epsilon)^{\frac{1}{2}}\left(1-\frac{\omega_{\mathrm{p}}^2}{\omega^2}\right)^{\frac{1}{2}}. \tag{1.54}$$

Hence as long as $\omega > \omega_{\mathrm{p}}$, the wave number is real. If it is real it has (by the rules of the game) no imaginary component; so the wave is not attenuated. This is quite interesting. By introducing a slight modification into our model, we may come to radically different conclusions. Assuming previously $J = \sigma\mathscr{E}$, we worked out that if any electrons are present at all the wave is bound to decay. Now we are saying that for sufficiently large $\omega\tau$ an electromagnetic wave may travel across our conductor without attenuation. Is this possible? It seems to contradict the empirical fact that radio waves cannot penetrate metals. True; but that's because radio waves haven't got high enough frequencies; let us try light waves. Can they penetrate a metal? No, they can't. It is another empirical fact that metals are not transparent. So we should try even higher frequencies. How high? Well, there's no need to go on guessing, we can work out the threshold frequency from eqn. (1.53). Taking the electron density in a typical metal as 6×10^{28} per m^3, we then get

$$\begin{aligned} f_{\mathrm{p}} &= \frac{1}{2\pi}\left(\frac{N_{\mathrm{e}}e^2}{m\epsilon_0}\right)^{\frac{1}{2}} \\ &= \frac{1}{2\pi}\left\{\frac{6\times10^{28}(1{\cdot}6\times10^{-19})^2}{9{\cdot}11\times10^{-31}\times8{\cdot}85\times10^{-12}}\right\}^{\frac{1}{2}} \\ &= 2{\cdot}2\ 10^{15}\ \mathrm{Hz}. \end{aligned} \tag{1.55}$$

At this frequency range you are probably more familiar with the wavelengths of electromagnetic waves. Converting the above calculated frequency into wavelength, we get

$$\lambda = \frac{c}{f_p} = \frac{3 \times 10^8}{2{\cdot}2 \times 10^{15}} = 136 \text{ nm} \tag{1.56}$$

Thus the threshold wavelength is below the edge of the visible region (400 nm). It is gratifying to note that our theory is in agreement with our everyday experience; metals are not transparent.

There is one more thing we need to check. Is the condition $\omega\tau \gg 1$ satisfied? For a typical metal at room temperature the value of τ is usually above 10^{-14} s, making $\omega\tau$ of the order of hundreds at the threshold frequency.

By making transmission experiments through a thin sheet of metal, the critical wavelength can be determined. The measured and calculated values are compared in Table 1.1. The agreement is not bad.

TABLE 1.1. *Threshold wavelengths in nm*

Metal	Observed wavelength	Calculated wavelength
Cs	440	360
Rb	360	320
K	315	290
Na	210	210
Li	205	150

Before going further I should like to say a little about the relationship of transmission, reflection, and absorption to each other. The concepts are simple and one can always invoke the principle of conservation of energy if in trouble.

Let us take the case when $\omega\tau \gg 1$; k is given by eqn. (1.54), and our conductor fills half the space, as shown in Fig. 1.4. What happens when an electromagnetic wave is incident from the left?

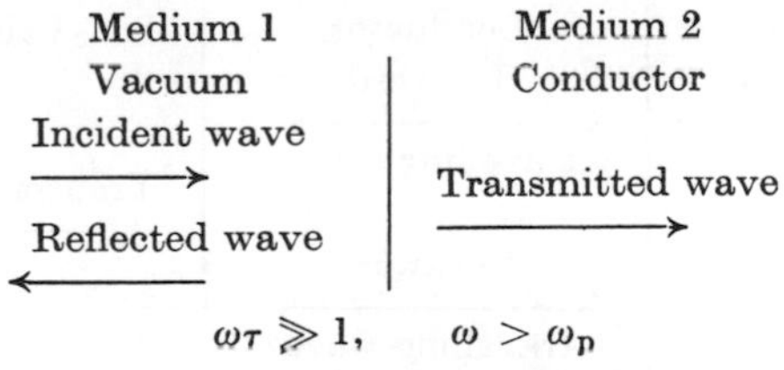

FIG. 1.4. Incident electromagnetic wave partly reflected and partly transmitted.

(i) $\omega > \omega_p$. The electromagnetic wave propagates in the conductor. There is also some reflection depending on the amount of mismatch.

Energy conservation says

energy in the incident wave = energy in the transmitted wave
+energy in the reflected wave.

Is there any absorption? No, because $\omega\tau \gg 1$.

(ii) $\omega < \omega_p$. In this case k is purely imaginary; the electromagnetic wave decays exponentially. Is there any absorption? No. Can the electromagnetic wave decay then? Yes, it can. Isn't this in contradiction with something or other? The correct answer may be obtained by writing out the energy balance. Since the wave decays and the conductor is infinitely long, no energy goes out at the right-hand side. So everything must go back. The electromagnetic wave is reflected, as shown in Fig. 1.5. The energy balance is

energy in the incident wave = energy in the reflected wave.

Medium 1 | Medium 2
Vacuum | Conductor
Incident wave →
No transmitted wave
Reflected wave ←

$\omega\tau \gg 1, \quad \omega < \omega_p$

FIG. 1.5. Incident electromagnetic wave reflected by the conductor.

(iii) Let us take now the case shown in Fig. 1.6 when our conductor is of finite dimension in the z-direction. What happens now if $\omega < \omega_p$? The wave now has a chance to get out at the other side, so there is a flow of energy (forwards and backwards) in the conductor. The wider the slab, the smaller is the amplitude of the wave that appears at the other side because the amplitude decays exponentially in the conductor.

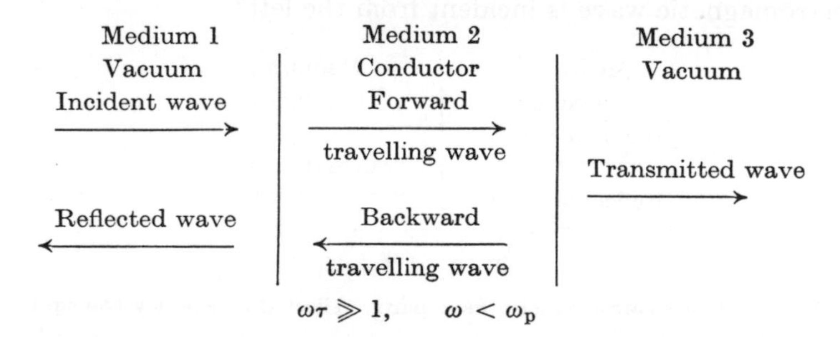

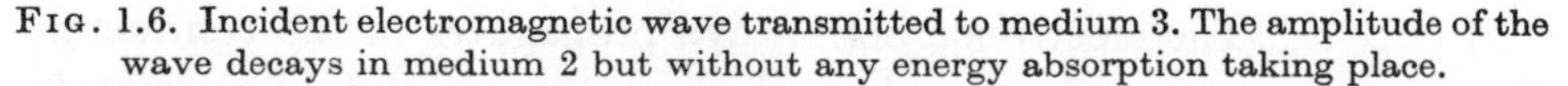

FIG. 1.6. Incident electromagnetic wave transmitted to medium 3. The amplitude of the wave decays in medium 2 but without any energy absorption taking place.

There is decay but no absorption. The amplitudes of the reflected and transmitted waves rearrange themselves in such a way as to conserve energy. If there is a smaller amount transmitted there will be a larger amount reflected.

If we choose a frequency such that $\omega\tau \ll 1$ then, of course, dissipative processes do occur and some of the energy of the electromagnetic wave is converted into heat. The energy balance in the most general case is

$$\text{energy in the incident wave} = \text{energy in the transmitted wave} \\ +\text{energy in the reflected wave} \\ +\text{energy absorbed}$$

A good example of the phenomena enumerated above is the reflection of radio waves from the ionosphere. The ionosphere is a layer which, as the name suggests, contains ions. There are free electrons and positively charged atoms, so our model should work. In a metal, atoms and electrons are closely packed; in the ionosphere the density is much smaller so that the critical frequency ω_p is also smaller. Its value is a few hundred megahertz. Thus, radio waves below this frequency are reflected by the ionosphere (this is why short radio waves can be used for long-distance communication) and those above this frequency are transmitted into space (and so can be used for space or satellite communication). The width of the ionosphere also comes into consideration but at the wavelengths used (it is the width in wavelengths that counts) it can well be regarded as infinitely wide.

1.6. Waves in the presence of an applied magnetic field

In the presence of a constant magnetic field the characteristics of electromagnetic waves will be modified but the solution can be obtained by exactly the same technique as before. The electromagnetic equations (1.22) and (1.23) are still valid for the a.c. quantities; the equation of motion should, however, contain the constant magnetic field, which we shall take in the positive z direction. Then comes the process of linearization when the cross-products of a.c. quantities are neglected. Thus we can again ignore the terms $\mathbf{v}\ \partial\mathbf{v}/\partial z$ and $\mathbf{v}\times\mathbf{B}$, where $\mathbf{B}$ is now the a.c. component of the magnetic field. But, be careful: $\mathbf{v}\times\mathbf{B}_0$ is *not* negligible; it is a first-order quantity. Thus the linearized equation of motion for this case is

$$m\left(\frac{\partial\mathbf{v}}{\partial t}+\frac{\mathbf{v}}{\tau}\right) = e(\mathscr{E}+\mathbf{v}\times\mathbf{B}_0) \tag{1.57}$$

In order to satisfy this vector equation we need both the v_x and v_y components. That means that the current density, and through that the electric and magnetic fields, will also have both x and y components.

Writing down all the equations is a little lengthy but the solution is not more difficult in principle. It may again be attempted in the exponential form, and $\partial/\partial z$ and $\partial/\partial t$ may again be replaced by ik and $-i\omega$ respectively. All the differential equations are then converted into algebraic equations, and by making the determinant of the coefficients zero we get the dispersion equation. I shall not go through the detailed derivation here because it would take up a great deal of time, and the resulting dispersion equation is hardly more complicated than eqn. (1.51). All that happens is that ω in the $\omega\tau$ term is replaced by $\omega \pm \omega_c$. Thus the dispersion equation for transverse electromagnetic waves in the presence of a longitudinal d.c. magnetic field is

$$k = \omega\sqrt{(\mu\epsilon)}\sqrt{\left(1+\frac{i\sigma}{\omega\epsilon\{1-i(\omega\pm\omega_c)\tau\}}\right)} \tag{1.58}$$

where

$$\omega_c = \frac{e}{m}B_0. \tag{1.59}$$

The plus and minus signs give circularly polarized electromagnetic waves rotating in opposite directions. I don't think it is worth finding out which is which unless you want to set up an experiment.

Let us see now what values of the parameters lead to waves propagating without attenuation. For that, the expression under the square root should become real and positive. Previously, we could achieve this aim by taking $\omega\tau$ large but now the same thing may occur for small values of ω as well, provided ω_c is large enough. Thus if

$$\omega_c\tau \gg 1 \quad \text{and} \quad \omega \ll \omega_c, \tag{1.60}$$

Eqn. (1.58) modifies to

$$k = \omega\sqrt{(\mu\epsilon)}\sqrt{\left(1\mp\frac{\sigma}{\omega\epsilon}\frac{1}{\omega_c\tau}\right)}. \tag{1.61}$$

The expression under the square root is now positive if σ is small enough; or, alternatively, it stays positive for any value of σ if the positive sign is taken. The second case looks interesting. It means that in a good conductor low-frequency waves may propagate without attenuation. This discovery was made by Aigrain about a decade ago. He called them *helicon* waves and the name stuck. They were found experimentally not long afterwards.

If σ is large enough the first term in eqn. (1.61) is negligible; using further eqn. (1.53) to bring in ω_p we may write the dispersion equation in the following form:

$$k = \omega\sqrt{\mu}\sqrt{\left(\frac{\epsilon\omega_p^2}{\omega\omega_c}\right)}. \tag{1.62}$$

We have an ordinary transverse electromagnetic wave but the dielectric constant is increased by a factor $\omega_p^2/\omega\omega_c$. How large is this factor? Let's take $\omega_p = 10^{16}$ radian s^{-1}, $\omega = 314$ radian s^{-1}, and $\omega_c = 10^{12}$ radian s^{-1}. Then the relative dielectric constant is

$$\epsilon_r = \frac{\omega_p^2}{\omega\omega_c} = 3{\cdot}18\times 10^{17} \tag{1.63}$$

It looks a bit odd but that's how it is. The relative dielectric constant is enormous. This means that both the wavelength and the wave velocity are reduced by a factor $\sqrt{3{\cdot}18\,.\,10^{17}} = 5{\cdot}65\,.\,10^8$. Thus we have an electromagnetic wave of frequency 50 Hz (the frequency of the mains!) which has a wavelength of about 10 mm and propagates with a velocity of 500 mm s^{-1}. This may actually happen in potassium at temperatures of a few kelvins. Why such a low temperature? Because we need $\omega_c\tau \gg 1$ and to reach such high values with the available magnetic fields the collision time must be long; and it is long enough only at low temperatures.

What is this good for? Are there any practical applications? Not at a few kelvins. A device needs to perform miracles to make it economic to work at such low temperatures. But the properties of helicon waves (namely the different propagation for different circular polarizations) have been used in isolators† working at about 30 MHz. They work at room temperature and use a semiconductor, indium antinomide.

1.7. Cyclotron resonance

Let us return to the general dispersion equation (1.58) and take the negative sign in the denominator (by this we chose a circularly polarized wave of definite rotation). To see more clearly what happens let us divide up the expression under the square root into its real and

† An isolator is a device that lets through a signal in one direction but not in the other.

imaginary parts. We get

$$k = \omega\sqrt{(\mu\epsilon)}\sqrt{\left(1-\frac{\omega_p^2\tau^2\left(1-\dfrac{\omega_c}{\omega}\right)}{1+(\omega-\omega_c)^2\tau^2}+i\frac{\omega_p^2\tau}{\omega}\frac{1}{1+(\omega-\omega_c)^2\tau^2}\right)} \tag{1.64}$$

This looks a bit complicated. In order to get a simple analytical expression let us confine our attention to semiconductors where ω_p is not too large and the applied magnetic field may be large enough to satisfy the conditions

$$\omega_c \gg \omega_p \quad \text{and} \quad \omega_c\tau \gg 1 \tag{1.65}$$

We intend to investigate now what happens when ω_c is close to ω. The second and third terms in eqn. (1.64) are then small in comparison with unity; so the square root may be expanded to give

$$k = \omega\sqrt{\mu\epsilon}\left(1+\frac{i}{2}\frac{\omega_p^2\tau}{\omega}\frac{1}{1+(\omega-\omega_c)^2\tau^2}\right), \tag{1.66}$$

The attenuation of the electromagnetic wave is given by the imaginary part of k. It may be seen that it has a maximum when $\omega_c = \omega$. Since ω_c is called the cyclotron† frequency this resonant absorption of electromagnetic waves is known as *cyclotron resonance*. The sharpness of the resonance depends strongly on the value of $\omega_c\tau$, as shown in Fig. 1.7 where Im k (normalized to its value at $\omega/\omega_c = 1$) is plotted against ω/ω_c. It may be seen that the resonance is hardly noticeable at $\omega_c\tau = 1$. The role of $\omega_c\tau$ is really analogous to that of Q in a resonant circuit. For good resonance we need a high value of $\omega_c\tau$.

The curves have been plotted using the approximate equation (1.66); nevertheless the conclusions are roughly valid for any value of ω_p. If you want more accurate resonance curves, use eqn. (1.64).

Why is there such a thing as cyclotron resonance? The calculation from the dispersion equation provides the figures but if we want the reasons, we should look at the following physical picture.

Suppose that at a certain point in space the a.c. electric field is at right angles to the constant magnetic field, B_0. The electron that happens to be at that point will experience a force at right angles to B_0 and will move along the arc of a circle. We can write a force equation. When the

† After an accelerating device called the cyclotron that works by accelerating particles in increasing radii in a fixed magnetic field.

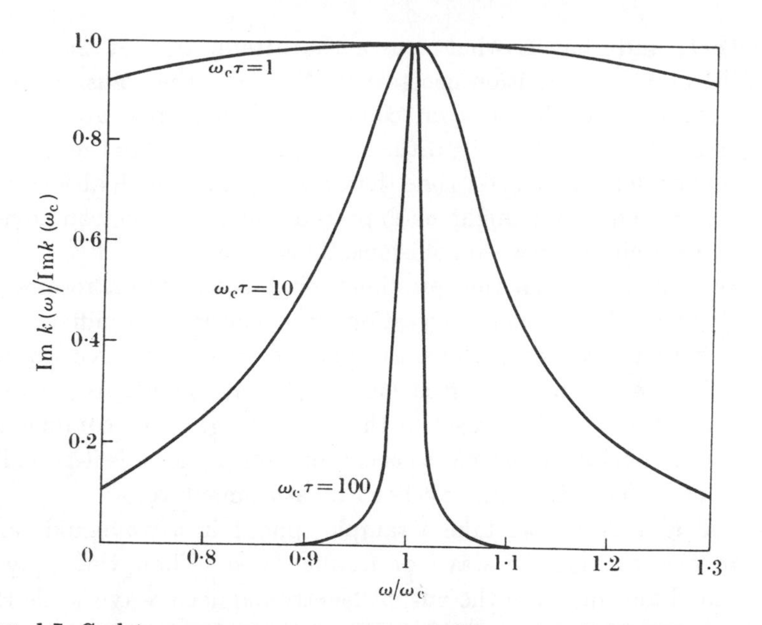

FIG. 1.7. Cyclotron resonance curves computed from eqn. (1.66). There is maximum absorption when the frequency of the electromagnetic wave agrees with the cyclotron frequency.

direction of motion is along the direction of $\mathscr{E}$ the magnetic and centrifugal forces are both at right angles to it, thus

$$B_0 ev = \frac{mv^2}{r}, \tag{1.67}$$

where r is the instantaneous radius of curvature of the electron's path. Thus the electron will move with an angular velocity

$$\omega_c = \frac{v}{r} = \frac{e}{m} B_0. \tag{1.68}$$

The orbits will *not* be circles, for superimposed on this motion is an acceleration varying with time in the direction of the electric field. Now if the frequency of the electric field ω and the cyclotron frequency ω_c are equal, the amplitude of the oscillation builds up. An electron that is accelerated north in one half-cycle will be ready to go south when the electric field reverses, and thus its speed will increase again. Notice that any increase in *speed* must come from the electric field; the acceleration produced by a magnetic field changes direction, not speed, since the force is always at right angles to the direction of motion. Thus, under resonance conditions, the electron will take up energy from the electric field; and that is what causes the attenuation of the wave. Why is the $\omega_c\tau > 1$ condition necessary? Well, τ is the collision time; $\tau = 1/\omega_c$ means that the electron collides with a lattice atom after going round one radian. Clearly, if the electron is exposed to the electric field for a considerably shorter time than a cycle, not much absorption can take place. The limit might be expected when the electron covers one radian of a circle between collisions. This gives $\omega_c\tau = 1$.

Now we may ask again the question: what is cyclotron resonance good for? There have been suggestions for making amplifiers and oscillators with the aid of cyclotron resonance where by clever means the sign of attenuation is reversed (turning it into gain). As far as I know none of these devices reached the ultimate glory of commercial exploitation. If cyclotron resonance is no good for devices, is it good for something else? Yes, it is an excellent measurement tool.

It is used as follows: we take a sample, put it in a waveguide and launch an electromagnetic wave of frequency ω. Then we apply a magnetic field and measure the output electromagnetic wave while the strength of magnetic field is varied. When the output is a minimum, the condition of cyclotron resonance is satisfied. We know ω so we know ω_c; we know the value of the magnetic field, B_0, so we can work out the mass of the electron from the formula

$$m = \frac{eB_0}{\omega} \tag{1.69}$$

But, you would say, what's the point in working out the mass of the electron? That's a fundamental constant, isn't it? Unfortunately, it isn't. When we put our electron in a crystal lattice its mass will be different. The actual value can be measured directly with the aid of cyclotron resonance. So once more, under the pressure of experimental results we have to modify our model. The bouncing billiard balls have variable mass. Luckily, the charge of the electron does remain a fundamental constant. We must be grateful for small favours.

1.8. Plasma waves

Electromagnetic waves are not the only type of waves that can propagate in a solid. There are sound waves and plasma waves as well. We know about sound waves; but what are plasma waves? In their simplest form they are density waves of charged particles in an electrically neutral medium. So they may exist in a solid that has some mobile carriers. The main difference between this case and the previously considered electromagnetic case is that now we permit the accumulation of space charge. At a certain point in space the local density of electrons may exceed the local density of positive carriers. Then an electric field arises owing to the repulsive forces between these 'unneutralized' electrons. The electric field tries to restore the equilibrium of positive and negative charges. It drives the electrons away from the regions where they accumulated. The result is, of course, that the electrons overshoot the mark, and some time later there will be a deficiency of electrons in the same region. An opposite electric field is then created which tries to draw back the electrons, etc. etc. This is the usual case of harmonic oscillation. Thus as far as an individual electron is concerned, it performs simple harmonic motion.

If we consider a one-dimensional model again where everything is the same in the transverse plane, then the resulting electric field has a longitudinal component only. A glance at eqn. (1.26), where $\nabla \times \mathscr{E}$ is worked out, will convince you that if the electric field has a z-component only then $\nabla \times \mathscr{E} = 0$, that is $\mathbf{B} = 0$. There is no magnetic field present; the interplay is solely between the charges and the electric field. For this reason these density waves are often referred to as electrostatic waves.

If $\mathbf{B} = 0$ then eqn. (1.22) takes the simple form

$$\mathbf{J} + \epsilon \frac{\partial \mathscr{E}}{\partial t} = 0 \tag{1.70}$$

We need the equation of motion, which for longitudinal motion will have exactly the same form as for transverse motion, namely

$$m \frac{\partial \mathbf{v}}{\partial t} = e\mathscr{E}, \tag{1.71}$$

where we have neglected the damping term $m\mathbf{v}/\tau$.†

Current density and velocity are related again by

$$\mathbf{J} = N_{e0} e v, \tag{1.72}$$

where N_{e0} is the equilibrium density of electrons.

Finally, there is the equation giving the relationship between the local density of electrons and the electric field, Poisson's equation,

$$\frac{\partial \mathscr{E}}{\partial z} = \frac{1}{\epsilon} N_e e \tag{1.73}$$

† Ignoring losses will considerably restrict the applicability of the formulae derived but our aim is here to show no more than the simplest possible case.

Substituting $\mathscr{E}$ from eqn. (1.71) into eqn. (1.70) and J from eqn. (1.72) into (1.70) we obtain

$$N_{e0}ev + \frac{\epsilon m}{e}\frac{\partial^2 v}{\partial t^2} = 0 \tag{1.74}$$

Following again our favourite method of replacing $\partial/\partial t$ by $-i\omega$, eqn. (1.74) reduces to

$$v\left\{N_{e0}e + \frac{\epsilon m}{e}(-\omega^2)\right\} = 0 \tag{1.75}$$

Since v must be finite, this means

$$N_{e0}e - \frac{\epsilon m}{e}\omega^2 = 0, \tag{1.76}$$

or, rearranging,

$$\omega^2 = \frac{N_{e0}e^2}{m\epsilon}. \tag{1.77}$$

This is our dispersion equation. It is a rather odd one because k does not appear in it. A relationship between k and ω gives the allowed values of k for a given ω. If k does not appear in the dispersion equation, *all* values of k are allowed. On the other hand, there is only a single value of ω allowed. Looking at it more carefully we may recognize that it is nothing else but ω_p, the frequency we met previously as the critical frequency of transparency for electromagnetic waves. Historically, it was first discovered in plasma oscillations (in gas discharges by Langmuir); so it is more usual to call it the 'plasma frequency', and that is where the subscript p comes from.

Summarizing, a lossless plasma wave may have any wavelength and may propagate at any velocity, but only at one single frequency, the plasma frequency.

What are plasmas good for? This is a very difficult question to answer. We have touched the simplest case only. Plasma physics covers a vast field with immense potentialities (for example fusion) but the number of devices based on some sort of charge imbalance is very limited.† This is because plasma waves are generally harmful. It is more the concern of physicists and engineers to *get rid* of plasma waves than to set them up.

† Microwave tubes belong to a special category. They do employ density waves but the electrons' space charge is not compensated.

1.9. Specific heat of metals

We have proposed, following the kinetic theory of gases, that the average energy per electron is $\frac{3}{2}kT$. Hence the specific heat contribution of the free electrons in a solid is

$$\frac{\partial}{\partial T}\left(\tfrac{3}{2}kT\right)N_e = \tfrac{3}{2}kN_e. \tag{1.78}$$

The lattice will also contribute to the specific heat a term† $3Nk$ (where N is the number of lattice atoms per unit volume). The lattice atoms also have three degrees of freedom, but as they vibrate their energy is equipartitioned between the kinetic and potential energy. Thus we should expect an alkali metal (where $N_e = N$) to have a 50% greater specific heat than an insulator having the same number of lattice atoms because of the electronic contribution. It turns out that metals and insulators have about the same specific heat. Metals behave as if the free electrons make practically no contribution to the specific heat. Our model fails again to explain the experimentally observed value. What shall we do? Modify our model. But how? Up to now the modifications have been fairly obvious. The 'wrong sign' of the Hall voltage could be explained by introducing positive carriers, and when cyclotron resonance measurements showed that the mass of an electron in a solid was different from the 'free' electron mass, we simply said: 'all right, the electron's mass is not a constant.' How should we modify our model now? There seems to be no simple way of doing so. An entirely new start is needed.

Examples

1. A 10 mm cube of germanium passes a current of 6·4 mA when 10 mV is applied between two of its parallel faces. Assuming that the charge carriers are electrons that have a mobility of 0·39 m^2 V^{-1} sec^{-1}, calculate the density of carriers. What is their collision time if the electron's effective mass in germanium is $0{\cdot}12m_0$ where m_0 is the free electron mass?

2. An electromagnetic wave of free space wavelength 0·5 mm propagates through a piece of indium antimonide that is placed in an axial magnetic field. There is resonant absorption of the electromagnetic wave at a magnetic field, $B = 0{\cdot}323$ weber/m^2.

(i) What is the effective mass of the particle in question?

(ii) Assume that the collision time is 15 times longer (true for electrons around liquid nitrogen temperatures) than in germanium in the previous example. Calculate the mobility.

† The formula is valid at room temperature but fails at low temperatures. In general, the lattice contribution to specific heat must also be explained with the aid of quantum theory but we shall not be able to do that in the present course.

(iii) Is the resonance sharp? What is your criterion?

3. If both electrons and holes are present the conductivities add. This is because under the effect of an applied electric field the holes and electrons flow in opposite directions, and a negative charge moving in the (say) $+z$ direction is equivalent to a positive charge moving in the $-z$ direction.

Assume that in a certain semiconductor the ratio of electronic mobility, μ_e, to hole mobility, μ_h, is equal to 10, the number of holes is $N_h = 10^{20}/m^3$, and the number of electrons is $N_e = 10^{19}/m^3$. The measured conductivity is $0{\cdot}455$ ohm^{-1} m^{-1}. Calculate the mobilities.

4. Measurements on sodium have provided the following data: resistivity $4{\cdot}7\ 10^{-8}$ ohm m, Hall coefficient $-2{\cdot}5\ 10^{-10}$ m^3 coulomb^{-1}, critical wavelength of transparency 210 nm, and density 971 kg/m^3.

Calculate (i) the density of electrons, (ii) the mobility, (iii) the effective mass, (iv) the collision time, (v) the number of electrons per atom available for conduction.

Electric conduction in sodium is caused by electrons. The number of atoms in a kg mole is $6{\cdot}02.\ 10^{26}$ and the atomic weight of sodium is 23.

5. The Hall coefficient is defined by the relationship $\mathscr{E}_H = RJB$, where $\mathscr{E}_H$ is the transverse electric field, J the longitudinal current density produced by the applied electric field, and B the applied magnetic field (as shown in Fig. 1.3). Derive an expression for R if both electrons and holes are present. The experimentally determined Hall coefficient is negative. Can you conclude that electrons are dominant charge carriers?

(Hint: Write down the equations of motion both for holes and electrons. Neglect the product of transverse velocity with the magnetic field because the transverse velocity of the carriers is much smaller than their velocity in the longitudinal direction (in the direction of the applied electric field). The transverse electric field may then be obtained from the condition that the transverse current is zero.)

6. An electromagnetic wave is incident from Medium 1 upon Medium 2 as shown in Figs. 1.4 and 1.5. Derive expressions for the reflected and transmitted power. Show that the transmitted electromagnetic power is finite when $\omega > \omega_p$ and zero when $\omega < \omega_p$.

(Hint: Solve Maxwell's equations separately in both media. Determine the constants by matching the electric and magnetic fields at the boundary. The power in the wave (per unit surface) is given by the Poynting vector.)

7. An electromagnetic wave is incident upon a medium of width d, as shown in Fig. 1.6. Derive expressions for the reflected and transmitted power. Calculate the transmitted power for the cases $d = 0{\cdot}25\ \mu$m and $d = 2{\cdot}5\ \mu$m when $\omega = 6{\cdot}28\ 10^{15}$ rad/sec, $\omega_p = 9\ 10^{15}$ rad/sec (take $\epsilon = \epsilon_0$ and $\mu = \mu_0$).

8. In a medium containing free charges the total current density may be written as $\mathbf{J}_{total} = \mathbf{J} + j\omega\epsilon\mathbf{E}$, where $\mathbf{J}$ is the particle current density, $\mathbf{E}$ electric field, ϵ dielectric constant, and ω frequency of excitation. For convenience the above expression is often written in the form $\mathbf{J}_{total} = j\omega\bar{\bar{\epsilon}}_{eqv}\mathbf{E}$, defining thereby an equivalent dielectric tensor $\bar{\bar{\epsilon}}_{eqv}$. Determine $\bar{\bar{\epsilon}}_{eqv}$ for a fully ionized electron-ion plasma to which a constant magnetic field B_0 is applied in the z direction.

2. The electron as a wave

The old order changeth, yielding place to new.

TENNYSON *The Idylls of the King*

Vezess uj utakra, Lucifer.

MADÁCH *Az Ember Tragediája*

2.1. Introduction

WE have considered the electron as a particle and managed to explain successfully a number of interesting phenomena. Can we explain the rest of electronics by gentle modifications of this model? Unfortunately (for students if not lecturers) the answer is no. The experimental results on specific heat have already warned us that something is wrong with our particles but the situation is, in fact, a lot worse. We find that the electron has wavelike properties too. The chief immigrant in this particular woodpile, the experiment that could not possibly be explained by a particle model, was the electron diffraction experiment of Davisson and Germer in 1927. The electrons behaved as waves.

We shall return to the experiment a little later; let us see first what the basic difference is between particle and wave behaviour. The difference can best be illustrated by the following 'thought' experiment. Suppose we were to fire bullets at a bulletproof screen with two slits in it (Fig. 2.1). We will suppose that the gun barrel is old and worn so that the bullets bespatter the screen around the slits uniformly after

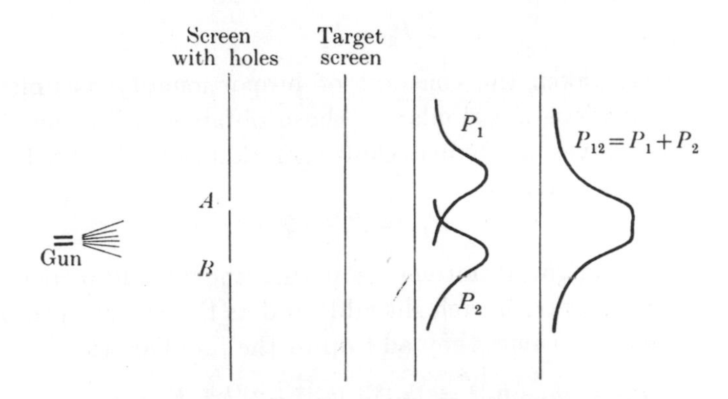

FIG. 2.1. An experiment with bullets.

a fairly large number of shots. If at first, slit B is closed by a bullet-proof cover, the bullets going through A make a probability pattern on the target screen, shown graphically in Fig. 2.1 as a plot of probability against distance from the gun nozzle–slit A axis. Calling this pattern P_1, we expect (and get) a similar but displaced pattern P_2 if slit B is opened and slit A is closed. Now if both slits are open, the combined pattern, P_{12} is simply

$$P_{12} = P_1 + P_2 \tag{2.1}$$

We will now think of a less dangerous and more familiar experiment, with waves in a ripple tank (Fig. 2.2). The gun is replaced by a vibrator

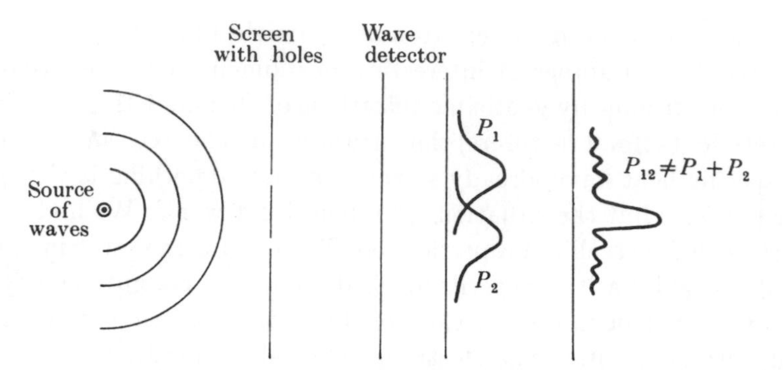

FIG. 2.2. An experiment with waves.

or ripple-generator, the slits are the same, and in the target plane there is a device to measure the ripple intensity, i.e. a quantity proportional to the square of the height of the waves produced. Then with one slit open we find

$$P_1 = h_1^2 \tag{2.2}$$

or with the other open

$$P_2 = h_2^2 \tag{2.3}$$

where we have taken the constant of proportionality as unity. The probability functions are similar to those obtained with one slit and bullets. So far waves and bullets show remarkable similarity. But with both slits open we find

$$P_{12} \neq P_1 + P_2 \tag{2.4}$$

Instead, as we might intuitively suppose, the instantaneous values of the wave heights from each slit add; and as the wavelength of each set of ripples is the same, they add up in the familiar way

$$P_{12} = |h_1 + h_2|^2 = |h_1|^2 + |h_2|^2 + 2\,|h_1|\,|h_2| \cos\delta \tag{2.5}$$

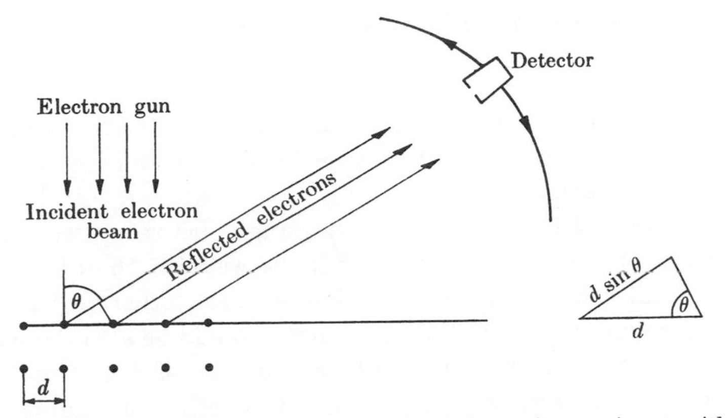

FIG. 2.3. Schematic representation of Davisson and Germer's experiment with low energy electrons. The electrons are effectively reflected by the surface layer of the crystal. The detector shows maximum intensity when the individual reflections add in phase.

where δ is the phase difference between the two interfering waves. Thus the crucial difference between waves and particles is that waves interfere but particles don't.

The big question in the 1920s was: are electrons like bullets or do they follow a theoretical prediction by L. de Broglie (in 1924)? According to de Broglie the electrons should have wavelike properties with a wavelength inversely proportional to particle momentum, viz.

$$\lambda = \frac{h}{mv}, \tag{2.6}$$

where h is Planck's constant† (not the height of the waves in the ripple tank) with a rather small numerical value, namely $6{\cdot}6 \times 10^{-34}$ J s, and m and v are the mass and speed of the electron.

To test de Broglie's hypothesis, Davisson and Germer fired a narrow beam of electrons at the surface of a single crystal of nickel (Fig. 2.3). The wavelike nature of the electron was conclusively demonstrated. The reflected beam displayed an interference pattern.

The arrangement is analogous to a reflection grating in optics; the grating is replaced by the regular array of atoms and the light waves are replaced by electron waves. Maximum response is obtained when the

† Planck introduced this quantity in 1901 in a theory to account for discrepancies encountered in the classical picture of radiation from hot bodies. He considered a radiator as an assembly of oscillators whose energy could not change continuously but must always increase or decrease by a quantum of energy, hf. This was the beginning of the twentieth century for science, and science has not been the same since. The confidence and assurance of nineteenth-century physicists disappeared, probably for ever. The most we can hope nowadays is that our latest models and theories go one step further in describing nature.

reflections add in phase, that is when the condition

$$n\lambda = d \sin \theta \tag{2.7}$$

is satisfied where n is an integral number, d is the lattice spacing, and λ is the wavelength to be determined as a function of electron-gun accelerating voltage.

From eqn. (2.7) the difference in angle between two successive maxima is of the order of λ/d. Thus if the wavelength of the radiation is too small the maxima lie too close to each other to be resolved. Hence for good resolution the wavelength should be about equal to the lattice spacing, which is typically a fraction of a nanometer. The electron velocity corresponding to a wavelength of 0·1 nm is

$$v = \frac{h}{m\lambda} = \frac{6{\cdot}6 \times 10^{-34}}{9{\cdot}1 \times 10^{-31} \times 10^{-10}} \text{ Js kg}^{-1}\text{ m}^{-1} = 7{\cdot}25\ 10^{6} \text{ m s}^{-1} \tag{2.8}$$

The accelerating voltage may be obtained from the condition of energy conservation

$$\tfrac{1}{2}mv^2 = eU$$

whence

$$U = \frac{mv^2}{2e} = \frac{9{\cdot}1 \times 10^{-31}\,(7{\cdot}25 \times 10^{6})^2}{2 \times 1{\cdot}6 \times 10^{-19}} \text{ kg m}^2\text{ s}^{-2}\text{ C}^{-1} = 150 \text{ V} \tag{2.9}$$

The voltages used by Davisson and Germer were of this order.

So electrons are waves. Are protons waves? Yes, they are; it can be shown experimentally. Are neutrons waves? Yes, they are; it can be shown experimentally. Are bullets waves? Well, they should be but there are some experimental difficulties in proving it. Take a bullet which has a mass of 10^{-3} kg and travels at a velocity 10^3 m s^{-1}. Then the bullet's wavelength is $6{\cdot}6 \times 10^{-34}$ m. Thus our reflecting agents or slits should be about 10^{-34} m apart to observe the diffraction of bullets, and that would not be easily realizable. Our bullets are obviously too fast. Perhaps with slower bullets we will get a diffraction pattern with slits a reasonable distance apart. Taking 10 mm for the distance between the slits, and requiring the same wavelength for the bullets, their velocity comes to 10^{-28} m s^{-1}; that is, the bullet would travel one metre in about 10^{21} years. Best modern estimates give the age of the universe as 10^{10} years so this way of doing the experiment runs again into practical difficulties.

The conclusion from this rather eccentric aside is of some importance. It seems to suggest that everything, absolutely everything, that we used to regard as particles may behave like waves if the right conditions are ensured. The essential difference between electrons and particles encountered in some other branches of engineering is merely one of size. Admittedly the factors involved are rather large. The bullet in our chosen example has a mass 10^{27} times the electron mass, so it is not entirely unreasonable that they behave differently.

2.2. The electron microscope

Particles are waves, waves are particles. This outcome of a few simple experiments mystifies the layman, delights the physicist, and provides the philosopher with material for a couple of treatises. What about the engineer? The engineer is supposed to ask the consequential (though grammatically slightly incorrect) question; what is this good for?

Well, one well-known practical effect of the wave nature of light is that the resolving power of a microscope is fundamentally limited by the wavelength of the light. If we want greater resolution we need a shorter wavelength. Let's use X-rays then. Yes, but they can't be easily focused; it is impossible to make good lenses for X-rays. Use electrons then; they have short enough wavelengths. An electron accelerated to a voltage of 150 V has a wavelength of 0·1 nm. This is already four thousand times shorter than the wavelength of violet light, and using higher voltages we can get even shorter wavelengths. Good, but can electrons be focused? Yes, they can. Very conveniently, just about the same time that Davisson and Germer proved the wave properties of electrons, Busch discovered that electric and magnetic fields of the right configuration can bring a diverging electron beam to a focus. So all we need is a fluorescent screen to make the incident electrons visible, and the electron microscope is ready (Fig. 2.4).

You know, of course, about the electron microscope, that it has a resolving power so great that it is possible to see large molecules with it. Our aim is mainly to emphasize the mental processes that lead from scientific discoveries to practical applications. But besides, there is one more interesting aspect of the electron microscope. It provides perhaps the best example for what is known as the 'duality of the electron'. To explain the operation of the electron microscope both the 'wave' and the 'particle' aspects of the electron are needed. The focusing is possible because the electron is a charged *particle*, and the great resolution is possible because it is a wave of extremely short wavelength.

In conclusion, it must be admitted that the resolving power of the electron microscope is not as large as would follow from the available

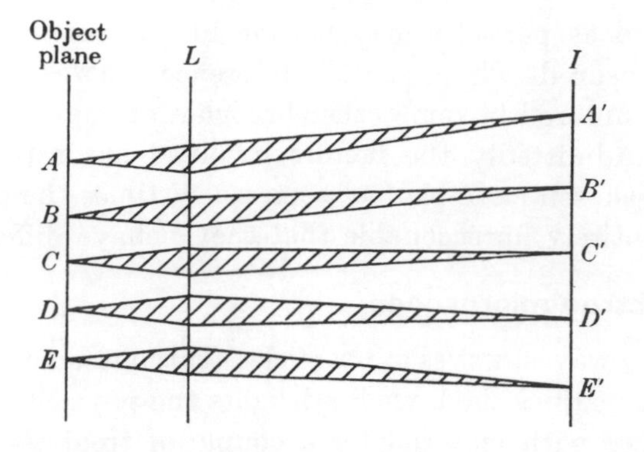

FIG. 2.4. Schematic representation of the salient features of an electron microscope. The points A, B, C, D, E are chosen in the plane of the object to be magnified. The electrons incident from the left leave a point on the object at small angles. The diverging electron beam is brought again to a focus in the image plane (I) by the lens in the plane L. The magnification is the ratio $A'E'$ to AE.

wavelengths of the electrons. The limitation in practice is caused by lens aberrations.

2.3. Some properties of waves

You are by now familiar with all sorts of waves, and you know that a wave of frequency ω and wave number k may be described by the formula

$$u = a \exp \mathrm{i}\varphi; \qquad \varphi = -(\omega t - kz) \tag{2.10}$$

where the positive z-axis is chosen as the direction of propagation.

The phase velocity may be defined as

$$v_\varphi = \left.\frac{\partial z}{\partial t}\right|_{\varphi=\text{constant}} = \frac{\omega}{k} = f\lambda \tag{2.11}$$

This is the velocity with which any part of the wave moves along. For a single frequency wave this is fairly obvious. One can easily imagine how the crest moves. But what happens when several waves are superimposed? The resultant wave is given by

$$u = \sum_n a_n \exp\{-\mathrm{i}(\omega_n t - k_n z)\} \tag{2.12}$$

where to each value of k_n belongs an a_n and an ω_n. Going over to the continuum case (when the number of components within an interval

Δk tends to infinity), we get

$$u = \int_{-\infty}^{\infty} a(k) \exp\{-\mathrm{i}(\omega t - kz)\}\, \mathrm{d}k \tag{2.13}$$

where $a(k)$ and ω are functions of k.

We shall return to the general case later; let us take for the time being, $t = 0$, then

$$u(z) = \int_{-\infty}^{\infty} a(k) \exp \mathrm{i}kz\, \mathrm{d}k \tag{2.14}$$

and investigate the relationship between k and z. We shall be interested in the case when the wave number and frequency of the waves do not spread out too far, that is $a(k)$ is zero everywhere with the exception of a narrow interval Δk. The simplest possible case is shown in Fig. 2.5

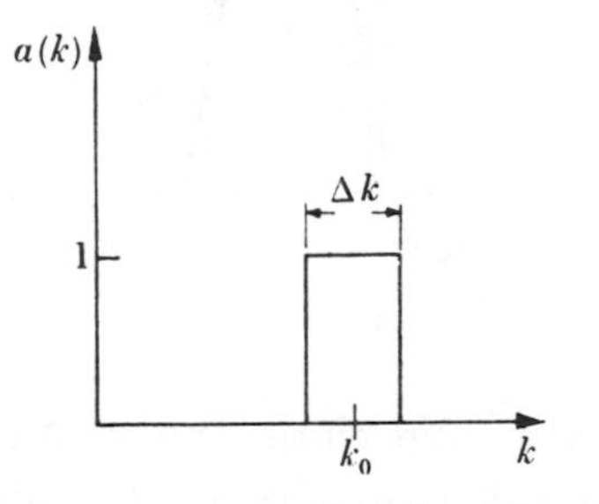

FIG. 2.5. The amplitude of the waves as a function of wave number, described by eqn. (2.15.)

where

$$a(k) = 1 \quad \text{for} \quad k_0 - \frac{\Delta k}{2} < k < k_0 + \frac{\Delta k}{2} \tag{2.15}$$

and

$$a(k) = 0$$

outside this interval. The integral (2.14) reduces then to

$$u(z) = \int_{k_0 - \Delta k/2}^{k_0 + \Delta k/2} \exp \mathrm{i}kz\, \mathrm{d}k \tag{2.16}$$

which can be easily integrated to give

$$u(z) = \Delta k \exp \mathrm{i}k_0 z \, \frac{\sin \dfrac{\Delta kz}{2}}{\dfrac{\Delta kz}{2}}. \tag{2.17}$$

We have here a wave whose *envelope* is given by the function

$$\frac{\sin \frac{\Delta kz}{2}}{\frac{\Delta kz}{2}}, \tag{2.18}$$

plotted in Fig. 2.6. It may be seen that the function is rapidly decreasing outside a certain interval Δz. We may say that the wave is essentially contained in this 'packet', and in the future we shall refer

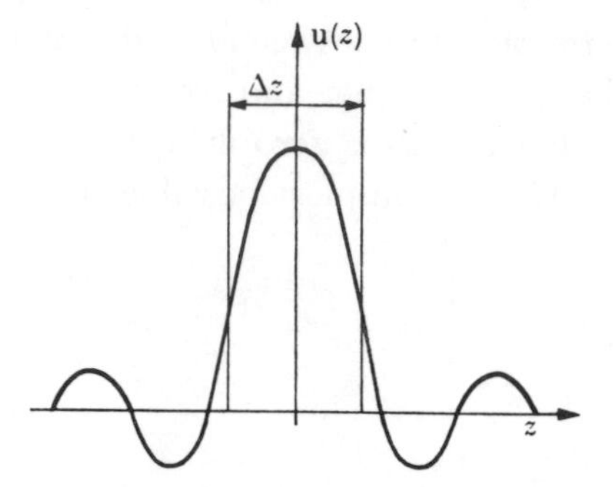

FIG. 2.6. The spatial variation of the amplitude of the wave packet of Fig. 2.5.

to it as a wave packet. We chose the width of the packet, rather arbitrarily, to be determined by the points where the amplitude drops to 0·63 of its maximum value, that is where

$$\frac{\Delta kz}{2} = \pm\frac{\pi}{2}. \tag{2.19}$$

Hence the relationship between the spread in wavenumber Δk, and the spread in space Δz, is as follows

$$\Delta k \Delta z = 2\pi. \tag{2.20}$$

An obvious consequence of this relationship is that by making Δk large, Δz must be small, and vice versa. This means that for having a narrow wave packet in space, we need a larger spread in wavenumber.

Let us return now to the time-varying case, still maintaining that $a(k)$ is essentially zero beyond the interval Δk. Eqn. (2.13) takes then the form

$$u = \int_{k_0-\Delta k/2}^{k_0+\Delta k/2} a(k)\exp\{-\mathrm{i}(\omega t - kz)\}\,\mathrm{d}k. \tag{2.21}$$

Let us now rewrite the above formula in the following form

$$u(z, t) = A(z, t)\exp\{-\mathrm{i}(\omega_0 t - k_0 z)\} \tag{2.22}$$

where

$$A(z, t) = \int_{k_0-\Delta k/2}^{k_0+\Delta k/2} u(k)\exp[-\mathrm{i}\{(\omega-\omega_0)t-(k-k_0)z\}]\,\mathrm{d}k \tag{2.23}$$

and ω_0 is the frequency at $k = k_0$.

We may now define two velocities. One is ω_0/k_0, which corresponds to the previously defined phase velocity, and is the velocity with which the individual components propagate. The other velocity may be defined by looking at the expression for A. Since A represents the envelope of the wave, we may say that the envelope has the same shape whenever

$$(\omega-\omega_0)t-(k-k_0)z = \text{constant}. \tag{2.24}$$

Hence we may define a velocity

$$v_\mathrm{g} = \frac{\partial z}{\partial t} = \frac{\omega-\omega_0}{k-k_0} \tag{2.25}$$

which, for sufficiently small Δk, reduces to

$$v_\mathrm{g} = \left(\frac{\partial \omega}{\partial k}\right)_{k=k_0} \tag{2.26}$$

v_g is called the *group velocity* because it gives the velocity of the wave packet.

2.4. Applications to electrons

We have discussed some properties of waves. It has been an exercise in mathematics. Now we take a deep plunge and will try to apply these properties to the particular case of the electrons. The first step is to identify the wave packet with an electron in your mental picture. This is not unreasonable. We are saying, in fact, that where the ripples are, *there* must be the electron. If the ripples are uniformly distributed in space (as is the case for a single frequency wave) the electron can be anywhere. If the ripples are concentrated in space in the form of a wave packet, the presence of an electron is indicated. Having identified the wave packet with an electron, we may identify the velocity of the wave packet with the electron velocity.

What can we say about the energy of the electron? We know that a photon of frequency ω has an energy

$$E = hf = \hbar\omega \tag{2.27}$$

where f is the frequency of the electromagnetic wave and $\hbar = h/2\pi$. Analogously, it may be suggested that the energy of an electron in a wave packet centred at the frequency ω is given by the same formula. Hence we may write down the energy of the electron (taking the potential energy as zero) in the form

$$\hbar\omega = \tfrac{1}{2}mv_g^2 \tag{2.28}$$

We can differentiate this partially with respect to k to get

$$\hbar\frac{\partial\omega}{\partial k} = mv_g\frac{\partial v_g}{\partial k} \tag{2.29}$$

which, with the aid of eqn. (2.26), reduces to

$$\hbar = m\frac{\partial v_g}{\partial k}. \tag{2.30}$$

Integrating, and taking the integration constant as zero, we get

$$\hbar k = mv_g \tag{2.31}$$

which can be expressed in terms of wavelength as

$$\lambda = \frac{h}{mv_g} \tag{2.32}$$

and this is nothing else but de Broglie's relationship. Thus if we assume the validity of the wave picture, identify the group velocity of a wave packet with the velocity of an electron, and assume that the centre frequency of the wave packet is related to the energy of the electron by Planck's constant, de Broglie's relationship automatically drops out.

This proves, of course, nothing. There are too many assumptions, too many identifications, representations, and interpretations; but, undeniably, the different pieces of the jigsaw puzzle do show some tendency to fit together. We have now established some connection between the wave and particle aspects, which seemed to be entirely distinct not long ago.

What can we say about the electron's position? Well, we identified the position of the electron with the position of the wave packet. So, wherever the wave packet is, there is the electron. But remember, the wave packet is not infinitely narrow; it has a width Δz, and there will thus be some uncertainty about the position of the electron.

Let us look again at eqn. (2.20). Taking note of the relationship expressed in eqn. (2.31) between wave number and momentum, eqn.

(2.20) may be rewritten as

$$\Delta p \, \Delta z = h. \tag{2.33}$$

This is the celebrated uncertainty relationship. It means that the uncertainty in the position of the electron is related to the uncertainty in the momentum of the electron. If we know the position of the electron with great precision, that is if Δz is very small, then the uncertainty in the velocity of the electron must be large. Let us put in a few figures to see the orders of magnitude involved. If we know the position of the electron with an accuracy of 10^{-9} m then the uncertainty in momentum is

$$\Delta p = 6{\cdot}6 \times 10^{-25} \text{ kg m s}^{-1}, \tag{2.34}$$

corresponding to

$$\Delta v \simeq 7 \times 10^{5} \text{ m s}^{-1}, \tag{2.35}$$

that is the uncertainty in velocity is quite appreciable.

Taking macroscopic dimensions, say 10^{-3} m for the uncertainty in position, and a bullet with a mass 10^{-3} kg, the uncertainty in velocity decreases to

$$\Delta v = 6{\cdot}6 \times 10^{-28} \text{ m s}^{-1} \tag{2.36}$$

which is something we can easily put up with in practice. Thus whenever we come to very small distances and very light particles, the uncertainty in velocity becomes appreciable, but with macroscopic objects and macroscopic distances the uncertainty in velocity is negligible. You can see that everything here depends on the value of h, which happens to be rather small in our universe. If it were larger by a factor of, say 10^{40}, the police would have considerable difficulty in enforcing the speed limit.

The uncertainty relationship has some fundamental importance. It did away (probably for ever) with the notion that distance and velocity can be simultaneously measured with arbitrary accuracy. It is applicable not only to position and velocity but to a number of other related pairs of physical quantities.† It may also help to explain qualitatively some complicated phenomena. We may, for example, ask the question why there is such a thing as a hydrogen atom consisting of a negatively charged electron and a positively charged proton. Why doesn't the electron eventually fall into the proton? Armed with our

† You might find it interesting to learn that electric and magnetic intensities are also subject to this law. They cannot be simultaneously measured to arbitrary accuracy.

knowledge of the uncertainty relationship, we can now say that this event is energetically unfavourable. If the electron is too near to the proton then the uncertainty in its velocity is high; so it may have quite a high velocity, which means high kinetic energy. Thus the electron's search for low potential energy (by moving near to the proton) is frustrated by the uncertainty principle, which assigns a large kinetic energy to it. The electron must compromise and stay at a certain distance from the proton.

2.5. Two analogies

The uncertainty relationship is characteristic of quantum physics. We would search in vain for anything similar in classical physics. The derivation is, however, based on certain mathematical formulae that also appear in some other problems. Thus, even if the phenomena are entirely different, the common mathematical formulation permits us to draw analogies.

Analogies may or may not be helpful. It depends to a certain extent on the man's imagination or lack of imagination and, of course, on familiarity or lack of familiarity with the analogue.

I believe in the use of analogies. I think they can help, both in memorizing a certain train of thought, and in arriving at new conclusions and new combinations. Even such a high-powered mathematician as Archimedes resorted to mechanical analogies when he wanted to convince himself of the truth of certain mathematical theorems. So this is quite a respectable method, and as I happen to know two closely related analogies, I shall describe them.

Notice first of all that $u(z)$ and $a(k)$ are related to each other by a Fourier integral in eqn. (2.14). In deriving eqn. (2.20) we made the sweeping assumption that $a(k)$ was constant within a certain interval but this is not necessary. We would get the same sort of final formula (with slightly different numerical constants) for any reasonable $a(k)$. The uncertainty relationship, as derived from the wave concept, is a consequence of the Fourier transform connection between $a(k)$ and $u(z)$. Thus whenever two functions are related in the same way, they can readily serve as analogues.

Do such functions appear in engineering practice? They do. The time variation of a signal and its frequency spectrum are connected by Fourier transform. A pulse of the length τ has a spectrum (Fig. 2.7) exactly like the envelope we encountered before. The width of the frequency spectrum (referred to as bandwidth in common language)

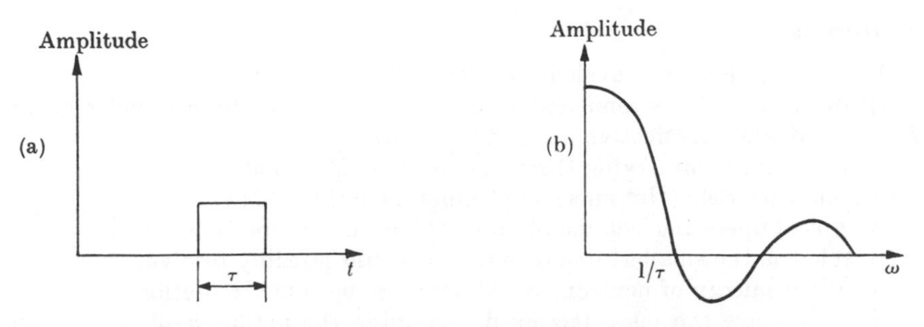

FIG. 2.7. A rectangular pulse and its frequency spectrum.

is related to the length of the pulse. All communication engineers know that the shorter the pulse the larger is the bandwidth to be transmitted. For television, for example, we need to transmit lots of pulses (the light intensity for some several hundred thousand spots twenty-five times per second), so the pulses must be short and the bandwidth large. This is why television works at much higher frequencies than radio broadcasting.

In the mathematical formulation, k and z of eqn. (2.20) are to be replaced by the frequency ω and time t. Hence the relationship for communication engineering takes the form

$$\Delta\omega \, \Delta t = 2\pi \tag{2.37}$$

The analogy is close indeed.

In the second analogue the size of an aerial and the sharpness of the radiation pattern are related. It is the same story. In order to obtain a sharp beam one needs a big aerial. So if you have ever wondered why radio astronomers use such giant aerials, here is the answer. They need narrow beams to be able to distinguish between the various radio stars, and they must pay for them by erecting (or excavating) big antennas.†

The mathematical relationship comes out as follows

$$\Delta\theta \, \Delta z = \lambda \tag{2.38}$$

where $\Delta\theta$ is the beamwidth, Δz is the linear dimension of the aerial and λ is the wavelength of the electromagnetic radiation (transmitted or received).

† Incidentally there is another reason why radio telescopes must be bigger than, say, radar aerials. They do a lot of work at a wavelength of 210 mm, which is seven times longer than the wavelength used by most radars. Hence for the same resolution an aerial seven times bigger is needed.

Examples

1. Find the de Broglie wavelength of the following particles:

(i) an electron in a semiconductor having average thermal velocity at $T = 300$ K and an effective mass of $m_e^* = am_0$,

(ii) a helium atom having thermal energy at $T = 300$ K,

(iii) an α-particle (He^4 nucleus) of kinetic energy 10 MeV.

2. A typical operating voltage of an electron microscope is 50 kV.

(i) What is the smallest distance that it could possibly resolve?

(ii) What energy of neutrons could achieve the same resolution?

(iii) What are the main factors determining the actual resolution of an electron microscope?

3. Electrons accelerated by a potential of 70 V are incident perpendicularly on the surface of a single crystal metal. The crystal planes are parallel to the metal surface and have a (cubic) lattice spacing of 0·352 nm. Sketch how the intensity of the scattered electron beam would vary with angle.

4. A beam of electrons of 10 keV energy passes perpendicularly through a very thin (of the order of a few nanometres) foil of our previous single crystal metal. Determine the diffraction pattern obtained on a photographic plate placed 0·1 m behind the specimen. How will the diffraction pattern be modified for a polycrystalline specimen? (Hint: Treat the lattice as a two dimensional array.)

5. Consider again an electron beam incident upon a thin metal foil but look upon the electrons as particles having a certain kinetic energy. In experiments with aluminium foils (J. Geiger and K. Wittmaack, *Zeitschrift für Physik*, **195,** 44, 1966) it was found that a certain fraction of the electrons passing through the metal had a loss of energy of 14·97 eV. We could explain this loss as being the creation of a particle of that much energy. But what particle? It cannot be a photon (a transverse electromagnetic wave in the wave picture) because an electron in motion sets up no transverse waves. It must be a particle that responds to a longitudinal electric field. So it might be a plasma wave of frequency ω_p which we could call a 'plasmon' in the particle picture. The energy of this particle would be $\hbar\omega_p$.

Calculate the value of $\hbar\omega_p$ for aluminium assuming three free electrons per atom. Compare it with the characteristic energy loss found.

The density of aluminium is 2700 kg/m^3 and its atomic weight is 27.

3. The electron

That's how it is, said Pooh.

A. A. MILNE *When we were six*

3.1. Introduction

WE have seen that some experimental results can be explained if we regard the electron as a particle, whereas the explanation of some other experiments is possible only if we look upon our electron as a wave. Now which is it? Is it a particle or is it a wave? It is neither, it is an electron.

An electron is an electron; this seems a somewhat tautological definition. What does it mean? I want to say by this that we don't have to regard the electron as something else, something we are already familiar with. It helps, of course, to know that the electron sometimes behaves as a particle because we have some intuitive idea of what particles are supposed to do. It is helpful to know that the electron may behave as a wave because we know a lot about waves. But we don't have to look at the electron as something else. It is sufficient to say that an electron is an electron as long as we have some means of predicting its properties.

How can we predict what an electron will do? Well, how can we predict any physical phenomena? We need some mathematical relationship between the variables. Prediction and mathematics are intimately connected in science—or are they? Can we make predictions without any mathematics at all? We can. Seeing, for example, dark heavy clouds gathering in the sky we may say that 'it's going to rain' and on a large number of occasions we'll be right. But this is not really a very profound and accurate prediction. We are unable to specify *how* dark the clouds should be for a certain amount of rain, and we would find it hard to guess the temporal variation of the positions of the clouds. So, as you know very well, meteorology is not yet an exact science.

In physics fairly good predictions are needed because otherwise it is difficult to get further money for research. In engineering the importance of predictions can hardly be overestimated. If the designer of a bridge or of a telephone exchange makes some wrong predictions, this mistake may bring upon him the full legal apparatus of the state or the frequent curses of the subscribers. Thus, for engineers, prediction is not a trifling matter.

Now what about the electron? Can we predict its properties? Yes, we can because we have an equation which describes the behaviour of the electron in mathematical terms. It is called Schrödinger's equation. Now I suppose you would like to know where Schrödinger's equation came from? It came from nowhere; or more correctly it came straight from Schrödinger's head, not unlike Pallas Athene who is reputed to have sprung out of Zeus' head (and in full armour too!). Schrödinger's equation is a product of Schrödinger's imagination; it cannot be derived from any set of physical assumptions. Schrödinger's equation is, of course, not unique in this respect. You have met similar cases before.

In the sixth form you learned Newton's equation. At the time you had just gained your first glances into the hidden mysteries of physics. You would not have dared to question your schoolmaster about the origin of Newton's equation. You were probably more reverent at the time, more willing to accept the word of authorities, and besides, Newton's equation looks so simple that one's credulity is not seriously tested. Force equals mass times acceleration; anyone is prepared to believe that much. And it seems to work in practice.

At the university you are naturally more inquisitive than in your schooldays; so you may have been a bit more reluctant to accept Maxwell's equations when you first met them. It must have been very disturbing to be asked to accept the equation

$$\nabla \times \mathscr{E} = -\frac{\partial \mathbf{B}}{\partial t} \tag{3.1}$$

as the truth and nothing but the truth. But then you were shown that this equation is really identical with the familiar induction law

$$U = -\frac{\partial \phi}{\partial t} \tag{3.2}$$

and the latter merely expresses the result of a simple experiment. Similarly

$$\nabla \times \mathbf{H} = \mathbf{J} \tag{3.3}$$

is only a rewriting of Ampere's law. So all is well again or rather all would be well if there wasn't another term on the right-hand side, the displacement current $\partial D/\partial t$. Now what is this term? Not many lecturers admit that it came into existence as a pure artifice. Maxwell felt there should be one more term there and that was it. True, Maxwell himself made an attempt to justify the introduction of displacement current by referring to the a.c. current in a capacitor, but very probably that was just a concession to the audience he had to communicate with. He must have been more concerned with refuting the theory of instantaneous action at a distance, and with deriving a velocity with which disturbances can travel.

The extra term had no experimental basis whatsoever. It was a brilliant hypothesis which enabled Maxwell to predict the existence of electromagnetic waves. When some years later Hertz managed to find these waves the hypothesis became a law. It was a momentous time in history though most history books keep silent about the event.†

I am telling you all this just to show that an equation which comes from nowhere in particular may represent physical reality. Of course, Schrödinger had good reasons for setting up his equation. He had immediate success in several directions. Whilst Maxwell's displacement current term explained no experimental observation, Schrödinger's equation could immediately account for the atomic spectrum of hydrogen, for the energy levels of the Planck oscillator, for the non-radiation of electronic currents in atoms, and for the shift of energy levels in strong transverse fields. He produced four papers in quick succession and noted at the end with quiet optimism:

'I hope and believe that the above attempt will turn out to be useful for explaining the magnetic properties of atoms and molecules, and also the electric current in the solid state.'

Schrödinger was right.‡ His equation turned out to be useful indeed. He was not *exactly* right, though. In order to explain all the properties of the solid state (including magnetism) two further requisites are needed: Pauli's principle and 'spin'. Fortunately, both of them can be stated in simple terms, so if we make ourselves familiar with Schrödinger's equation the rest is relatively easy.

† According to most historians' definition, an event is important if it affects a large number of people to a considerable extent for a long time. If historians were faithful to this definition they should write a lot about Maxwell and Hertz because by predicting and proving the existence of electromagnetic waves, Maxwell and Hertz had more influence on the life of ordinary people nowadays than any nineteenth-century general, statesman, or philosopher (with the possible exception of Karl Marx).

‡ In the above discussion the role of Schrödinger in setting up modern quantum physics was very much exaggerated. There were a number of others (Heisenberg, Born, Dirac, Pauli to name a few) who made comparable contributions but since this is not a course in the history of science and the Schrödinger formulation is adequate for our purpose we shall not discuss these contributions.

3.2. Schrödinger's equation

After such a lengthy introduction, let us have now the celebrated equation itself. In the usual notation

$$-\frac{\hbar^2}{2m}\nabla^2\Psi+V\Psi = i\hbar\frac{\partial\Psi}{\partial t} \tag{3.4}$$

where m is the mass of the electron and V is the potential in which the electron moves.

We have a partial differential equation in Ψ. But what is Ψ? It is called the wavefunction and

$$|\Psi(x, y, z; t)|^2 \, dx \, dy \, dz \tag{3.5}$$

gives the probability that the electron can be found at time t in the volume element $dx\,dy\,dz$ in the immediate vicinity of the point x, y, z. To show the significance of this function better, $|\Psi|^2$ is plotted in Fig. 3.1 for a hypothetical case where $|\Psi|^2$ is independent of time and varies

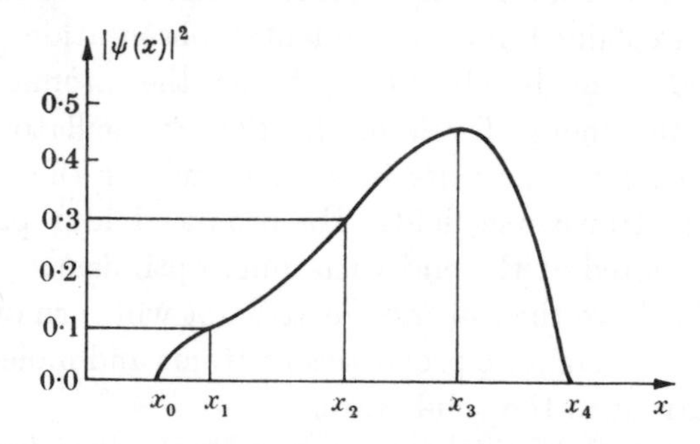

FIG. 3.1. Introducing the concept of the wavefunction. $|\psi(x)|^2\,dx$ is proportional to the probability that the electron may be found in the interval dx at the point x.

only in one dimension. If we make many measurements on this system we shall find that the electron is always between x_0 and x_4 (the probability of being outside this region is zero), that it is most likely to be found in the interval dx around x_3, and it is three times as probable to find the electron at x_2 than at x_1. Since the electron must be somewhere, the probability of finding it between x_0 and x_4 must be unity, that is

$$\int_{x_0}^{x_4} |\Psi(x)|^2 \, dx = 1. \tag{3.6}$$

The above example does not claim to represent any physical situation. It is shown only to illustrate the meaning of $|\Psi|^2$.

The physical content of eqn. (3.4) will be clearer when we shall treat more practical problems but there is one thing we can say immediately. Schrödinger's equation does not tell us the *position* of the electron, only the *probability* that it will be found in the vicinity of a certain point.

The description of the electron's behaviour is statistical but there is nothing particularly new in this. After all, you met statistical descriptions before in gasdynamics for example, and there was considerably less fuss about it.

The main difference is that in classical mechanics we use statistical methods in order to simplify the calculations. We are too lazy to write up 10^{27} differential equations to describe the motion of all the gas molecules in a vessel so we rely instead on a few macroscopic quantities like pressure, temperature, average velocity, etc. We use statistical methods because we elect to do so. It is merely a question of convenience. This is not so in quantum mechanics. The statistical description of the electron is inherent in quantum theory. That is the best we can do. We cannot say much about an electron at a given time. We can only say what happens on the average when we make many observations on one system or we can predict the statistical outcome of simultaneous measurements on identical systems. It may be sufficient to make one single measurement (specific heat or electrical conductivity) when the phenomenon is caused by the collective interaction of a large number of electrons.

We cannot even say how an electron moves as a function of time. We cannot say this because the position and the momentum of an electron cannot be simultaneously determined. The limiting accuracy is given by the uncertainty relationship, eqn. (2.33).

3.3. Solutions of Schrödinger's equation

Let us separate the variables and attempt a solution in the following form

$$\Psi(\mathbf{r}, t) = \psi(\mathbf{r})w(t) \tag{3.7}$$

where $\mathbf{r}$ represents now all the spatial variables. Substituting eqn. (3.7) into eqn. (3.4), and dividing by ψw we get

$$-\frac{\hbar^2}{2m}\frac{\nabla^2\psi}{\psi}+V = \mathrm{i}\hbar\frac{1}{w}\frac{\partial w}{\partial t}. \tag{3.8}$$

Since the left-hand side is a function of $\mathbf{r}$ and the right-hand side is a function of t, they can be equal only if they are both separately equal to a constant which we shall call E, that is we obtain two differential equations as follows

$$\mathrm{i}\hbar\,\frac{\partial w}{\partial t} = Ew \tag{3.9}$$

and

$$-\frac{\hbar^2}{2m}\,\nabla^2\psi + V\psi = E\psi. \tag{3.10}$$

The solution of eqn. (3.9) is simple enough. We can immediately integrate and get

$$w = \exp\left(-\mathrm{i}\,\frac{E}{\hbar}\,t\right) \tag{3.11}$$

and this is nothing else but our good old wave solution (at least as a function of time) if we equate

$$E = \hbar\omega. \tag{3.12}$$

This is actually something we have suggested before (eqn. (2.27)) by recourse to Planck's formula. So we may call E the energy of the electron. However, before making such an important decision let us investigate eqn. (3.10) which also contains E. We could rewrite eqn. (3.10) in the form

$$\left(-\frac{\hbar^2}{2m}\,\nabla^2 + V\right)\psi = E\psi. \tag{3.13}$$

The second term in the bracket is potential energy so we are at least in good company. The first term contains ∇^2, the differential operator you will have met many times in electrodynamics. Writing it symbolically in the form

$$-\frac{\hbar^2}{2m}\,\nabla^2 = \frac{1}{2m}\,(-\mathrm{i}\hbar\,\nabla)^2 \tag{3.14}$$

we can immediately see that by introducing the new notation

$$\mathbf{p} = -\mathrm{i}\hbar\,\nabla \tag{3.15}$$

and calling it the 'momentum operator' we may arrive at an old familiar relationship

$$\frac{\mathbf{p}^2}{2m} = \text{Kinetic energy}. \tag{3.16}$$

Thus on the left-hand side of eqn. (3.13) we have the sum of kinetic and potential energies in operator form and on the right-hand side we have a constant E having the dimensions of energy. Hence we may, with good conscience, interpret E as the total energy of the electron.

You might be a little bewildered by these definitions and interpretations but you must be patient. You cannot expect to unravel the mysteries of quantum mechanics at the first attempt. The fundamental difficulty is that first steps in quantum mechanics are not guided by intuition. You cannot have any intuitive feelings because the laws of quantum mechanics are not directly experienced in everyday life. The most satisfactory way (at least for the few who are mathematically inclined) is to plunge into the full mathematical treatment and leave the physical interpretation to a later stage. Unfortunately this method is lengthy and far too abstract for an engineer. So the best we can do is to digest alternately a little physics and a little mathematics and hope that the two will meet.

3.4. The electron as a wave

Let us look at the simplest case when $V = 0$ and the electron can move only in one dimension. Then eqn. (3.13) (which is often called the time independent Schrödinger equation) reduces to

$$\frac{\hbar^2}{2m}\frac{\partial^2\psi}{\partial z^2}+E\psi = 0. \tag{3.17}$$

The solution of this differential equation is a wave in space. Hence the general solution of Schrödinger's equation for the present problem is

$$\Psi = \exp\left(-\mathrm{i}\frac{E}{\hbar}t\right)\{A\exp(\mathrm{i}kz)+B\exp(-\mathrm{i}kz)\} \tag{3.18}$$

where A and B are constants representing the amplitudes of the forward and backward travelling waves and k is related to E by

$$E = \frac{\hbar^2k^2}{2m}. \tag{3.19}$$

In this example we have chosen the potential energy of the electron as zero, thus eqn. (3.19) must represent the kinetic energy. Hence we may conclude that $\hbar k$ must be equal to the momentum of the electron. We have come to this conclusion before, heuristically, on the basis of the wave picture but now we have the full authority of Schrödinger's equation behind us.

You may notice too that $p = \hbar k$ is an alternative expression of de Broglie's relationship, thus we have obtained from Schrödinger's equation both the wave behaviour and the correct wavelength.

What can we say about the position of the electron? Take $B = 0$ for simplicity, then we have a forward travelling wave with a definite value for k. The probability of finding the electron at any particular point is given by $|\psi(z)|^2$ which according to eqn. (3.18) is unity, independently of z. This means physically that there is an equal probability of the electron being at any point on the z-axis. The electron can be anywhere, that is the uncertainty in the electron's position is infinite. This is only to be expected. If the value of k is given then the momentum is known, so the uncertainty in the momentum of the electron is zero; hence the uncertainty in position must be infinitely great.

3.5. The electron as a particle

Eqn. (3.17) is a linear differential equation, hence the sum of the solutions is still a solution. We are therefore permitted to add up as many waves as we like, that is a wave packet (as constructed in Chapter 2) is also a solution of Schrödinger's equation.

We can now be a little more rigorous than before. A wave packet represents an electron because $|\psi(z)|^2$ is appreciably different from zero only within the packet. With the choice $a(k) = 1$ in the interval Δk, it follows from eqns. (2.16) and (2.17) that the probability of finding the electron is given by†

$$|\psi(z)|^2 = K\left(\frac{\sin\dfrac{\Delta kz}{2}}{\dfrac{\Delta kz}{2}}\right)^2. \tag{3.20}$$

3.6. The electron meeting a potential barrier

Consider again a problem where the motion of the electron is constrained in one dimension and the potential energy is assumed to take the form shown in Fig. 3.2.

You are familiar with the classical problem where the electron starts somewhere on the negative z-axis (say at $-z_0$) in the positive direction with a definite velocity. The solution may be obtained purely from energetic considerations. If the kinetic energy of the electron E is smaller than V_2, the electron is turned back by the potential barrier

† The constant K may be determined by the normalization condition

$$\int_{-\infty}^{\infty} |\psi(z)|^2 = 1.$$

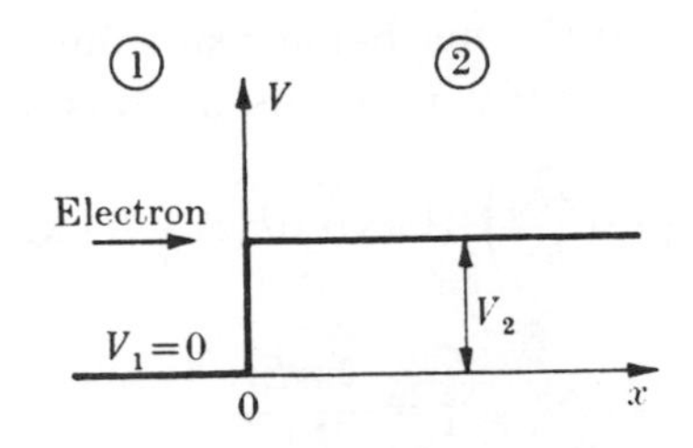

FIG. 3.2. An electron incident upon a potential barrier.

at $x = 0$. If $E > V_2$ the electron slows down but carries on regardless.

How should we formulate the equivalent quantum mechanical problem? We should represent our electron by a wave packet centred in space on $-z_0$ and should describe its momentum with an uncertainty Δp. We should use the wave function obtained as initial condition at $t = 0$ and should solve the time dependent Schrödinger equation. This would be a very illuminating exercise, alas much too difficult mathematically.

We have to be satisfied by solving a related problem. We shall give our electron a definite energy, that is, a definite momentum, and we shall put up with the concomitant uncertainty in position. We shall not be able to say anything about the electron's progress towards the potential barrier but we shall have a statistical solution which will give the probability of finding the electron on either side of this potential barrier.

Specifying the momentum and not caring about the position of the electron is not so unphysical as you might think. The conditions stated may be approximated in practice by shooting a sufficiently sparse† electron beam towards the potential barrier with a well defined velocity. We are not concerned then with the positions of individual electrons only with their spatial distribution on the *average* which we call the macroscopic charge density. Hence we may identify $e\,|\psi|^2$ with the charge density.

This is not true in general. What is always true is that $|\psi|^2$ gives the *probability* of an electron being found at z. Be careful, $|\psi(z)|^2$ does *not* give the fraction of the electron's charge residing at z. The charge of the electron is *not* smoothed out. *When the electron is found, the whole electron is there.* If, however, a large number of electrons behave identically then $|\psi|^2$ may be justifiably regarded to be proportional to the charge density.

† So that the interaction of the electrons can be neglected.

Let us proceed now to the mathematical solution. In region 1 where $V_1 = 0$ the solution is already available in eqns. (3.18) and (3.19)

$$\Psi_1 = \exp\left(-\mathrm{i}\,\frac{E}{\hbar}\,t\right)\{A \exp(\mathrm{i}k_1 z)+B\exp(-\mathrm{i}k_1 z)\} \tag{3.21}$$

and

$$k_1^2 = \frac{2mE}{\hbar^2}. \tag{3.22}$$

In region 2 the equation to be solved is as follows (the time dependent part of the solution remains the same because E is specified)

$$\frac{\hbar^2}{2m}\frac{\partial^2\psi}{\partial z^2}+(E-V_2)\psi = 0 \tag{3.23}$$

with the general solution

$$\Psi_2 = \exp\left(-\mathrm{i}\,\frac{E}{\hbar}\,t\right)\{C\exp(\mathrm{i}k_2 z)+D\exp(-\mathrm{i}k_2 z)\}, \tag{3.24}$$

where

$$k_2^2 = \frac{2m}{\hbar^2}\,(E-V_2). \tag{3.25}$$

Now we shall ask the question, depending on the relative magnitudes of E and V_2, what is the probability that the electron can be found in regions 1 and 2 respectively.

It is actually easier to speak about this problem in wave language because then the form of the solution is automatically suggested. Whenever a wave is incident on some sort of discontinuity, there is a reflected wave and there is a transmitted wave. Since no wave is incident from region 2 we can immediately decide that D must be zero.

In order to determine the remaining constants we have to match the two solutions at $z = 0$, requiring that both ψ and $\partial\psi/\partial z$ should be continuous. From eqns. (3.21) and (3.24) the above conditions lead to the algebraic equations

$$A+B = C \tag{3.26}$$

and

$$\mathrm{i}k_1(A-B) = \mathrm{i}k_2 C \tag{3.27}$$

whence

$$\frac{B}{A} = \frac{k_1-k_2}{k_1+k_2}, \qquad \frac{C}{A} = \frac{2k_1}{k_1+k_2}. \tag{3.28}$$

Let us distinguish now two cases: (i) $E > V_2$. In this case $k_2^2 > 0$, k_2 is real which means an oscillatory solution in region 2. The values of k_2

and k_1 are however, different. Thus B/A is finite, that is there is a finite amount of reflection. In contrast to the classical solution there is some probability that the electron is turned back by the potential discontinuity. (ii) $E < V_2$. In this case $k_2^2 < 0$, k_2 is imaginary, that is it declines exponentially in region 2.† Since $|C/A| > 0$, there is a finite (though declining) probability of finding the electron at $z > 0$. Classically an electron has no chance of getting inside region 2. Under the laws of quantum mechanics the electron may penetrate the potential barrier.

A third case of interest is when the potential profile is as shown in Fig. 3.3, and $E < V_2$. Then k_2 is imaginary and k_3 is real. Hence one

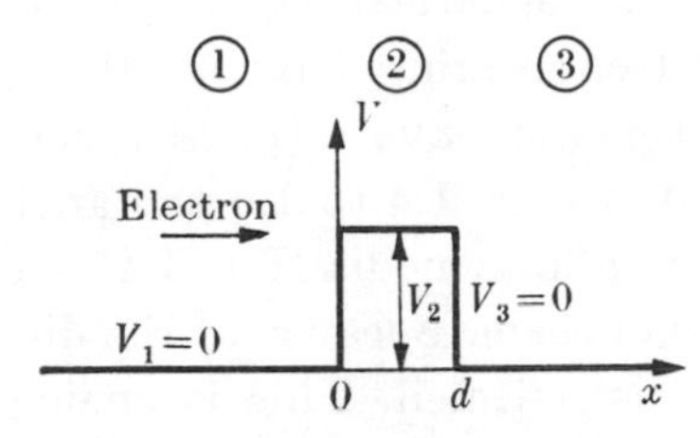

FIG. 3.3. An electron incident upon a narrow potential barrier.

may expect that $|\psi|^2$ declines in region 2 and is constant in region 3. The interesting thing is that $|\psi|^2$ in region 3 is not zero. Thus there is a finite probability that the electron crosses the potential barrier and appears at the other side with energy unchanged. Since there is an exponential decline in region 2, it is necessary that that region should be narrow, to obtain any appreciable probability in region 3. If we are thinking in terms of the incident electron beam, we may say that a certain fraction of the electrons will get across the potential barrier. This is called the tunnel effect or simply tunnelling.

As you will see later it is an important effect which we shall often invoke to explain phenomena as different as the bonding of the hydrogen molecule or the operation of the tunnel diode.‡

3.7. Two analogies

Without the help of Schrödinger's equation we could not have guessed how electrons behaved when meeting a potential barrier. But having found the solutions in the form of propagating and exponentially

† The exponentially increasing solution cannot be present for physical reasons.

‡ A more mundane example of tunnelling occurs every time we switch on an electric light. The contacts are always covered with an oxide film, that in bulk would be an insulator. But it is rubbed down to a few molecules thickness by the mechanical action of the switch, and the tunnelling is so efficient that we do not notice it.

decaying waves, a physical picture, I hope, is emerging. There is always a physical picture if you are willing to think in terms of waves. Then it is quite natural that discontinuities cause reflections and that the waves might be attenuated.

The concepts are not appreciably more difficult than those needed to describe the motion of classical electrons, but you need time to make yourself familiar with them. 'Familiarity breeds contempt' may very well apply to arts subjects but in most branches of science the saying should be reformulated as 'lack of familiarity breeds bewilderment'.

Assuming that you have already developed some familiarity with waves it may help to stress the analogy further. If we went a little more deeply into the mathematical relationships we would find that the problem of an electron meeting a potential barrier is entirely analogous to an electromagnetic wave meeting a new medium. Recalling the situations depicted in Figs. 1.4 to 1.6 the analogies are as follows.

(i) There are two semi-infinite media (Fig. 1.4); electromagnetic waves propagate in both of them. Because of the discontinuity, a certain part of the wave is reflected. This is analogous to the electron meeting a potential barrier (Fig. 3.2) with an energy $E > V_2$. Some electrons are reflected because of the presence of a discontinuity in potential energy.

(ii) There are two semi-infinite media (Fig. 1.5); electromagnetic waves may propagate in the first one but not in the second one. The field intensities are, however, finite in medium 2 because the electromagnetic wave penetrates to a certain extent. This is analogous to the electron meeting a potential barrier (Fig. 3.2) with an energy $E < V_2$. In spite of not having sufficient energy some electrons may penetrate into region 2.

(iii) There are two semi-infinite media separated from each other by a third medium (Fig. 1.6); electromagnetic waves may propagate in media 1 and 3 but not in the middle one. The wave incident from medium 1 declines in medium 2 but a finite amount arrives and can propagate in medium 3. This is analogous to the electron meeting a potential profile shown in Fig. 3.3, with an energy $E < V_2$. In spite of not having sufficient energy some electrons may cross region 2 and may appear and continue their journey in region 3.

Instead of taking plane waves propagating in infinite media one might make the analogy physically more realizable (though mathematically less perfect) by employing waveguides. Discontinuities can then be represented by joining two waveguides of different cross-sections, and the exponentially decaying wave may be obtained by using a cut-off waveguide (dimension smaller than half free-space wavelength). Then all the above phenomena can be easily demonstrated in the laboratory.

3.8. The electron in a potential well

In our previous examples the electron was free to roam in the one-dimensional space. Now we shall make an attempt to trap it by presenting it with a region of low potential energy which is commonly called a potential well. The potential profile assumed is shown in Fig. 3.4.

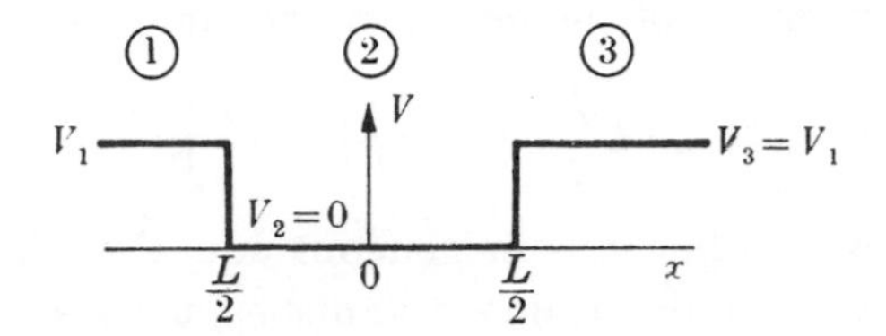

FIG. 3.4. An electron in a potential well.

If $E > V_1$ the solutions are very similar to those discussed before but when $E < V_1$ a new situation arises.

We have by now sufficient experience in solving Schrödinger's equation for a constant potential so we shall write down the solutions without further discussions.

In region 3 there is only an exponentially decaying solution

$$\psi_3 = C\exp(-\gamma z) \tag{3.29}$$

where

$$\gamma^2 = \frac{2m}{\hbar^2}(V_1 - E). \tag{3.30}$$

In region 2 the potential is zero. The solution is either symmetric or antisymmetric.† Accordingly

$$\psi_{2s} = A\cos kz \tag{3.31}$$

$$\psi_{2a} = A\sin kz \tag{3.32}$$

where

$$k^2 = \frac{2m}{\hbar^2}E. \tag{3.33}$$

In region 1 the solution must decay again, this time towards negative infinity. If we wish to satisfy the symmetry requirement as well, the

† This is something we have not proved. It is true (the proof can be obtained fairly easily from Schrödinger's equation) in general that if the potential function is symmetric the solution must be either symmetric or antisymmetric.

wave function must look like

$$\psi_1 = \pm C \exp \gamma z. \tag{3.34}$$

Let us investigate the symmetric solution first. We have then eqns. (3.29) and (3.31), and eqn. (3.34) with the positive sign. The conditions to be satisfied are the continuity of ψ and $\partial\psi/\partial z$ at $L/2$ and $-L/2$ but owing to the symmetry it is sufficient to do the matching at (say) $L/2$. From the continuity of the wave function

$$A \cos k\frac{L}{2} - C \exp\left(-\gamma\frac{L}{2}\right) = 0. \tag{3.35}$$

From the continuity of the derivative of the wave function

$$Ak \sin k\frac{L}{2} - C\gamma \exp\left(-\gamma\frac{L}{2}\right) = 0. \tag{3.36}$$

We have now two linear homogeneous equations in A and C which are soluble only if the determinant vanishes, that is

$$\begin{vmatrix} \cos k\dfrac{L}{2} & -\exp\left(-\gamma\dfrac{L}{2}\right) \\ k \sin k\dfrac{L}{2} & -\gamma \exp\left(-\gamma\dfrac{L}{2}\right) \end{vmatrix} = 0 \tag{3.37}$$

leading to

$$k \tan k\frac{L}{2} = \gamma \tag{3.38}$$

Thus k and γ are related by eqn. (3.38). Substituting their values from eqns. (3.30) and (3.33) respectively we get

$$E^{\frac{1}{2}} \tan\left(\frac{2m}{\hbar^2} E \frac{L^2}{4}\right)^{\frac{1}{2}} = (V_1 - E)^{\frac{1}{2}} \tag{3.39}$$

which is an algebraic equation to be solved for E. Nowadays one feeds this sort of equation into a computer and has the results printed in a few seconds. But let us be old-fashioned and solve the equation graphically by plotting the left-hand side and the right-hand side separately. Putting in the numerical values we know

$$m = 9{\cdot}1 \times 10^{-31}\ \text{kg}, \qquad \hbar = 1{\cdot}05 \times 10^{-34}\ \text{J s}$$

and we shall take

$$\frac{L}{2} = 5 \times 10^{-10}\ \text{m}, \qquad V_1 = 1{\cdot}6 \times 10^{-18}\ \text{J}.$$

As may be seen in Fig. 3.5 the curves intersect each other in three points; so there are three solutions and that is the lot. If $E > V_1$ the

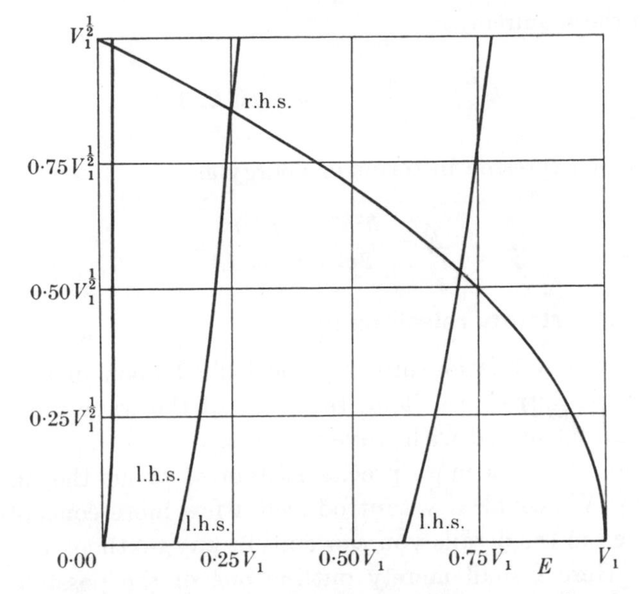

FIG. 3.5. A plot of the two sides of eqn. (3.39) against E for $L/2 = 5\times10^{-10}$ m and $V_1 = 1{\cdot}6\times10^{-18}$ J.

electron can have any energy it likes, but if $E < V_1$ there are only three possible energy levels. To be correct there are three energy levels for the symmetric solution and a few more for the antisymmetric solution.

We have at last arrived at the solution of the first quantum-mechanical problem which deserves literally the name quantum mechanical. Energy is no longer continuous, it cannot take arbitrary values. Only certain discrete energy levels are permitted. In the usual jargon of quantum mechanics it is said: energy is quantized.

We may generalize further from the above example. The discrete energy levels obtained are not a coincidence. It is true in general that whenever we try to confine the electron the solution consists of a discrete set of wave functions and energy levels.

3.9. The potential well with a rigid wall

We call the potential wall rigid when the electron cannot penetrate even quantum-mechanically. This happens when $V_1 = \infty$. We shall briefly investigate this case because we shall need the solution later.

Eqns (3.31) and (3.32) are still valid in the zero potential region but now (since the electron cannot penetrate the potential barrier) the continuity condition is

$$\psi\left(\pm\frac{L}{2}\right) = 0 \tag{3.40}$$

leading to the requirement

$$k\frac{L}{2} = \frac{n\pi}{2}, \qquad n = 1, 2, 3\ldots \tag{3.41}$$

which may be expressed in terms of energy as

$$E = \frac{\hbar^2 k^2}{2m} = \frac{h^2}{8m}\frac{n^2}{L^2}. \tag{3.42}$$

3.10. The uncertainty relationship

The uncertainty relationship may be looked upon in a number of ways. We have introduced it on the basis of the wave picture where electrons were identified with wave packets.

Can we make now a more precise statement about the uncertainty relationship? We could, if we introduced a few more concepts. If you are interested in the details you can consult any textbook on quantum mechanics. Here I shall merely outline one of the possible ways of deriving the uncertainty relationship.

First of all, the average value of a physically measurable quantity (called an observable) is defined in quantum mechanics as

$$\langle A \rangle = \frac{\int_V \psi^* A \psi \, \mathrm{d(volume)}}{\int_V |\psi|^2 \, \mathrm{d(volume)}} \tag{3.43}$$

where the integration is over the volume of interest (wherever ψ is defined). A is in general an operator; it is $-i\hbar\nabla$ for the momentum and simply $\mathbf{r}$, the radius vector, for the position. Assuming that Schrödinger's equation is solved for a particular case, we know the wave function ψ and hence with the aid of eqn. (3.43) we can work out the average and r.m.s. values of both the electron's position and of its momentum. Identifying Δz and Δp with

$$\{\langle (z - \langle z \rangle)^2 \rangle\}^{\frac{1}{2}} \quad \text{and} \quad \{\langle (p - \langle p \rangle)^2 \rangle\}^{\frac{1}{2}}$$

respectively, we get

$$\Delta z \, \Delta p \geqslant h. \tag{3.44}$$

There is actually another often-used form of the uncertainty relationship

$$\Delta E \, \Delta t \geqslant h \tag{3.45}$$

which may be derived from relativistic quantum theory (where time is on equal footing with the spatial coordinates) and interpreted in the following way. Assume that an electron sits in a higher energy state of a system (e.g. in a potential well). It may fall to the lowest energy state by emitting a photon of energy $\hbar\omega$. So if we know the energy of the lowest state, we could work out the energy of that particular higher state by measuring the frequency of the emitted photon. But if the electron spends only a time Δt in the higher state then the energy of the state can be determined with an accuracy not greater than $\Delta E = h/\Delta t$. This is borne out by the measurements. The emitted radiation is not monochromatic; it covers a finite range of frequencies.

3.11. Philosophical implications

The advent of quantum mechanics brought problems to the physicist which previously belonged to the sacred domain of philosophy. The engineer can still afford to ignore the philosophical implications but by a narrow margin only. In another decade philosophical considerations might be relevant in the discussion of devices, so I'll try to give you a foretaste of the things which might come.

To illustrate the sort of questions philosophers are asking, take the following one: We see a tree in the quad, so the tree must be there. We have the evidence of our senses (the eye in this particular case) that the tree exists. But what happens when we don't look at the tree, when no one looks at the tree at all; does the tree still exist? It's a good question.

Had philosophers been content *asking* this and similar questions, the history of philosophy would be an easier subject to study. Unfortunately, driven by usual human passions (curiosity, vivid imagination, vanity, ambition, the desire to be cleverer than the next man, craving for fame, etc.) philosophers did try to answer the questions. To the modern scientist most of their answers and debates don't seem to be terribly edifying. I just want to mention Berkeley who maintained that matter would cease to exist if unobserved, but luckily there is God who perceives everything, so matter may exist after all. This view was attacked by Ronald Knox in the following limerick:

> There was a young man who said, 'God
> Must think it exceedingly odd
>   If he finds that this tree
>   Continues to be
> When there's no one about in the Quad.'

Berkeley replied in kind:

> Dear Sir:
> Your astonishment's odd:
> *I* am always about in the Quad.
> And that's why the tree
> Will continue to be,
> Since observed by
> Yours faithfully,
> GOD.

You do, I hope, realize that only a minority of philosophical arguments were ever conducted in the form of limericks, and the above examples are not typical. I mention them partly for entertainment and partly to emphasize the problem of the tree in the quad a little more.

In the light of quantum mechanics we should look at the problem from a slightly different angle: The question is not so much what happens while the tree is unobserved but rather what happens while the tree is unobservable. The tree can leave the quad because for a brief enough time it can have a high enough energy at its disposal, and no experimenter has any means of knowing about it. We are prevented by the uncertainty relationship ($\Delta E \Delta t \simeq h$) from ever learning whether the tree did leave the quad or not.

You may say that this is against common sense. It is but the essential point is whether or not it violates the Laws of Nature (as we know them today). Apparently it doesn't. You may maintain that for that critical Δt interval the tree stays where it always has stood. Yes, it's a possible view. You may also maintain that the tree went over for a friendly visit to the quad of another college and came back. Yes, that's another possible view.

Is there any advantage in imagining that the tree did make that brief excursion? I can't see any, so I would opt for regarding the tree as being in the quad at all times.

But the problem remains, and becomes of more practical interest when considering particles of small size. A free electron travelling with a velocity 10^6 m s^{-1} has an energy of 2·84 eV. Assume that it wants to 'borrow' the same amount of energy again. It may borrow that much energy for an interval $\Delta t = h/2{\cdot}84\ \text{eV} \simeq 1{\cdot}5 \times 10^{-15}$ s. Now, you may ask what can an electron do in a time interval as short as $1{\cdot}5 \times 10^{-15}$ s? Quite a lot; it can get comfortably from one atom to the next. And remember all this on borrowed energy. So if there was a barrier of (say) five electron volts, our electron could easily move *over* it, and having scaled the barrier it could return the borrowed energy, and no one would be able to find out how the electron made the journey. Thus from a purely philosophical argument we could make up an alternative picture of tunnelling. We may say that tunnelling across potential barriers comes about because the electron can borrow energy for a limited time. Is this a correct description of what happens? I don't know but it is a *possible* description. Is it useful? I suppose it is always useful to have various ways of describing the same event; that always improves understanding. But the crucial test is whether this way of thinking will help in arriving at new conclusions which can be experimentally tested. For an engineer the criterion is even clearer; if an engineer can think up a new device using these sorts of arguments (e.g. violating energy conservation for a limited time) and the device works (or even better it can be sold for ready money) then the method is vindicated. The end justifies the means, as Machiavelli said.

Theoretical physicists, I believe, do use these methods. In a purely particle description of Nature, for example, the Coulomb force between two electrons is attributed to the following cause. One of the electrons borrows some energy to create a photon which goes dutifully to the other electron where it is absorbed returning thereby the energy borrowed. The farther the two electrons are from each other, the less energy can be borrowed, and therefore lower frequency photons are emitted and absorbed. Carrying on these arguments (if you are interested about more details ask a theoretical physicist) they do manage to get correctly the forces between electrons. So there are already some people who find it useful to play around with these concepts.

This is about as much as I want to say about the philosophical role of the electron. There are, incidentally, a number of other points where philosophy and quantum mechanics meet (e.g. the assertion of quantum mechanics that no event can be predicted with certainty, merely with a certain probability) but I think we may have already gone beyond that what is absolutely necessary for the education of an engineer.

Examples

1. An electron confined by a rigid one-dimensional potential well (Fig. 3.4 with $V_1 = \infty$) may be anywhere within the interval $2a$. So the uncertainty in its position is $\Delta x = 2a$. There must be a corresponding uncertainty in the momentum of the electron and hence it must have a certain kinetic energy. Calculate this energy from the uncertainty relationship and compare it with the value obtained from eqn. (3.42) for the ground state.

2. The wavefunction for a rigid potential well is given by eqns. (3.31) and (3.32) and the permissible values of k by eqn. (3.42). Calculate the average values of

$$z, \qquad (z-\langle z\rangle)^2, \qquad p, \qquad (p-\langle p\rangle)^2$$

(Hint: Use eqn. (3.43). The momentum operator in this one-dimensional case is $-i\hbar\,\partial/\partial z$.)

3. The classical equivalent of the potential well is a particle bouncing between two perfectly elastic walls with uniform velocity.

(i) Calculate the classical average values of the quantities enumerated in the previous example.

(ii) Show that for high-enough energies the quantum mechanical solution tends to the classical solution.

4. In electromagnetic theory the conservation of charge is represented by the continuity equation

$$\nabla.\mathbf{J} = -e\frac{\partial N}{\partial t}$$

where $\mathbf{J}$ = current density and N = density of electrons.

Assume that $\Psi(x, t)$ is a solution of Schrödinger's equation in a one-dimensional problem. Show that by defining the current density as

$$J(x) = -\frac{i\hbar e}{2m}\left[\Psi^*\frac{\partial\Psi}{\partial x} - \Psi\frac{\partial\Psi^*}{\partial x}\right]$$

the continuity equation is satisfied.

5. The time-independent Schrödinger equation for the one-dimensional potential shown in Fig. 3.2 is solved in Section 3.6. Using the definition of current density given in the example above, derive expressions for the reflected and transmitted currents. Show that the transmitted current is finite when $E > V_2$ and zero when $E < V_2$. Comment on the analogy with example 1.6.

6. Solve the time-independent Schrödinger equation for the one-dimensional potential shown in Fig. 3.3;

$$V(x) = 0, \qquad x < 0,$$

$$V(x) = V_2, \qquad 0 < x < d,$$

$$V(x) = 0, \qquad x > d.$$

Assume that an electron beam is incident from the $x < 0$ region with an energy E. Derive expressions for the reflected and transmitted current. Calculate the transmitted current when $V_2 = 2{\cdot}5$ eV, $E = 0{\cdot}5$ eV, $d = 2$ Å and $d = 20$ Å.

7. Solve the time-independent Schrödinger equation for the one-dimensional potential well shown in Fig. 3.6, restricting the analysis to even functions of ψ only. The solution may be expressed in determinant form. Without expanding the determinant explain what its roots represent.

8. Show that the differential equation for the electric field of a plane electro-

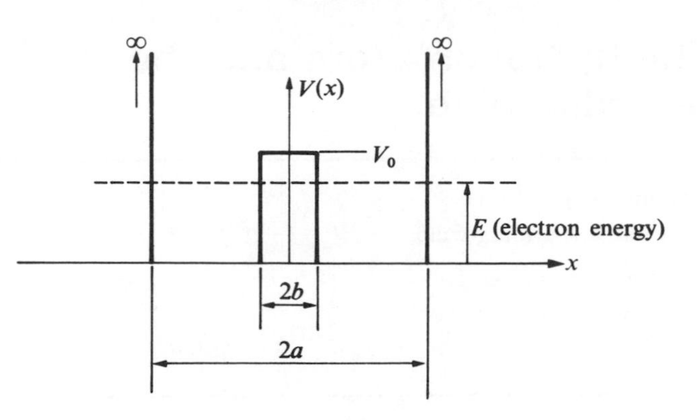

FIG. 3.6. A one-dimensional potential well.

magnetic wave, assuming $\exp(-i\omega t)$ time dependence has the same form as the time-independent Schrödinger equation for constant potential. Show further that the expression for the Poynting vector of the electromagnetic wave is of the same functional form as that for the quantum mechanical current.

9. The potential energy of a classical harmonic oscillator is given as

$$V(x) = \tfrac{1}{2}m\omega_0^2 x^2.$$

We get the 'quantum' harmonic oscillator by putting the above potential function into Schrödinger's equation.

The solutions for the four lowest states are as follows

$$\psi(\zeta) = H_n(\zeta)\exp(-\tfrac{1}{2}\zeta^2),$$

where

$$\zeta = \alpha x, \qquad \alpha^2 = \frac{m\omega_0}{\hbar},$$

$$H_0 = 1, \qquad H_1 = 2\zeta, \qquad H_2 = 4\zeta^2-2, \qquad \text{and} \qquad H_3 = 8\zeta^3-12\zeta.$$

Find the corresponding energies. Compare them with the energies Planck postulated for photons.

4. The hydrogen atom and the periodic table

I see the atoms, free and fine,
That bubble like a sparkling wine;
I hear the songs electrons sing,
Jumping from ring to outer ring;

LISTER *The Physicist*

4.1. The hydrogen atom

UP to now we have been concerned with rather artificial problems. We said: let's assume that the potential energy of our electron varies as a function of distance this way or that way without specifying the actual physical mechanism responsible for it. It wasn't a waste of time. It gave an opportunity of becoming acquainted with Schrödinger's equation, and of developing the first traces of a physical picture based, perhaps paradoxically, on the mathematical solution.

It would, however, be nice to try our newly acquired technique on a more physical situation where the potential is caused by the presence of some other physical 'object'. The simplest 'object' would be a proton, which, as we know, becomes a hydrogen atom if joined by an electron.

We are going to ask the following questions: (i) What is the probability that the electron is found at a distance r from the proton? (ii) What are the allowed energy levels?

The answers are again provided by Schrödinger's equation. All we have to do is to put in the potential energy due to the presence of a proton and solve the equation.

The wave function is a function of time, and one might want to solve problems where the conditions are given at $t = 0$ and one is interested in the temporal variation of the system. These problems are complicated and of little general interest. What we should like to know is how a hydrogen atom behaves on the average and for that purpose the solution given in eqn. (3.7) is adequate. We may then forget about the temporal variation, because

$$|w(t)|^2 = 1, \qquad (4.1)$$

and solve eqn. (3.13), the time-independent Schrödinger equation.

The proton, we know, is much heavier than the electron; so let us regard it as infinitely heavy (that is immobile) and place it at the origin of our coordinate system.

The potential energy of the electron at a distance r from the proton is known from electrostatics:

$$V(r) = -\frac{e^2}{4\pi\epsilon_0 r}. \tag{4.2}$$

Thus the differential equation to be solved is

$$\frac{\hbar^2}{2m}\nabla^2\psi + \left(\frac{e^2}{4\pi\epsilon_0 r} + E\right)\psi = 0. \tag{4.3}$$

It would be hard to imagine a physical configuration much simpler than that of a proton and an electron, and yet it is difficult to solve the corresponding differential equation. It is difficult because the $1/r$ term does not lend itself readily to analytical solutions. Thanks to the arduous efforts of nineteenth-century mathematicians the general solution is known, but it would probably mean very little to you. Unless you have a certain familiarity with the properties of associated Legendre functions, it won't make you much happier if you learn that associated Legendre functions happen to be involved. So I won't quote the general solution because that would be meaningless, nor shall I derive it because that would be boring. But just to give an idea of the mathematical operations needed I shall show the derivation for the simplest possible case when the solution is spherically symmetric, and even then only for the lowest energy.

The potential energy of the electron depends only on the distance r; it therefore seems advantageous to solve eqn. (4.3) in the spherical coordinates r, θ, ϕ (Fig. 4.1). If we restrict our attention to the spherically symmetrical case when ψ depends neither on ϕ nor on θ but only on r, then we can transform eqn. (4.3) without too much trouble. We shall need the following partial derivatives

$$\frac{\partial\psi}{\partial x} = \frac{\partial r}{\partial x}\frac{\partial\psi}{\partial r} \tag{4.4}$$

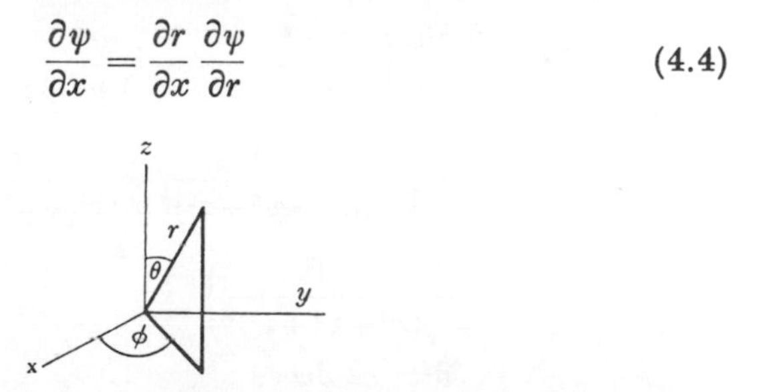

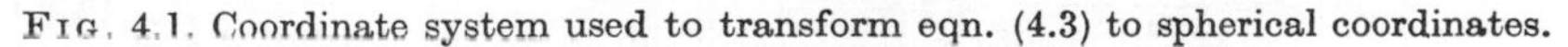
Fig. 4.1. Coordinate system used to transform eqn. (4.3) to spherical coordinates.

and

$$\frac{\partial^2\psi}{\partial x^2} = \frac{\partial}{\partial x}\left(\frac{\partial\psi}{\partial r}\frac{\partial r}{\partial x}\right). \tag{4.5}$$

When we differentiate eqn. (4.5) we have to remember that $\partial\psi/\partial r$ is a function of r and $\partial r/\partial x$ is still a function of x; therefore

$$\begin{aligned}\frac{\partial}{\partial x}\left(\frac{\partial\psi}{\partial r}\frac{\partial r}{\partial x}\right) &= \frac{\partial r}{\partial x}\frac{\partial}{\partial r}\left(\frac{\partial\psi}{\partial r}\right)\frac{\partial r}{\partial x}+\frac{\partial\psi}{\partial r}\frac{\partial^2 r}{\partial x^2}\\ &= \frac{\partial^2\psi}{\partial r^2}\left(\frac{\partial r}{\partial x}\right)^2+\frac{\partial\psi}{\partial r}\frac{\partial^2 r}{\partial x^2}.\end{aligned} \tag{4.6}$$

Obtaining the derivatives in respect with y and z in an analogous manner, we get finally

$$\begin{aligned}\nabla^2\psi &= \frac{\partial^2\psi}{\partial x^2}+\frac{\partial^2\psi}{\partial y^2}+\frac{\partial^2\psi}{\partial z^2}\\ &= \frac{\partial^2\psi}{\partial r^2}\left\{\left(\frac{\partial r}{\partial x}\right)^2+\left(\frac{\partial r}{\partial y}\right)^2+\left(\frac{\partial r}{\partial z}\right)^2\right\}+\\ &\quad+\frac{\partial\psi}{\partial r}\left(\frac{\partial^2 r}{\partial x^2}+\frac{\partial^2 r}{\partial y^2}+\frac{\partial^2 r}{\partial z^2}\right).\end{aligned} \tag{4.7}$$

We now have to work out the partial derivatives of r. Since

$$r = (x^2+y^2+z^2)^{\frac{1}{2}}, \tag{4.8}$$

we get

$$\frac{\partial r}{\partial x} = \frac{x}{(x^2+y^2+z^2)^{\frac{1}{2}}} \tag{4.9}$$

and

$$\frac{\partial^2 r}{\partial x^2} = \frac{1}{(x^2+y^2+z^2)^{\frac{1}{2}}}-\frac{x^2}{(x^2+y^2+z^2)^{\frac{3}{2}}}, \tag{4.10}$$

and similar results for the derivatives by y and z. Substituting all of them in eqn. (4.7) we get

$$\begin{aligned}\nabla^2\psi &= \frac{\partial^2\psi}{\partial r^2}\left(\frac{x^2}{x^2+y^2+z^2}+\frac{y^2}{x^2+y^2+z^2}+\frac{z^2}{x^2+y^2+z^2}\right)+\\ &\quad+\frac{\partial\psi}{\partial r}\left\{\frac{3}{(x^2+y^2+z^2)^{\frac{1}{2}}}-\frac{x^2}{(x^2+y^2+z^2)^{\frac{3}{2}}}-\frac{y^2}{(x^2+y^2+z^2)^{\frac{3}{2}}}-\right.\\ &\quad\left.-\frac{z^2}{(x^2+y^2+z^2)^{\frac{3}{2}}}\right\}\\ &= \frac{\partial^2\psi}{\partial r^2}+\frac{2}{r}\frac{\partial\psi}{\partial r}.\end{aligned} \tag{4.11}$$

Thus for the spherically symmetrical case of the hydrogen atom the Schrödinger equation takes the form

$$\frac{\hbar^2}{2m}\left(\frac{\partial^2\psi}{\partial r^2}+\frac{2}{r}\frac{\partial\psi}{\partial r}\right)+\left(E+\frac{e^2}{4\pi\epsilon_0 r}\right)\psi = 0. \tag{4.12}$$

It may be seen by inspection that a solution of this differential equation is

$$\psi = \mathrm{e}^{-c_0 r}. \tag{4.13}$$

The constant c_0 can be determined by substituting eqn. (4.13) in eqn. (4.12)

$$\frac{\hbar^2}{2m}\left\{c_0^2\mathrm{e}^{-c_0 r}+\frac{2}{r}(-c_0\mathrm{e}^{-c_0 r})\right\}+\left(E+\frac{e^2}{4\pi\epsilon_0 r}\right)\mathrm{e}^{-c_0 r} = 0. \tag{4.14}$$

The above equation must be valid for every value of r, that is the coefficients of $\exp(-c_0 r)$ and that of $\exp(-c_0 r)/r$ must vanish. This condition is satisfied if

$$E = -\frac{\hbar^2 c_0^2}{2m} \tag{4.15}$$

and

$$\frac{\hbar^2 c_0}{m} = \frac{e^2}{4\pi\epsilon_0}. \tag{4.16}$$

From eqn. (4.16)

$$c_0 = \frac{e^2 m}{4\pi\hbar^2\epsilon_0}, \tag{4.17}$$

which substituted in eqn. (4.15) gives

$$E = -\frac{me^4}{8\epsilon_0^2 h^2}. \tag{4.18}$$

Thus the wave function assumed in eqn. (4.13) is a solution of the differential equation (4.12), provided that c_0 takes the value prescribed by eqn. (4.17). Having obtained the value of c_0 the energy is determined as well. It can take only one single value satisfying eqn. (4.18).

The negative sign of the energy means only that the energy of this state is below our chosen zero point. (By writing the Coulomb potential in the form of eqn. (4.2) we tacitly took the energy as zero when the electron is at infinity.)

Let us work out now the energy obtained above numerically. Putting in the constants we get

$$E = -\frac{(9{\cdot}1\times10^{-31})(1{\cdot}6\times10^{-19})^4}{8(8{\cdot}85\times10^{-12})^2(6{\cdot}62\times10^{-34})^2}\frac{\text{kg C}^4}{\text{F}^2\,\text{m}^{-2}\,\text{J}^2\,\text{s}^2}$$

$$= -2{\cdot}18\times10^{-18}\,\text{J}. \tag{4.19}$$

Expressed in joules this number is rather small. Since in most of the subsequent investigations this is the order of energy we shall be concerned with, and since there is a strong human temptation to use numbers only between 0·01 and 100, we abandon with regret the MKS unit of energy and use instead the electron volt, which is the energy of an electron when accelerated to 1 volt. Since

$$1\,\text{eV} = 1{\cdot}6\times10^{-19}\,\text{J} \tag{4.20}$$

the above energy in the new unit comes to the more reasonable-looking numerical value

$$E = 13{\cdot}6\,\text{eV}. \tag{4.21}$$

From experimental studies of the spectrum of hydrogen it was known well before the discovery of quantum mechanics that the lowest energy level of hydrogen must be $-13{\cdot}6$ eV, and it was a great success of Schrödinger's theory that the same figure could be deduced from a respectable-looking differential equation.

What can we say about the electron's position? As we have discussed many times before, the probability that an electron can be found in an elementary volume (at the point r, θ, ϕ) is proportional to $|\psi|^2$, that is in the present case to $\exp(-2c_0r)$. The highest probability is at the origin and it decreases exponentially to zero as r tends to infinity. We could, however, ask a slightly different question: what is the probability that the electron can be found in the spherical shell between r and $r+\mathrm{d}r$? Then the probability distribution is proportional to

$$r^2\,|\psi|^2 = r^2\mathrm{e}^{-2c_0r}, \tag{4.22}$$

which has now a maximum, as can be seen in Fig. 4.2. The numerical value of the maximum can be determined by differentiating eqn. (4.22)

$$\frac{\mathrm{d}}{\mathrm{d}r}(r^2\mathrm{e}^{-2c_0r}) = 0 = \mathrm{e}^{-2c_0r}(2r-2c_0r^2), \tag{4.23}$$

whence

$$r = \frac{1}{c_0} = \frac{4\pi\hbar^2\epsilon_0}{e^2m} = 0{\cdot}0528\,\text{nm}. \tag{4.24}$$

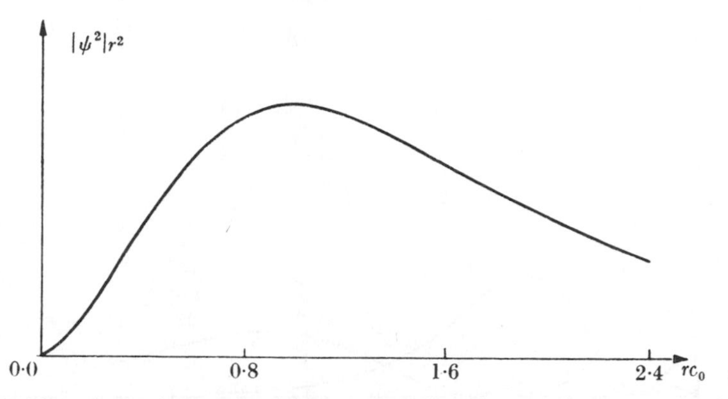

FIG. 4.2. Plot of eqn. (4.22), showing the probability that an electron (occupying the lowest energy state) may be found in the spherical shell between r and $r+dr$.

This radius was again known in pre-quantum-mechanical times and was called the radius of the first Bohr orbit (where electrons can orbit without radiating). Thus in quantum theory the Bohr orbit appears as the most probable position of the electron.

We have squeezed out about as much information from our one meagre solution as is possible; we should look now at the other solutions which I shall give without any proof. Sticking for the moment to the spherically symmetrical case the wavefunction is

$$\psi_n(r) = \mathrm{e}^{-c_n r} L_n(r), \tag{4.25}$$

where L_n is a polynomial, and the corresponding energies are (in electron volts)

$$E_n = -13{\cdot}6\,\frac{1}{n^2}, \qquad n = 1, 2, 3.... \tag{4.26}$$

The solution we obtained before was for $n = 1$. It gives the lowest energy and it is therefore usually referred to as the *ground state*.

If we have a large number of hydrogen atoms, most of them are in their ground state but some of them will be in excited states, which are given by $n > 1$. The probability distributions for the higher excited states have maxima farther from the origin as shown in Fig. 4.3 for $n = 1, 2, 3$. This is fair enough; for $n > 1$ the energy of the electron is nearer to zero (which is the energy of a free electron); so it is less strongly bound to the proton. If it is less strongly bound, it can wander farther away; so the radius corresponding to maximum probability increases.

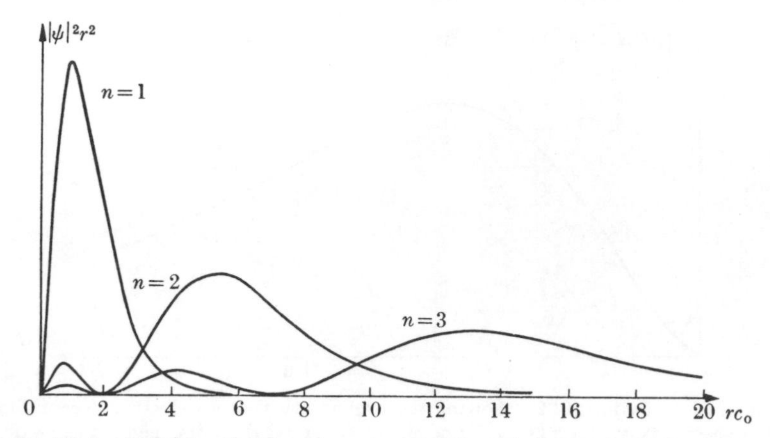

FIG. 4.3. Plots of $\psi_n^2 r^2$ for the three lowest energy ($n = 1, 2, 3$) spherically symmetrical solutions. The curves are normalized so that the total probabilities (the area under the curves) are equal.

4.2. Quantum numbers

So much about spherically symmetrical solutions. The general solution includes, of course, our previously obtained solutions (denoted by $R(r)$ from here on) but shows variations in θ and ϕ as well. It can be written as

$$\psi_{n,l,m_l}(r, \theta, \phi) = R_{nl}(r)Y_l^{m_l}(\theta, \phi). \tag{4.27}$$

We have met n before; l and m_l represent two more discrete sets of constants which ensure that the solutions have physical meaning. These discrete sets of constants always appear in the solutions of partial differential equations; you may remember them from the problems of the vibrating string or of the vibrating membrane. They are generally called *eigenvalues*; in quantum mechanics they are referred to as quantum numbers.

It may be shown (alas, not by simple means) from the original differential equation (eqn. (4.12)) that the quantum numbers must satisfy the following relationships

$$\begin{aligned} n &= 1, 2, 3 \ldots \\ l &= 0, 1, 2 \ldots n-1 \\ m_l &= 0, \pm 1, \pm 2 \ldots \pm l. \end{aligned} \tag{4.28}$$

For $n = 1$ there is only one possibility: $l = 0$ and $m_l = 0$, and the corresponding wavefunction is the one we guessed in eqn. (4.13). For the spherically symmetrical case the wavefunctions have already been plotted for $n = 1, 2, 3$; now let us see a wavefunction which is dependent on direction. Choosing $n = 2$, $l = 1$, $m_l = 0$, the corresponding wavefunction is

$$\begin{aligned}\psi_{210} &= R_{21}(r)\,Y_1^0(\theta,\,\phi)\\ &= r\mathrm{e}^{-c_0r/2}\cos\theta. \end{aligned} \tag{4.29}$$

This equation tells us how the probability of finding the electron varies as a function of r and θ. Thus the equal-probability surfaces may be determined. The spherical symmetry has gone but there is still cylindrical symmetry (no dependence on ϕ). It is therefore sufficient to plot the curves in, say, the xz-plane. This is done in Fig. 4.4, where the unit

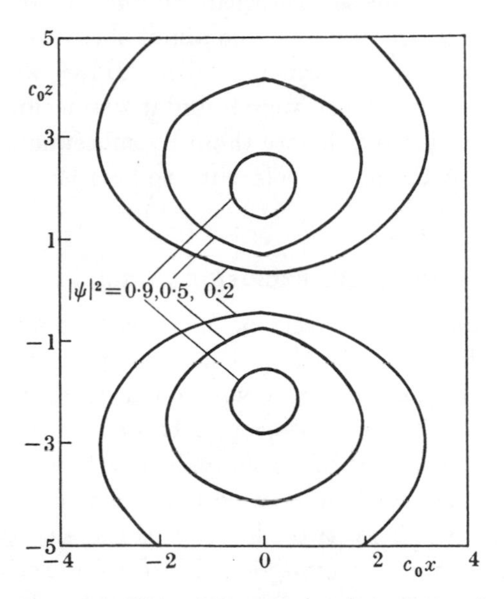

FIG. 4.4. Plots of constant $|\psi_{210}|^2$ in the xz-plane.

of distance is taken as one Bohr radius and the maximum probability (at $x = 0$ and $z = \pm 2/c_0$) is normalized to unity. It can be clearly seen that the $\theta = 0$ and $\theta = \pi$ directions are preferred (which of course follows from eqn. (4.29) directly); there is a higher probability of finding the electron in those directions.

With $n = 2$ and $l = 1$ there are two more states but they give nothing new. The preferential directions in those cases are in the direction of the $\pm x$ and $\pm y$-axes respectively.

For higher values of n and l the equal-probability curves look more and more complicated. Since at this level of treatment they won't add much to our picture of the hydrogen atom we can safely omit them.

I have to add a few words about notations. However convenient the parameters n and l might appear they are never (or at least very rarely) used in that form. The usual notation is a number (equal to the value of n) followed by a letter, which is related to l by the following rule

$$\begin{array}{lcccccccc} l = & 0 & 1 & 2 & 3 & 4 & 5 & 6 & 7 \\ & \mathrm{s} & \mathrm{p} & \mathrm{d} & \mathrm{f} & \mathrm{g} & \mathrm{h} & \mathrm{i} & \mathrm{k}. \end{array}$$

Thus if you wish to refer to the states with $n = 3$ and $l = 1$, you call them the 3p states or the 3p configuration. The reason for this rather illogical notation is of course historical. In the old days when only spectroscopic information was available about these energy levels they were called s for sharp, p for principal, d for diffuse and f for fundamental. When more energy levels were found it was decided to introduce some semblance of order and denote them by subsequent letters of the alphabet. That is how the next levels came to bear the letters g, h, i, k, etc.

4.3. Electron spin and Pauli's exclusion principle

The quantum numbers n, l, and m_l have been obtained from the solution of Schrödinger's equation. Unfortunately, as I mentioned before, they do not represent the whole truth; there is one more quantum number to be taken into account. It is called the spin quantum number, denoted by s, and it takes the values $\pm\frac{1}{2}$.

Historically, spin had to be introduced to account for certain spectroscopic measurements where two closely spaced energy levels were observed when only one was expected. These were explained in 1925 by Uhlenbeck and Goudsmit by assuming that the electron can spin about its own axis. This classical description is very much out of fashion nowadays but the name spin stuck and has been universally used ever since. Today the spin is looked upon just as another quantum number obtainable from a more complicated theory which includes relativistic effects as well.

So we have now four quantum numbers: n, l, m_l, and s. Any permissible combination of these quantum numbers (eqn. (4.28) shows what is permissible) gives a state; the wavefunction is determined, the electron's energy is determined, everything is determined. But what happens when we have more than one electron? How many of them can occupy the same state? One, said Pauli. *There can be no more than one electron in any given state.* This is Pauli's exclusion principle. We shall use it as a separate assumption though it can be derived from a relativistic quantum theory.

Although both the spin and the exclusion principle are products of rather involved theories, both of them can be explained in simple terms. So even if you don't learn where they come from, you can easily remember them.

4.4. The periodic table

We have so far tackled the simplest configuration when there are only two particles: one electron and one proton. How should we attempt the solution for a more complicated case; for helium, for example, which has two protons and two electrons? (Helium has two neutrons as well, but since they are neutral they have no effect on the electrons; thus when discussing the energy levels of electrons neutrons can be disregarded.)

The answer is still contained in Schrödinger's equation but the form of the equations is more complicated. The differential operator ∇ operated on the coordinate of the electron. If we have two electrons we need two differential operators. Thus

$$\nabla^2\psi$$

is replaced by

$$\nabla_1^2\psi+\nabla_2^2\psi,$$

where the indices 1 and 2 refer to electrons 1 and 2 respectively. We may take the protons† infinitely heavy again and put them at the origin of the coordinate system. Thus the potential energy of electron 1 at a distance r_1 from the protons is $-2e^2/4\pi\epsilon_0 r_1$, and similarly for electron 2. There is, however, one more term in the expression for potential energy: the potential energy due to the two electrons. If the distance between them is r_{12} then this potential energy is $e^2/4\pi\epsilon_0 r_{12}$. It is of positive sign because the two electrons repel each other. We can now write down Schrödinger's equation for two protons and two electrons:

$$-\frac{\hbar^2}{2m}(\nabla_1^2\psi+\nabla_2^2\psi)+\frac{1}{4\pi\epsilon_0}\left(-\frac{2e^2}{r_1}-\frac{2e^2}{r_2}+\frac{e^2}{r_{12}}\right)\psi = E\psi. \tag{4.30}$$

Note that the wave function ψ now depends on six variables, namely on the three spatial coordinates of each electron.

† It is a separate story how positively charged protons and neutrons can peacefully coexist in the nucleus, and the answer is still only partly known.

Can this differential equation be solved? The answer, unfortunately, is no. No analytical solutions have been found. So we are up against mathematical difficulties even with helium. Imagine then the trouble we should have with tin. A tin atom has 50 protons and 50 electrons; the corresponding differential equation has 150 independent variables and 1275 terms in the expression for potential energy. This is annoying. We have the correct equation but we can't solve it because our mathematical apparatus is inadequate. What shall we do? Well, if we can't get exact solutions we can try to find approximate solutions. This is fortunately possible. Several techniques have been developed for solving the problem of individual atoms by successive approximations. The mathematical techniques are not particularly interesting, and so I shall mention only the simplest physical model that leads to the simplest mathematical solution.

In this model we assume that there are Z positively charged protons in the nucleus, and the Z electrons floating around the nucleus are unaware of each other. If the electrons are independent of each other, then the solution for each of them is the same as for the hydrogen atom provided that the charge at the centre is taken as Ze. This means putting Ze^2 instead of e^2 into eqn. (4.2) and Z^2e^4 instead of e^4 into eqn. (4.18). Thus we can rewrite all the formulae used for the hydrogen atom, and in particular the formula for energy, which now stands as

$$E_n = -13{\cdot}6\,\frac{Z^2}{n^2}. \tag{4.31}$$

That is, the energy of the electrons decreases with increasing Z. In other words, the energy is below zero by a larger amount; that is more energy is needed to liberate an electron. This is fairly easy to understand; a larger positive charge in the nucleus will bind the electron more strongly.

The model of entirely independent electrons is rather crude but it can go a long way towards a qualitative explanation of the chemical properties of the elements. We shall see how Mendeleev's periodic table can be built up with the aid of the quantum-mechanical solution of the hydrogen atom.

There are two important points to realize:

(i) We have our set of quantum numbers n, l, m_l, and s, each one specifying a state with a definite energy. The energy depends on n only, but several states exist for every value of n.

(ii) Pauli's exclusion principle must be obeyed. Each state can be occupied by one electron only.

Hence we shall start by taking the lowest energy level, count the number of states, fill them up one by one with electrons, and then proceed to the next energy level; and so on.

According to eqn. (4.31) the lowest energy level is obtained with $n = 1$. Then $l = 0$, $m = 0$, and there are two possible states of spin, $s = \pm\frac{1}{2}$. Thus the lowest energy level may be occupied by two electrons. Putting in one electron we get hydrogen, putting in two electrons we get helium, putting in three electrons ... No, we can't do that; if we want an element with three electrons, then the third electron must go into a higher energy level.

With helium the $n = 1$ 'shell' is closed, and this fact determines the chemical properties of helium. If the helium atom happens to meet other electrons (in events officially termed *collisions*) it can offer only high energy states. Since all electrons look for low energy states they generally decline the invitation. They manifest no desire to become attached to a helium atom.

If the probability of attracting an electron is small, can the helium atom give away one of its electrons? This is not very likely. It can offer to its own electrons comfortable low-energy states. The electrons are quite satisfied and stay. Thus the helium atom neither takes up nor gives away electrons. Helium is chemically inert.

We now have to start the next energy shell with $n = 2$. The first element there is lithium, containing two electrons with $n = 1$, $l = 0$ and one electron with $n = 2$, $l = 0$. Adopting the usual notations we may say that lithium has two 1s electrons and one 2s electron. Since the 2s electron has higher energy it can easily be tempted away. Lithium is chemically active.

The next element is beryllium with two 1s and two 2s electrons; then comes boron with two 1s, two 2s, and one 2p electrons, which, incidentally, can be denoted in an even more condensed manner as $1s^2$, $2s^2$, $2p^1$. Employing this new notation the six electrons of carbon appear as $1s^2$, $2s^2$, $2p^2$, the seven electrons of nitrogen as $1s^2$, $2s^2$, $2p^3$, the eight electrons of oxygen as $1s^2$, $2s^2$, $2p^4$, and the nine electrons of fluorine as $1s^2$, $2s^2$, $2p^5$.

Let us pause here for a moment. Recall that a 2p state means $n = 2$ and $l = 1$ which according to eqn. (4.28) can have three states ($m_l = 0$ and $m_l = \pm 1$) or, taking account of spin as well, six states altogether. In the case of fluorine five of them are occupied, leaving one empty low-energy state to be offered to outside electrons. The offer is often taken up, and so fluorine is chemically active.

Lithium and fluorine are at the opposite ends, the former having one *extra* electron, the latter *needing* one more electron to complete the shell. So it seems quite reasonable that when they are together the extra electron of lithium will occupy the empty state of the fluorine atom, making up the compound LiF. A chemical bond is born, a chemist would say.

We shall discuss bonds later in more detail. Let us return meanwhile to the rather protracted list of the elements. After fluorine comes neon. The $n = 2$ shell is completed: no propensity to take up or give away electrons. Neon is chemically inert like helium.

The $n = 3$ shell starts with sodium, which has just one 3s electron and should therefore behave chemically like lithium. A second electron fills the 3s shell in magnesium. Then come aluminium, silicon, phosphorus, sulphur, and chlorine with one, two, three, four, and five 3p electrons respectively. Chlorine is again short of one electron to fill the 3p shell, and so behaves like fluorine. The 3p shell is completed in argon, which is again inert.

So far everything has gone regularly, and by the rules of the game the next electron should go into the 3d shell. It doesn't. Why? Well, why should it? The electrons in potassium are under no obligation to follow the energy hierarchy of the hydrogen atom like sheep. They arrange themselves in such a way as to have the lowest energy. If there were *no* interaction between the electrons, the energy levels of the elements would differ only by the factor Z^2, conforming otherwise to that of the hydrogen atom. If the interaction between the electrons mattered a lot, we should completely abandon the classification based on the energy levels of the hydrogen atom. As it happens, the electron interactions are responsible for small quantitative† changes that cause qualitative change in potassium—and in the next few elements, called the *transition elements*. First the 4s shell is filled, and only after that are the 3d states occupied. The balance between the two shells remains, however, delicate. After vanadium (with three 3d and two 4s electrons) one electron is withdrawn from the 4s shell; hence chromium has five 3d electrons but only one 4s electron. The same thing happens later

† In the hydrogen-type solutions the energy depends only on n, whereas taking account of electron interactions the energy increases with increasing values of l. It just happens that in potassium the energy of the 3d level ($n = 3, l = 2$) is higher than that of the 4s level ($n = 4, l = 0$).

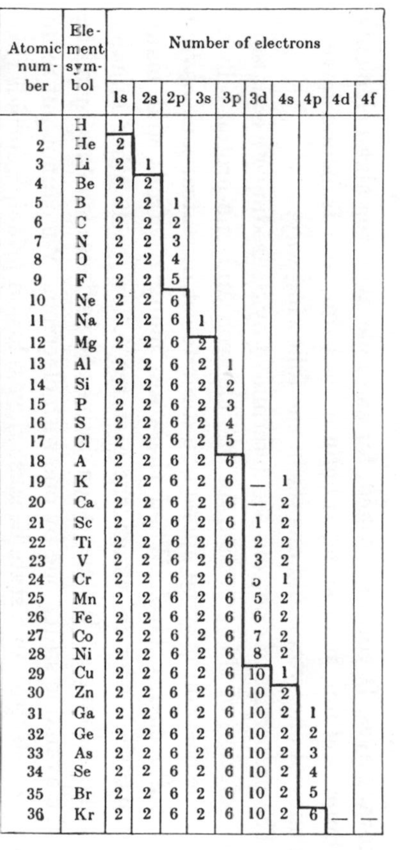

Atomic number	Element symbol	Number of electrons									
		1s	2s	2p	3s	3p	3d	4s	4p	4d	4f
1	H	1									
2	He	2									
3	Li	2	1								
4	Be	2	2								
5	B	2	2	1							
6	C	2	2	2							
7	N	2	2	3							
8	O	2	2	4							
9	F	2	2	5							
10	Ne	2	2	6							
11	Na	2	2	6	1						
12	Mg	2	2	6	2						
13	Al	2	2	6	2	1					
14	Si	2	2	6	2	2					
15	P	2	2	6	2	3					
16	S	2	2	6	2	4					
17	Cl	2	2	6	2	5					
18	A	2	2	6	2	6					
19	K	2	2	6	2	6	—	1			
20	Ca	2	2	6	2	6	—	2			
21	Sc	2	2	6	2	6	1	2			
22	Ti	2	2	6	2	6	2	2			
23	V	2	2	6	2	6	3	2			
24	Cr	2	2	6	2	6	5	1			
25	Mn	2	2	6	2	6	5	2			
26	Fe	2	2	6	2	6	6	2			
27	Co	2	2	6	2	6	7	2			
28	Ni	2	2	6	2	6	8	2			
29	Cu	2	2	6	2	6	10	1			
30	Zn	2	2	6	2	6	10	2			
31	Ga	2	2	6	2	6	10	2	1		
32	Ge	2	2	6	2	6	10	2	2		
33	As	2	2	6	2	6	10	2	3		
34	Se	2	2	6	2	6	10	2	4		
35	Br	2	2	6	2	6	10	2	5		
36	Kr	2	2	6	2	6	10	2	6	—	—

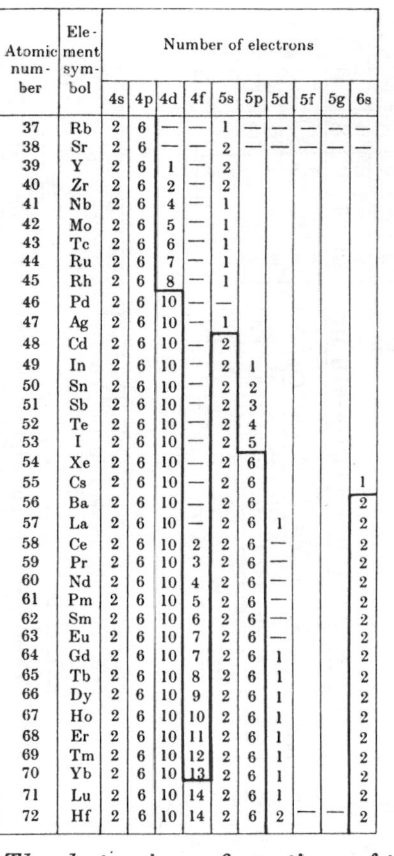

Atomic number	Element symbol	Number of electrons									
		4s	4p	4d	4f	5s	5p	5d	5f	5g	6s
37	Rb	2	6	—	—	1	—	—	—	—	—
38	Sr	2	6	—	—	2	—	—	—	—	—
39	Y	2	6	1	—	2					
40	Zr	2	6	2	—	2					
41	Nb	2	6	4	—	1					
42	Mo	2	6	5	—	1					
43	Tc	2	6	6	—	1					
44	Ru	2	6	7	—	1					
45	Rh	2	6	8	—	1					
46	Pd	2	6	10	—	—					
47	Ag	2	6	10	—	1					
48	Cd	2	6	10	—	2					
49	In	2	6	10	—	2	1				
50	Sn	2	6	10	—	2	2				
51	Sb	2	6	10	—	2	3				
52	Te	2	6	10	—	2	4				
53	I	2	6	10	—	2	5				
54	Xe	2	6	10	—	2	6				
55	Cs	2	6	10	—	2	6				1
56	Ba	2	6	10	—	2	6				2
57	La	2	6	10	—	2	6	1			2
58	Ce	2	6	10	2	2	6	—			2
59	Pr	2	6	10	3	2	6	—			2
60	Nd	2	6	10	4	2	6	—			2
61	Pm	2	6	10	5	2	6	—			2
62	Sm	2	6	10	6	2	6	—			2
63	Eu	2	6	10	7	2	6	—			2
64	Gd	2	6	10	7	2	6	1			2
65	Tb	2	6	10	8	2	6	1			2
66	Dy	2	6	10	9	2	6	1			2
67	Ho	2	6	10	10	2	6	1			2
68	Er	2	6	10	11	2	6	1			2
69	Tm	2	6	10	12	2	6	1			2
70	Yb	2	6	10	13	2	6	1			2
71	Lu	2	6	10	14	2	6	1			2
72	Hf	2	6	10	14	2	6	2	—	—	2

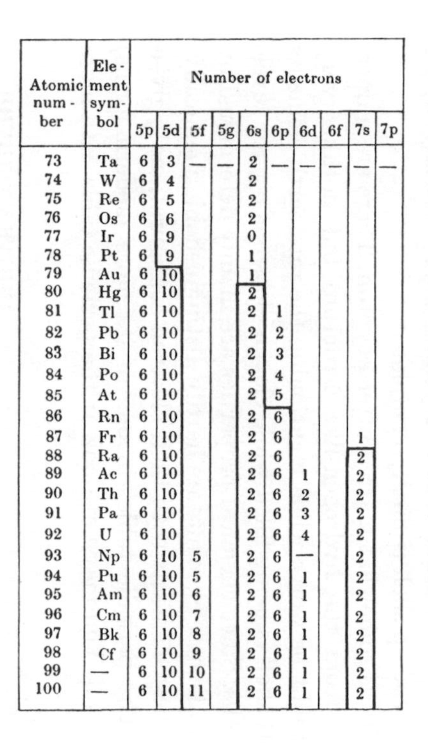

Atomic number	Element symbol	Number of electrons									
		5p	5d	5f	5g	6s	6p	6d	6f	7s	7p
73	Ta	6	3	—	—	2	—	—	—	—	—
74	W	6	4			2					
75	Re	6	5			2					
76	Os	6	6			2					
77	Ir	6	9			0					
78	Pt	6	9			1					
79	Au	6	10			1					
80	Hg	6	10			2					
81	Tl	6	10			2	1				
82	Pb	6	10			2	2				
83	Bi	6	10			2	3				
84	Po	6	10			2	4				
85	At	6	10			2	5				
86	Rn	6	10			2	6				
87	Fr	6	10			2	6			1	
88	Ra	6	10			2	6			2	
89	Ac	6	10			2	6	1		2	
90	Th	6	10			2	6	2		2	
91	Pa	6	10			2	6	3		2	
92	U	6	10			2	6	4		2	
93	Np	6	10	5		2	6	—		2	
94	Pu	6	10	5		2	6	1		2	
95	Am	6	10	6		2	6	1		2	
96	Cm	6	10	7		2	6	1		2	
97	Bk	6	10	8		2	6	1		2	
98	Cf	6	10	9		2	6	1		2	
99	—	6	10	10		2	6	1		2	
100	—	6	10	11		2	6	1		2	

TABLE 4.1. *The electronic configurations of the elements.*

with copper, but apart from that everything goes smoothly up to krypton, where the 4p shell is finally completed.

The regularity is somewhat marred after krypton. There are numerous deviations from the hydrogen-like structure but nothing very dramatic. It might be worth while mentioning the rare earth elements in which the 4f shell is being filled while eleven electrons occupy levels in the outer shells. Since chemical properties are mainly determined by the outer shells all these elements are hardly distinguishable chemically.

A list of all these elements with their electron configurations is given in Table 4.1. The periodic table (in one of its more modern forms) is given in Fig. 4.5. You may now look at the periodic table with more

IA	IIA	IIIA	IVA	VA	VIA	VIIA	VIII			IB	IIB	IIIB	IVB	VB	VIB	VIIB	
		1 H	2 He														
3 Li	4 Be											5 B	6 C	7 N	8 O	9 F	10 Ne
11 Na	12 Mg											13 Al	14 Si	15 P	16 S	17 Cl	18 Ar
19 K	20 Ca	21 Sc	22 Ti	23 V	24 Cr	25 Mn	26 Fe	27 Co	28 Ni	29 Cu	30 Zn	31 Ga	32 Ge	33 As	34 Se	35 Br	36 Kr
37 Rb	38 Sr	39 Y	40 Zr	41 Nb	42 Mo	43 Tc	44 Ru	45 Rh	46 Pd	47 Ag	48 Cd	49 In	50 Sn	51 Sb	52 Te	53 I	54 Xe
55 Cs	56 Ba	57 La	72 Hf	73 Ta	74 W	75 Re	76 Os	77 Ir	78 Pt	79 Au	80 Hg	81 Tl	82 Pb	83 Bi	84 Po	85 At	86 Rn
87 Fr	88 Ra	89 Ac															

58 Ce	59 Pr	60 Nd	61 Pm	62 Sm	63 Eu	64 Gd	65 Tb	66 Dy	67 Ho	68 Er	69 Tm	70 Yb	71 Lu	IIIA
90 Th	91 Pa	92 U	93 Np	94 Pu	95 Am	96 Cm	97 Bk	98 Cf	99 Es	100 Fm	101 Md	102 No	103	

FIG. 4.5. The periodic table of the elements.

knowing eyes. If you were asked, for example, why the alkali elements lithium, sodium, potassium, rubidium, caesium, and francium have a valency of one, you could answer in the following way.

The properties of electrons are determined by Schrödinger's equation. The solution of this equation for one electron and one proton tells us that the electron may be in one of a discrete set of states, each having a definite energy level. When there are many electrons and many protons, the order in which these states follow each other remains roughly unchanged. We may then derive the various elements by filling up the available states one by one with electrons. We cannot put more than one electron in a state because the exclusion principle forbids this.

The energy of the states varies in steps. Within a 'shell' there is a slow variation in energy but a larger energy difference between shells.

Hence whenever a new shell is initiated there is one electron with considerably higher energy than the rest. Since all electrons strive for lower energy this electron can easily be lost to another element.

All the alkali elements start new shells. Therefore each of them may lose an electron; each of them may contribute one unit to a new chemical configuration; and each of them has a valency of one.

We may pause here for a moment. You have had the first taste of the power of Schrödinger's equation. You can see now that the solution of all the basic problems that have haunted the chemists for centuries is provided by a modest-looking differential equation. The chaos prevailing before has been cleared and a sturdy monument has been erected in its stead. If you look at it carefully you will find that it possesses all the requisites of artistic creation. It is like a Greek temple. You can see in the background the stern regularity of the columns but the statues placed between them are all different.

Examples

1. Calculate the wavelength of electromagnetic waves needed to excite a hydrogen atom from the 1s into the 2s state.
2. Electromagnetic radiation of wavelength 20 nm is incident on atomic hydrogen. What is the maximum velocity at which an ionized electron may leave the atom?
3. An excited argon ion in a gas discharge radiates a spectral line of wavelength 450 nm. The transition from the excited to the ground state that produces this radiation takes an average time of 10^{-8} sec. What is the inherent width of the spectral line? What other factors in practice will cause the line width to be much greater than the value calculated from these data?
4. What is the shortest time in which an energy transition in an atom can take place if it results in the emission of a spectral line of 1000 nm wavelength?
5. Determine the most probable orbiting radius of the electron in a hydrogen atom from the following very crude considerations. The electron tries to move as near as possible to the nucleus in order to lower its potential energy. But if the electron is somewhere within the region 0 to r_{m} (i.e. we know its position with an uncertainty, r_{m}) the uncertainty in its momentum must be $\Delta p \simeq \hbar/r_{\mathrm{m}}$. So the kinetic energy of the electron is roughly $\hbar^2/2mr_{\mathrm{m}}^2$.

   Determine r_{m} from the condition of minimum energy. Compare the radius obtained with that of the first Bohr orbit.
6. Solve Schrödinger's equation for the ground state of helium neglecting the potential term between the two electrons. What is the energy of the ground state calculated this way? The measured value is $-24{\cdot}6$ eV. What do you think the difference is caused by? Give an explanation in physical terms.
7. Write down the time-independent Schrödinger equation for lithium.

5. Bonds

Striking the electric chain wherewith we are darkly bound

BYRON *Childe Harold's Pilgrimage*

Those whom God has joined together,
Let no man put asunder.

MARRIAGE SERVICE

5.1. Introduction

As we have seen, an electron and a proton may strike up a companionship, the result being a hydrogen atom. We have found that the energy of the electron is a negative number, that is, the electron in the vicinity of the proton has a lower energy than it would have if it were an infinite distance away (which corresponds to zero energy). The minimum comes about as some sort of compromise between the kinetic and potential energy but the important thing is that a minimum exists. The electron comes closer to find lower energy.

Can we say the same thing about two hydrogen atoms? Would they too come close to each other in order to reduce the total energy? Yes, they come close; they combine and make up a hydrogen molecule. This combination between atoms is called a *chemical bond*, and the discipline that is concerned with these combinations is chemistry.

You may justifiably ask why we talk of chemistry in a course on electrical properties of materials. Well, in a sense, chemistry is just a branch of the electrical properties of materials. The only way to explain chemical bonds is to use electrical and some specific quantum-mechanical properties. And not only chemistry is relegated to this position: metallurgy too. When a large number of atoms conglomerate and make up a solid, the reason is again to be sought in the behaviour of electrons. Thus all the mechanical properties of solids, including their very solidity, spring from the nature of their electrical components.

This is true in principle but not quite true in practice. We know the fundamental laws, and so we could work out everything (the outcome of all chemical reactions, the strength of all materials) if only the mathematical problems could be overcome. Bigger computers and improved techniques of numerical analysis might one day make such calculations feasible, but for the moment it is not practicable to go back to first principles. So we are not going to solve the problems of chemistry and metallurgy here. Nevertheless, we need to understand the nature of the chemical bond to proceed further. The bond between hydrogen atoms leads to the bond between germanium atoms, which leads to the energy structure of germanium and to the possibility of doping, and eventually to the transistor. You might think that this is too high a price to pay for understanding transistors and a dozen more electronic devices. You may be right. Unless you find these glimpses behind the scenes fascinating in themselves, the eventual understanding of electronic devices is not sufficient compensation for the labour expended. But try not to think in too narrow terms. Learning something about the foundations may help you later when confronting wider problems.

5.2. General mechanical properties of bonds

Before classifying and discussing particular bond types we can make a few common-sense deductions about what sort of forces must be involved in a bond. First of all there must be an attractive force. An obvious candidate for this role is the Coulomb attraction between unlike charges, which we have all met many times, giving a force proportional to r^{-2} where r is the separation.

We know that sodium easily mislays its outer-ring (often called valence) electron, becoming Na^+, and that chlorine is an avid collector of a spare electron. So, just as we mentioned earlier with lithium and fluorine, the excess electron of sodium will fill up the energy shell of chlorine, creating a positively charged sodium ion with a negatively charged chlorine ion. These two ions will attract each other; that is obvious. What is less obvious, however, is that NaCl crystallizes into a very definite structure with the Na and Cl ions 0·28 nm apart. What stops them getting closer? Surely the Coulomb forces are great at 0·28 nm. Yes, they are great, but they are not the only forces acting. When the ions are very close to each other and start becoming distorted, new forces arise that tend to re-establish the original undistorted separate state of the ions. These repulsive forces are of short range. They come into play only when the interatomic distance becomes comparable with the atomic radius. Thus we have two opposing forces that balance each other at the equilibrium separation, r_0.

It is possible to put this argument into graphical and mathematical form. If we plot the total energy of two atoms against their separation,

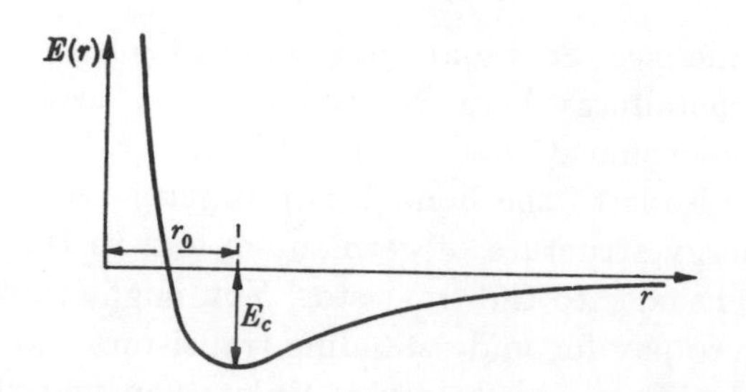

FIG. 5.1. The essential general appearance of the energy versus separation curve if two atoms are to bond together. The equilibrium separation is r_0 and the bond energy is E_c.

the graph must look something like Fig. 5.1. The 'common-sense' points about this diagram are as follows:

(i) The energy tends to zero at large distances—in other words we define zero energy as the energy in the absence of interaction.

(ii) At large distances the energy is negative and increases with increasing distance. This means that from infinity down to the point r_0 the atoms attract each other.

(iii) At very small distances the energy is rising rapidly, that is, the atoms repel each other up to the point r_0.

(iv) The curve has a minimum value at r_0 corresponding to an equilibrium position. Here the attractive and repulsive forces just balance each other.

In the above discussion we have regarded r as the distance between two atoms, and r_0 as the equilibrium distance. The same argument applies, however, if we think of a solid that crystallizes in a cubic structure. We may then interpret r as the interatomic distance in the solid.

Let us now see what happens when we compress the crystal, that is, when we change the interatomic distance by brute force. According to our model the energy will increase, but when the external influence is removed the crystal will return to its equilibrium position. In some other branches of engineering this phenomenon is known as elasticity. So if we manage to obtain the $E(r)$ curve we can calculate all the elastic properties of the solid. Let us work out as an example the bulk elastic constant. We shall take a cubical piece of material of side a (Fig. 5.2) and calculate the energy changes under isotropic compression.

Regarding $E(r)$ as the energy per atom, the total energy of the material is $N_a a^3 E(r_0)$ in equilibrium, where N_a is the number of atoms per unit volume. If the cube is uniformly compressed the interatomic distance will decrease by Δr, and the total energy will increase to

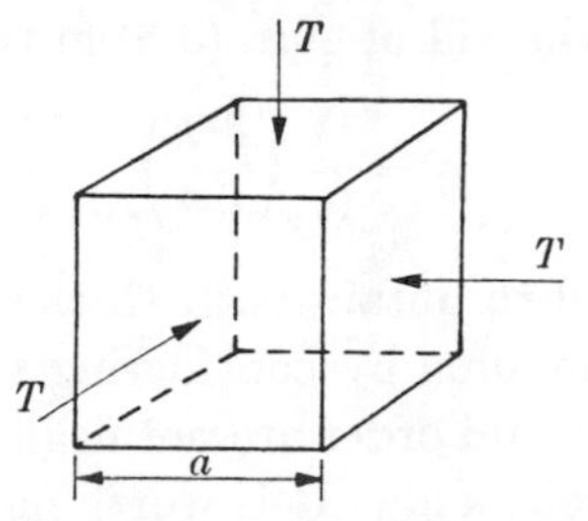

FIG. 5.2. A cube of material, side a, isotropically compressed.

$N_a a^3 E(r_0 - \Delta r)$. Expanding $E(r_0 - \Delta r)$ into a Taylor series and noting that $(\partial E/\partial r)_{r=r_0} = 0$, we get

$$E(r_0 - \Delta r) = E(r_0) + \frac{1}{2}\left(\frac{\partial^2 E}{\partial r^2}\right)_{r=r_0} (\Delta r)^2 + \dots . \tag{5.1}$$

Hence the net increase in energy is equal to

$$\frac{1}{2} N_a a^3 \left(\frac{\partial^2 E}{\partial r^2}\right)_{r=r_0} (\Delta r)^2. \tag{5.2}$$

This increase in energy is due to the work done by moving the six faces of the cube. The total change in linear dimension is $(a/r_0)\Delta r$; thus we may say that each face has moved by a distance $(a/2r_0)\Delta r$. Hence, while the stress is increasing from 0 to T, the total work done on the piece of material is

$$6\frac{1}{2} T a^2 \frac{a \Delta r}{2r_0}. \tag{5.3}$$

From the equality of eqns (5.2) and (5.3) we get

$$\frac{3}{2} T \frac{a^3}{r_0} \Delta r = \frac{1}{2} \frac{a^3}{r_0^3} \left(\frac{\partial^2 E}{\partial r^2}\right)_{r=r_0} (\Delta r)^2, \tag{5.4}$$

whence

$$T = \frac{1}{3r_0} \left(\frac{\partial^2 E}{\partial r^2}\right)_{r=r_0} \frac{\Delta r}{r_0}. \tag{5.5}$$

Defining the bulk elastic modulus by the relationship of stress to the volume-change caused, that is

$$\begin{aligned} T &= c\frac{\Delta a^3}{a^3} \\ &\simeq c\frac{3\Delta a}{a} \\ &= c\frac{3\Delta r}{r_0}, \end{aligned} \tag{5.6}$$

we can obtain c with the aid of eqn. (5.5) in the form

$$c = \frac{1}{9r_0}\left(\frac{\partial^2 E}{\partial r^2}\right)_{r=r_0} \tag{5.7}$$

So we have managed to obtain both Hooke's law and an expression for the bulk elastic modulus by considering the interaction of atoms. If terms higher than second order are not negligible, we have a material that does *not* obey Hooke's law. It is worth noting that most materials do obey this 'law' for small deformations, but not for large ones. This is in line with the assumptions we have made in the derivation.

For the purpose of making some rough calculations, the characteristic curve of Fig. 5.1 may be approximated by the following simple mathematical expression

$$E(r) = \frac{A}{r^n} - \frac{B}{r^m}, \tag{5.8}$$

where the first term on the right-hand side represents repulsion and the second term attraction. By differentiating eqn. (5.8) we can get E_c, the minimum of the $E(r)$ curve at the equilibrium distance $r = r_0$, in the form

$$E_c = \frac{B}{r_0^m}\left(\frac{m}{n} - 1\right). \tag{5.9}$$

For a stable bond, $E_c < 0$, which can be satisfied only if

$$m < n. \tag{5.10}$$

That is, the repulsive force has a higher index than the attractive one.

5.3. Bond types

There is no sharp distinction between the different types of bonds. For most bonds, however, we may say that one or the other mechanism dominates. Thus a classification is possible; the four main types are: (i) ionic, (ii) metallic, (iii) covalent, and (iv) Van der Waals.

5.4. Ionic bonds

A typical representative of an *ionic* crystal is NaCl, which we have already discussed in some detail. The crystal structure is regular and looks exactly like the one shown in Fig. 1.1. We have negatively charged Cl ions and positively charged Na ions. We may now ask the question, what is the cohesive energy of this crystal? Cohesive energy is what we have denoted by E_c in Fig. 5.1, that is the energy needed to take the crystal apart. How could we calculate this? If the binding is due mainly to electrostatic forces, then all we need to do is to sum the electrostatic energy due to pairs of ions.

Let us start with an arbitrary Na ion. It will have six Cl ions at a distance a giving the energy

$$-\frac{e^2}{4\pi\epsilon_0}\frac{6}{a}. \tag{5.11}$$

There are then 12 Na ions at a distance $a\sqrt{2}$ contributing to the energy by the amount

$$\frac{e^2}{4\pi\epsilon_0}\frac{12}{a\sqrt{2}}. \tag{5.12}$$

Next come eight chlorine atoms at a distance $a\sqrt{3}$, and so on. Adding up the contributions from all other ions, we have an infinite sum (well, practically infinite) of the form

$$-\frac{e^2}{4\pi\epsilon_0}\left(\frac{6}{a}-\frac{12}{a\sqrt{2}}+\frac{8}{a\sqrt{3}}\cdots\right). \tag{5.13}$$

We have to add together sums such as eqn. (5.13) for every Na and Cl ion to get the cohesive energy. It would actually be twice the cohesive energy, because we counted each pair twice or we may say that it is the cohesive energy per NaCl unit.

The infinite summations look a bit awkward but fortunately there are mathematicians who are fond of problems of this sort; they have somehow managed to sum up all these series, not only for the cubical structure of NaCl, but for the more complicated structures of some other ionic crystals as well. Their labour brought forth the formula

$$\text{Electrostatic energy} = -B\frac{e^2}{4\pi\epsilon_0 a}, \tag{5.14}$$

where B is called the *Madelung constant.* For a simple cubic structure its value is 1·748. Taking $a = 0{\cdot}28$ nm and putting the constants into eqn. (5.14) we get for the cohesive energy

$$E = 8{\cdot}94 \text{ eV}, \tag{5.15}$$

which is about 10% above the experimentally observed value. There are other types of energies involved as well (as, for example, the energy due to the slight deformation of the atoms) but, as the numerical results show, they must be of lesser significance. We have thus confirmed our starting-point that NaCl may be regarded as an ionic bond.

5.5. Metallic bonds

Having studied the construction of atoms we are now in a somewhat better position to talk about metals. Conceptually, the simplest metal is a monovalent alkali metal where each atom contributes one valence electron to the common pool of electrons. So we are, in fact, back to our very first model when we regarded a conductor as made up of lattice ions and charged billiard balls bouncing around.

We may now ask the question: how is a piece of metal kept together? 'By electrostatic forces', is the simplest, though not quite accurate, answer. Thus the *metallic* bond is similar to the ionic bond in the sense that the main role is played by electrostatic forces, but there is a difference as far as the positions of the charges are concerned. In metals the carriers of the negative charge are highly mobile; thus we may expect a bond of somewhat different properties. Since electrons whizz around and visit every little part of the metal, the electrostatic forces are ubiquitous and come from all directions. So we may regard the electrons as a glue that holds the lattice together. It is quite natural, then, that a small deformation does not cause fracture. Whether we compress or try to pull apart a piece of metal the cohesive forces are still there and acting vigorously. This is why metals are so outstandingly ductile and malleable.

5.6. The covalent bond

So far we have discussed two bonds, which depend on the fact that unlike charges attract—a familiar, old but nevertheless true, idea. But why should atoms like carbon or silicon hang together? It is possible to purify silicon so that its resistivity is several ohm metres—there can be no question of a lot of free electrons swarming around, nor is there an ionic bond. Carbon in its diamond form is the hardest material known. Not only must it form strong bonds but they must also be exceptionally precise and directional to achieve this hardness.

The properties of the covalent bond (also called the *valence* or *homopolar* bond on occasions) is the most important single topic in chemistry. Yet its mechanism was completely inexplicable before the rise of quantum mechanics.

The exact mathematical description is immensely difficult, even for people with degrees, so in an undergraduate course we must be modest. The most we can hope for is to get a good physical picture of the bond mechanism, and perhaps an inkling of how a theoretical physicist would start solving the problem.

The simplest example of the covalent bond is the hydrogen molecule, where two protons are kept together by two electrons. The bond comes about because both electrons orbit around both atoms. Another way of describing the bond is to appeal to the atoms' desire to have filled shells. A hydrogen atom needs two electrons (of opposite spin) to fill the 1s shell, and lacking any better source of electrons it will consider snatching that extra electron from a fellow hydrogen atom. Naturally the other hydrogen atom will resist, and at the end they come to a compromise and share both their electrons. It is as if two men, each anxious to secure two wives for himself, were to agree to share wives.

Another example is chlorine, which has five 3p electrons and is eagerly awaiting one more electron to fill the shell. The problem is again solved by sharing an electron pair with another chlorine atom. Thus each chlorine atom for some time has the illusion that it has managed to fill its outer shell.

Good examples of covalent bonds in solids are carbon, silicon, and germanium. Their electron configurations may be obtained from Table 4.1. They are as follows:

$$\text{C}: \quad 1s^2, 2s^2, 2p^2$$

$$\text{Si}: \quad 1s^2, 2s^2, 2p^6, 3s^2, 3p^2$$

$$\text{Ge}: \quad 1s^2, 2s^2, 2p^6, 3s^2, 3p^6, 3d^{10}, 4s^2, 4p^2$$

It can be easily seen that the common feature is two s and two p electrons in the outer ring. The s shells (2s, 3s, 4s respectively) are filled; so one may expect all three substances to be divalent, since they have two extra electrons in the p shells. Alas, all of them are tetravalent. The reason is that because of interaction (which occurs when several atoms are brought close together) the spherical symmetry of the outer s electrons is broken up and they are persuaded to join the p electrons in forming the bonds. Hence, for the purpose of bonding, the atoms of carbon, silicon, and germanium may be visualized with four dangling electrons at the outside. When the atoms are brought close to each other, these electrons establish the bonds by pairing up. The four electrons are arranged symmetrically in space, and the bonds must therefore be tetrahedral, as shown in Fig. 5.3.

In covalent bonds all the available electrons pair up and orbit around a pair of atoms; none of them can wander away to conduct electricity. This is why carbon in the form of diamond is an insulator. (Graphite, which has a different type of bond, is a fairly good conductor.) The covalent bonds are weaker in silicon and germanium, and some of the

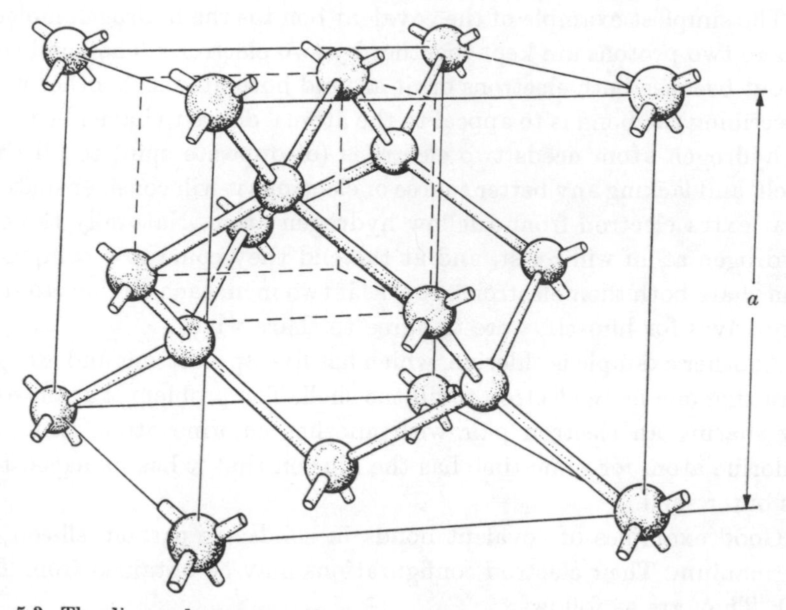

FIG. 5.3. The diamond structure. Notice that each atom is symmetrically surrounded in an imaginary cube by its four nearest neighbours. These are covalently bonded, indicated by tubular connections in the figure.

electrons might be 'shaken off' by the thermal vibrations of the crystal. This makes them able to conduct electricity to a certain extent. They are not conductors; we call them *semiconductors*.

5.7. The Van der Waals bond

If the outer shell is not filled, atoms will exert themselves to gain some extra electrons and they become bonded in the process. But what happens when the shell is already filled and there are no electrostatic forces either, as for example in argon? How will argon solidify? For an explanation some quantum-mechanical arguments are needed again.

We have described the atoms as consisting of a positive nucleus and the electrons around the nucleus as having certain probabilities of being in certain places. Since the electrons are sometimes here and sometimes there, there is no reason why the centres of positive and negative charge should always be coincident. Thus we could regard atoms as fluctuating dipoles. If atom A has a dipole moment then it will induce an opposite dipole moment on atom B. On average there will be an attractive force, since the tendency described leads always to attraction, never to repulsion.

This is called a *Van der Waals* bond. Such bonds are responsible for the formation of organic crystals. The forces are fairly weak (and may be shown to vary with the seventh power of distance); consequently these materials have low melting and boiling points.

5.8. Feynman's coupled mode approach

We are now going to discuss a more mathematical theory of the covalent bond, or rather of its simplest case, the bonding of the hydrogen molecule. We shall do this with the aid of Feynman's coupled modes. This approach proved amazingly powerful in Feynman's hands, enabling him to explain besides the hydrogen molecule such diverse phenomena as the nuclear potential between a proton and a neutron, and the change of the K° particle into its own antiparticle. There is in fact hardly a problem in quantum mechanics that Feynman could not treat by the technique of coupled modes. Of necessity we shall be much less ambitious and discuss only a few relatively simple phenomena.

I should really start by defining the term 'coupled mode'. But to define is to restrict, to put a phenomenon or a method into a neat little box in contradistinction to other neat little boxes. I'm a little reluctant to do so in the present instance because I am sure I would then exclude many actual or potential applications. Not being certain of the limitations of the approach, I would rather give you a vague description, just a general idea of the concepts involved.

The coupled mode approach is concerned with the properties of coupled oscillating systems like mechanical oscillators (e.g. pendulums), electric circuits, acoustic systems, molecular vibrations, and a number of other things you might not immediately recognize as oscillating systems. The approach was quite probably familiar to the better physicists of the last century but has become fashionable only recently. Its essence is to divide the system up into its components, investigate the properties of the individual components in isolation, and then reach conclusions about the whole system by assuming that the components are weakly coupled to each other. Mathematicians would call it a perturbation solution because the system is perturbed by introducing the coupling between the elements.

First of all we should derive the equations. These, not unexpectedly, turn out to be coupled linear differential equations. Let us start again with Schrödinger's equation (eqn. (3.4)) but put it in the operator form of eqn. (3.13), that is

$$\left(-\frac{\hbar^2}{2m}\nabla^2+V\right)\Psi = i\hbar\frac{\partial\Psi}{\partial t}. \qquad (5.16)$$

The operator in parentheses is usually called the *Hamiltonian operator* and denoted by H. So we may write Schrödinger's equation in the simple and elegant form

$$H\Psi = i\hbar\frac{\partial\Psi}{\partial t}. \tag{5.17}$$

We have attempted (eqn. (3.7)) the solution of this partial differential equation before by separating the variables

$$\Psi = w(t)\psi(\mathbf{r}). \tag{5.18}$$

Let us try to do the same thing again but in the more general form

$$\Psi = \sum_j w_j(t)\psi_j(\mathbf{r}), \tag{5.19}$$

where a number of solutions (not necessarily finite) are superimposed.

Up to now we have given all our attention to the spatial variation of the wavefunction. We have said that if an electron is in a certain state it turns up in various places with certain probabilities. Now we are going to change the emphasis. We shall not enquire into the spatial variation of the probability at all. We shall be satisfied with asking the much more limited question: what is the probability that the electron (or more generally a set of particles) is in state j at time t? We don't care what happens to the electron in state j as long as it is in state j. We are interested only in the *temporal* variation, that is, we shall confine our attention to the function $\omega(t)$.

We shall get rid of the spatial variation in the following way. Let us substitute eqn. (5.19) into eqn. (5.17)

$$\sum_j w_j H\psi_j = i\hbar\sum_j \psi_j\frac{dw_j}{dt}. \tag{5.20}$$

Then multiply both sides by ψ_k and integrate over the volume. We then obtain

$$\sum_j w_j \int \psi_k H\psi_j \, dv = i\hbar\sum_j \frac{dw_j}{dt}\int \psi_j\psi_k \, dv, \tag{5.21}$$

where dv is the volume element.

Now ψ_j and ψ_k are two solutions of the time-independent Schrödinger equation and they have the remarkable property (I have to ask you to believe this) of being *orthogonal* to each other. You may have met simple examples of orthogonality of functions before, if at no other place than in the derivation of the coefficients of a Fourier series. The condition can be simply stated in the following form

$$\int \psi_k \psi_j \, dv = \begin{matrix} C_{kj} \\ 0 \end{matrix} \quad \text{if} \quad \begin{matrix} k = j \\ k \neq j \end{matrix}. \tag{5.22}$$

Multiplying the wave functions with judiciously chosen constants, C_{kj} can be made unity and then the wave functions are called *orthonormal*. Assuming that this is the case and introducing the notation

$$H_{kj} = \int \psi_k H \psi_j \, dv \tag{5.23}$$

we get the following differential equations†

$$i\hbar \frac{dw_k}{dt} = \sum_j H_{kj} w_j \tag{5.24}$$

for each value of k.

This is the equation we sought. It is independent of the spatial variables, and depends only on time. It is therefore eminently suitable for telling us how the probability of being in a certain state varies with time.

You may quite justifiably worry at this point about how you can find the wave functions, how you can make them orthonormal, and how you can evaluate integrals looking as complex as eqn. (5.23). The beauty of Feynman's approach is that neither the wave function nor H_{kj} need be calculated. It will suffice to guess H_{kj} on purely physical grounds.

We have not so far said anything about the summation. How many wavefunctions (that is states) are we going to have? We may have an infinite number, as for the electron in a rigid potential well, or it may be finite. If, for example, only the spin of the electron matters then we have two states and no more. The summation should run through $j = 1$ and $j = 2$. Two is of course the minimum number. In order to have coupling one needs at least two components, and it turns out that two components are enough to reach some quite general conclusions about the properties of coupled systems. So the differential equations we are

† The derivation would be analogous if, instead of one electron, a set of particles was involved. Schrödinger's equation would then be written in terms of a set of spatial variables and there would be multiple integrals instead of the single integral here. The integrations would be more difficult to perform but the final form would still be that of eqn. (5.24).

going to investigate look as follows:

$$i\hbar\frac{dw_1}{dt} = H_{11}w_1+H_{12}w_2, \tag{5.25}$$

$$i\hbar\frac{dw_2}{dt} = H_{21}w_1+H_{22}w_2. \tag{5.26}$$

If $H_{12} = H_{21} = 0$ the two states are not coupled. Then the differential equation for state (1) is

$$i\hbar\frac{dw_1}{dt} = H_{11}w_1, \tag{5.27}$$

which has a solution

$$w_1 = K_1 \exp\left(-i\frac{H_{11}}{\hbar}t\right). \tag{5.28}$$

The probability of being in state (1) is thus

$$|w_1(t)|^2 = |K_1|^2. \tag{5.29}$$

This is not a very exciting solution but it is at least consistent. If there is no coupling between the states then the probability of being in state (1) does not vary with time. Once in state (1), always in state (1). The same is true of course for state (2). In the absence of coupling nothing changes.

Before solving the coupled differential equations let us briefly discuss the physical concepts of uncoupled states and the meaning of coupling. What do we mean exactly by coupling? We can explain this with our chosen example, the hydrogen molecule, or better still the even simpler case, the hydrogen molecular ion.

The hydrogen molecular ion consists of a hydrogen atom to which a proton is attached. We may then imagine our uncoupled states as shown in Fig. 5.4. We choose for state (1) the state when the electron is

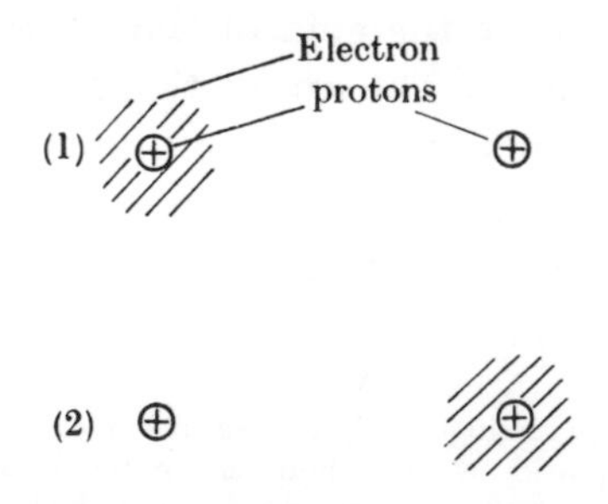

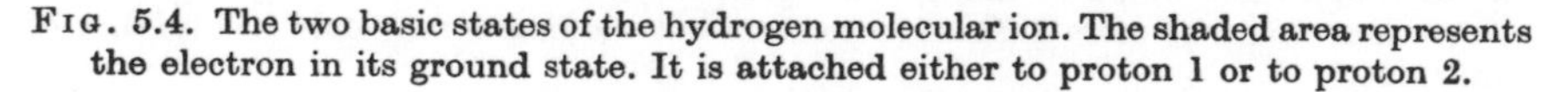
FIG. 5.4. The two basic states of the hydrogen molecular ion. The shaded area represents the electron in its ground state. It is attached either to proton 1 or to proton 2.

in the vicinity of proton 1 and occupying the lowest energy (ground) level, and proton 2 is just alone with no electron of its own. State (2) represents the alternative arrangement when the electron is attached to proton 2 and proton 1 is bare.

When we say that we consider only these two states, we are not denying the existence of other possible states. The electron could be in any of its excited states around the proton, and the whole configuration of three particles may vibrate, rotate, or move in some direction. We are going to ignore all these complications. We say that as far as our problem is concerned only the two states mentioned above are of any significance.

What do we mean when we say that these two states are uncoupled? We mean that if the electron is at proton 1 in the beginning, it will always stay there. Similarly, if the electron is at proton 2 in the beginning, it will always stay at proton 2. Is this complete separation likely? Yes, if the protons are far from each other this is the only thing that can happen. What can we expect when the protons are brought closer to each other? Classically, the electron that is in the vicinity of proton 1 should still remain with proton 1 because this is energetically more favourable. The electron cannot leave proton 1 because it faces an adverse potential barrier. According to the laws of quantum mechanics this is no obstacle, however. The electron may tunnel through the potential barrier and arrive at proton 2 with energy unchanged. Thus as the two protons approach each other there is an increasing probability that the electron jumps over from proton 1 to proton 2 and vice versa. And this is what we mean by coupling. The two states are not entirely separate. When the electron jumps from one proton to the other proton it introduces coupling between the two states.

What do we mean by *weak* coupling? It means that even in the presence of coupling it is still meaningful to talk about one or the other state. The states influence each other but may preserve their separate entities.

Let us return now to the solution of eqns. (5.25) and (5.26). As we are going to investigate symmetric cases only we may introduce the simplifications

$$H_{11} = H_{22} = E_0, \qquad H_{12} = H_{21} = -A, \tag{5.30}$$

leading to

$$\mathrm{i}\hbar\frac{\mathrm{d}w_1}{\mathrm{d}t} = E_0 w_1 - A w_2, \tag{5.31}$$

$$\mathrm{i}\hbar\frac{\mathrm{d}w_2}{\mathrm{d}t} = -A w_1 + E_0 w_2. \tag{5.32}$$

Following the usual recipe the solution may be attempted in the form

$$w_1 = K_1 \exp\left(-\mathrm{i}\frac{E}{\hbar}t\right), \qquad w_2 = K_2 \exp\left(-\mathrm{i}\frac{E}{\hbar}t\right). \tag{5.33}$$

Substituting eqn. (5.33) into eqns. (5.31) and (5.32) we get

$$K_1E = E_0K_1 - AK_2 \tag{5.34}$$

$$K_2E = -AK_1 + E_0K_2 \tag{5.35}$$

which has a solution only if

$$\begin{vmatrix} E_0 - E & -A \\ -A & E_0 - E \end{vmatrix} = 0. \tag{5.36}$$

Expanding the determinant we get

$$(E_0 - E)^2 = A^2 \tag{5.37}$$

whence

$$E = E_0 \pm A. \tag{5.38}$$

If there is no coupling between the two states, then $E = E_0$; that is, both states have the same energy. If there is coupling, the energy level is split. There are two new energy levels $E_0 + A$ and $E_0 - A$. This is a very important phenomenon that you will meet again and again. Whenever there is coupling, the energy splits.

The energies $E_0 \pm A$ may be defined as the energies of so-called stationary states obtainable from linear combinations of the original states. For our purpose it will suffice to know that we can have states with energies $E_0 + A$ and $E_0 - A$.

How will these energies vary with d, the distance between the protons? What is A anyway? A has come into our equations as a coupling term. The larger A, the larger the coupling and the larger the split in energy. Hence A must be related to the tunnelling probability that the electron may get through the potential barrier between the protons. Since tunnelling probabilities vary exponentially with distance (we have talked about this before when solving Schrödinger's equation for a tunnelling problem), A must vary roughly in the way shown in Fig. 5.5.

Now what is E_0? It is the energy of the states shown in Fig. 5.4. It consists of the potential and kinetic energies of the electron and of the potential energies of the protons (assumed immobile again). When the two protons are far away their potential energies are practically zero, and the electron's energy (since it is bound to a proton) is a

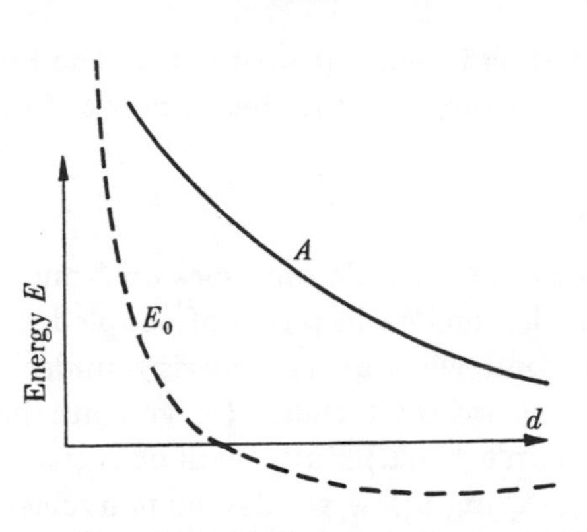

FIG. 5.5. The variation of E_0 and A with the interproton separation, d. E_0 is the energy when the states shown in Fig. 5.4 are uncoupled. A is the coupling term.

negative quantity. Thus E_0 is negative for large interproton distances but rises rapidly when the separation of the two protons is less than the average distance of the fluctuating electron from the protons. A plot of E_0 against d is also shown in Fig. 5.5.

We may now obtain the energy of our states by forming the combinations $E_0 \pm A$. Plotting these in Fig. 5.6 we see that $E_0 - A$ has a minimum, that is, at that particular value of d a stable configuration exists. We may also argue in terms of forces. Decreasing energy means an attractive force. Thus, when the protons are far away and we consider the state with the energy $E_0 - A$ there is an attractive force between the protons. This will be eventually balanced by the Coulomb repulsion between the protons, and an equilibrium will be reached.

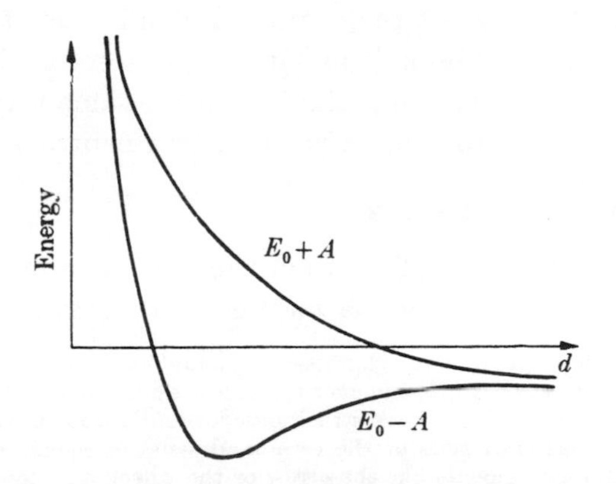

FIG. 5.6. Summing the quantities in Fig. 5.5 to get $E_0 + A$ and $E_0 - A$. The latter function displays all the characteristics of a bonding curve.

Thus in order to explain semi-quantitatively the hydrogen molecular ion we have had to introduce a number of new or fairly new quantum-mechanical ideas.

5.9. Nuclear forces

Feynman in his *Lectures on Physics* goes on from here and discusses a large number of phenomena in terms of coupled modes. Most of the phenomena are beyond what an engineering undergraduate needs to know; so with regret we omit them. (If you are interested you can always read Feynman's book.) But I cannot resist the temptation to follow Feynman in saying a few words about nuclear forces. With the treatment of the hydrogen molecular ion behind us we can really acquire some understanding of how forces between protons and neutrons arise.

It is essentially the same idea that we encountered before. A hydrogen atom and a proton are held together owing to the good services of an electron. The electron jumps from the hydrogen atom to the proton, converting the latter into a hydrogen atom. Thus when a reaction

$$\mathrm{H, p \rightarrow p, H} \tag{5.39}$$

takes place a bond is formed.

Yukawa† proposed in the middle of the 1930s that the forces between nucleons may have the same origin. Let us take the combination of a proton and a neutron. We may say again that a reaction

$$\mathrm{p, n \rightarrow n, p} \tag{5.40}$$

takes place and a bond is formed. 'Something' goes over from the proton to the neutron which causes the change, and this 'something' is called a positively-charged π-meson. Thus, just as an electron holds together two protons in a hydrogen molecular ion, in the same way a positively-charged π-meson holds together a proton and a neutron in the nucleus.

5.10. The hydrogen molecule

The hydrogen molecule differs from the hydrogen molecular ion by having one more electron. So we may choose our states as shown in

† If you permit me a digression in a digression, I should like to point out that Yukawa, a Japanese, was the first non-European ever to make a significant contribution to theoretical physics. Many civilizations have struck independently upon the same ideas, as for example the virgin births of gods or the commandments of social conduct, invented independently useful instruments like the arrow or the wheel, and developed independently similar judicial procedures and constitutions, but, interestingly, no civilization other than the European one bothered about theoretical physics.

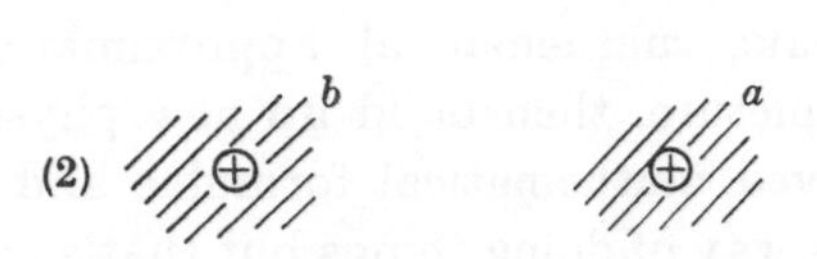

FIG. 5.7. The two basic states of the hydrogen molecule. Each electron can be attached to either proton leading to a coupling between the states.

Fig. 5.7. State (1) is when electron *a* is with proton 1 and electron *b* with proton 2, and state (2) is obtained when the electrons change places.

How do we know which electron is which? Aren't they indistinguishable? Yes, they are but we may distinguish them by assigning opposite spins to them.

We may now explain the bond of the hydrogen molecule in a manner analogous to that of the hydrogen molecular ion, but instead of a single electron jumping to and fro we have now two electrons changing places. Thus we may argue again that owing to symmetry the energies of the two states are identical. The coupling between the states due to the exchange of electrons splits the energy levels, one becoming somewhat higher, the other somewhat lower. Having the chance to lower the energy results in an attractive force which is eventually balanced by the repulsive force between the protons. And that is the reason why the hydrogen molecule exists.

It is interesting to compare this picture with the purely intuitive one described earlier, based on the atoms' 'desire' to fill the energy shells. In the present explanation we are saying that the bond is due to the *exchange* of electrons; previously we said the bond was due to *sharing* of the electrons. Which is it? Is it sharing or swapping? It is neither. Both explanations are no more than physical pictures to help the imagination.

We could equally well have said that the hydrogen molecule exists because it comes out mathematically from our basic premisses, that is the spin and Pauli's principle added to Schrödinger's equation. The problem is a purely mathematical one that can be solved by approximate methods. There is no need whatsoever for a physical picture. This argument would hold its ground if mathematical approximations were always available. But they are *not* available. A mathematician without some physical (or chemical) intuition would be completely lost. The mathematical techniques are inadequate without the help of physics. So we must strive to build up physical pictures from mathematical solutions, then make mathematical approximations based on the acquired physical picture, then build up new physical pictures based on the newly derived mathematical formulae and so on, and so on. It seems a tortuous way of doing things but that's how it is.

It is a lot easier in classical physics. Our physical picture is readily acquired in conjunction with our other faculties. We don't need to be taught that two bricks can't occupy the same place: we know they can't.

In studying phenomena concerned with extremely small things beyond the powers of direct observation, the situation is different. The picture of an atom with filled and unfilled energy shells is not a picture acquired through personal experience. It has come about by solving a differential equation. But once the solution is obtained, a physical picture starts emerging. We may visualize little boxes, or concentric spheres, or rows of seats in the House of Commons filling up slowly with MPs. The essential thing is that we *do* form some kind of picture of the energy shells. And once the shell picture is accepted it helps us find an explanation for the next problem, the bond between the atoms.

So you should not be unduly surprised that many alternative explanations are possible. They reflect attempts to develop intuition in a discipline where intuition does not come in a natural way.

Whenever confronted with new problems one selects from this store of physical pictures the ones likely to be applicable. If one of the physical pictures does turn out to be applicable, it is a triumph both for the picture and for the man who applied it. If all attempts fail, then either a new physical picture or a better man is needed to tackle the problem.

5.11. An analogy

One of the most important conclusions of the foregoing discussion was that 'whenever there is coupling the energy levels split'. This is a very important relationship in quantum mechanics but it could also be regarded as a simple mathematical consequence of the mathematical formulation. If we have coupled differential equations something will always get split somewhere. The example we are all familiar with is that of coupled electric resonant circuits shown in Fig. 5.8. If the two circuits

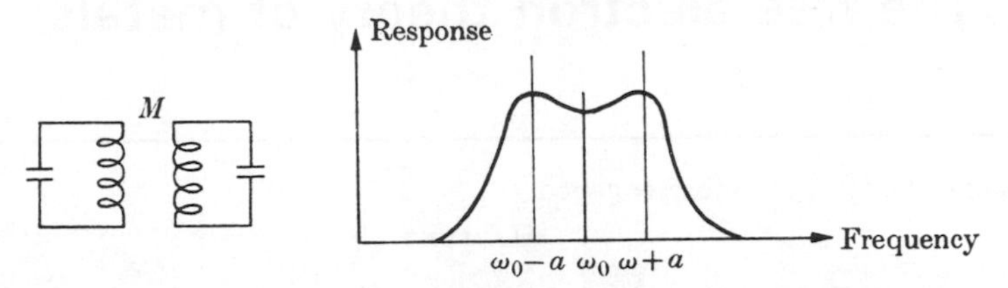

FIG. 5.8. The coupled circuit analogy. Two resonant circuits tuned to ω_0 when far apart (no coupling between them), have their resonant frequency split to $\omega_0 \pm a$ (cf. $E_0 \pm A$) when coupled.

are far away from each other (that is they are uncoupled) both of them have resonant frequencies ω_0. When the circuits are coupled there are two resonant frequencies $\omega_0 \pm a$, that is, we may say the resonant frequencies are split.

Examples

1. Discuss qualitatively the various mechanisms of bonding. Give examples of materials for each type of bond and also materials that do not have a clear single bond type.
2. Show that the force between two aligned permanent dipoles a distance r apart is attractive and varies as r^{-4}.

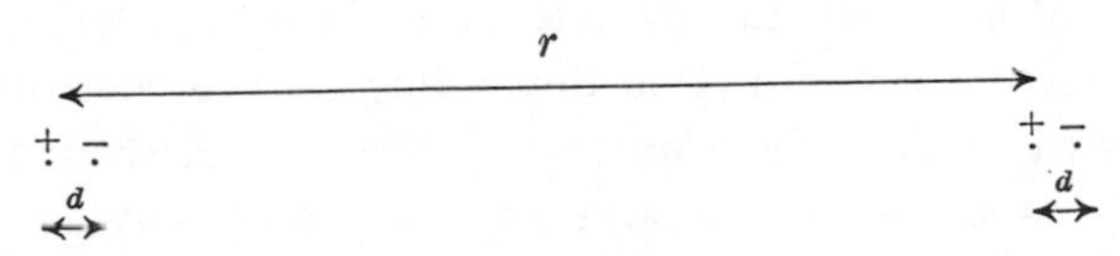

3. The interaction energy between two atoms may be phenomenologically described by eqn. (5.8). Show that the molecule will break up when the atoms are pulled apart to a distance

$$r_b = \left(\frac{n+1}{m+1}\right)^{1/(n-m)} r_0$$

where r_0 is the equilibrium distance between the atoms. Discuss the criterion of breaking used to get the above result.
4. For the KCl crystal the variation of energy may also be described by eqn. (5.8) but now r means the interatomic distance in the cubic crystal. Take $m = 1$, $n = 9$, $B = 1{\cdot}75e^2/4\pi\epsilon_0$. The bulk modulus of elasticity is $1{\cdot}88\ 10^{10}$ newton m^{-2}. Calculate the separation of the K^+–Cl^- ions in the ionic solid.
5. For a symmetrical coupled system the decrease in energy (in respect to the uncoupled case) is A as shown by eqn. (5.38). Show that, for an unsymmetrical system ($H_{11} \neq H_{22}$) with the same coupling ($H_{12}H_{21} = A^2$), the decrease in energy is less than A.

6. The free electron theory of metals

Struggling to be free, art more engaged

HAMLET

Much have I travelled in the realms of gold,
And many goodly states and kingdoms seen.

KEATS *On First Looking into Chapman's Homer.*

6.1. Free electrons

THE electric and magnetic properties of solids are mainly determined by the properties of electrons in them. Protons can usually be relegated to subordinate roles like ensuring charge neutrality. Neutrons may sometimes need to be considered, as for example in some superconducting materials where the critical temperature depends on the total mass of the nucleus, but on the whole the energy levels of electrons hold the key to the properties of solids.

The mathematical problem is not unlike the one we met in the case of individual atoms. How can we determine the energy levels of electrons in a solid? Take a wavefunction depending on the coordinates of 10^{25} electrons; write down the Coulomb potential between each pair of electrons, between electrons and protons; and solve Schrödinger's equation. This is an approach which, as you have probably guessed, we are not going to try. But what can we do instead? We can take a much simpler model, which is mathematically soluble, and hope that the solution will make sense.

Let us start our search for a simple model by taking a piece of metal and noting the empirical fact (true at room temperature) that there are no electrons beyond the boundaries of the metal. So there is some mechanism keeping the electrons inside. What is it? It might be an infinite potential barrier at the boundaries. And what about inside? How will the potential energy of an electron vary in the presence of that enormous number of nuclei and other electrons? Let us say it will be uniform. You may regard this a sweeping assumption (and, of course, you are absolutely right) but it works. It was introduced by Sommerfeld in 1928, and has been known as the 'free electron' model of a metal. The electrons inside the metal (more correctly the valence electrons which occupy the outer ring) are entirely free to roam around but they are not allowed to leave the metal.

You may recognize that the model is nothing else but the potential well we met before. There we obtained the solution for the one-dimensional case in the following form:

$$E = \frac{\hbar^2 k^2}{2m}$$

$$= \frac{h^2}{8m}\frac{n^2}{L^2}. \tag{6.1}$$

If we imagine a cube of side L containing the electrons then we get for the energy in the same manner

$$E = \frac{\hbar^2}{2m}(k_x^2+k_y^2+k_z^2)$$

$$= \frac{h^2}{8mL^2}(n_x^2+n_y^2+n_z^2), \tag{6.2}$$

where n_x, n_y, n_z are integers.

6.2. The density of states and the Fermi–Dirac distribution

The allowed energy, according to eqn. (6.2) is an integral multiple of $h^2/8mL^2$. For a volume of 10^{-6} m^3 this unit of energy is

$$E_{\text{unit}} = \frac{(6{\cdot}62\times10^{-34})^2}{8\times9{\cdot}1\times10^{-31}\times10^{-4}} = 0{\cdot}6\times10^{-33}\text{ J} = 3{\cdot}74\times10^{-15}\text{ eV}. \tag{6.3}$$

This is the energy difference between the first and second levels, but since the squares of the integers are involved the difference between neighbouring energy levels increases at higher energies. Let us anticipate the result obtained in the next section and take for the maximum energy $E = 3$ eV, which is a typical figure. Taking $n_x^2 = n_y^2 = n_z^2$ this maximum energy corresponds to a value of $n_x \cong 1{\cdot}64\times10^7$. Now an energy level just below the maximum energy can be obtained by taking the integers n_x-1, n_x, n_x. We get for the energy difference

$$\Delta E \cong 1{\cdot}22\times10^{-7}\text{ eV}, \tag{6.4}$$

that is, even at the highest energy the difference between neighbouring energy levels is as small as 10^{-7} eV. We can therefore say that in a macroscopically small energy interval dE there are still many discrete energy levels. So we can introduce the concept of density of states, which will simplify our calculations considerably.

The next question we ask is how many states are there between the energy levels E and $E+\text{d}E$. It is convenient to introduce for this purpose the new variable n with the relationship

$$n^2 = n_x^2 + n_y^2 + n_z^2. \tag{6.5}$$

Thus n represents a vector to a point n_x, n_y, n_z in three-dimensional space. In this space every integer specifies a state, that is a unit cube contains exactly one state. Hence the number of states in any volume is just equal to the numerical value of the volume. Thus in a sphere of radius n the number of states is

$$\frac{4n^3\pi}{3}. \tag{6.6}$$

Since n and E are related this is equivalent to saying that the number of states having energies less than E is

$$\frac{4n^3\pi}{3} = \frac{4\pi}{3}K^{\frac{3}{2}}E^{\frac{3}{2}} \quad \text{with} \quad K = \frac{8mL^2}{h^2}. \tag{6.7}$$

Similarly, the number of states having energies less than $E + \mathrm{d}E$ is

$$\frac{4\pi}{3}K^{\frac{3}{2}}(E + \mathrm{d}E)^{\frac{3}{2}}. \tag{6.8}$$

So the number of states having energies between E and $E + \mathrm{d}E$ is equal to

$$\begin{aligned} Z(E)\,\mathrm{d}E &= \frac{4\pi}{3}K^{\frac{3}{2}}\{(E + \mathrm{d}E)^{\frac{3}{2}} - E^{\frac{3}{2}}\} \\ &= 2\pi K^{\frac{3}{2}}E^{\frac{1}{2}}\,\mathrm{d}E. \end{aligned} \tag{6.9}$$

This is not the end yet. We have to note that only positive values of n_x, n_y, n_z are permissible; therefore we have to divide by a factor 8. Allowing further for the two values of spin we have to multiply by a factor 2. We get finally

$$Z(E)\,\mathrm{d}E = CE^{\frac{1}{2}}\,\mathrm{d}E \quad \text{with} \quad C = 4\pi L^3(2m)^{\frac{3}{2}}/h^3. \tag{6.10}$$

Eqn. (6.10) gives us the number of states but we would also like to know the number of *occupied* states, that is, the number of states that contain electrons. For that we need to know the probability of occupation, $F(E)$. This function can be obtained by a not-too-laborious exercise in statistical mechanics. One starts with the Pauli principle (that no state can be occupied by more than one electron) and works out the most probable distribution on the condition that the total energy and the total number of particles are given. The result is the so-called

Fermi–Dirac distribution

$$F(E) = \frac{1}{\{\exp(E - E_{\mathrm{F}})/kT\}+1}. \tag{6.11}$$

where E_{F} is a parameter called the Fermi level. It has the easily memorized property that at

$$E = E_{\mathrm{F}}, \qquad F(E) = \tfrac{1}{2}, \tag{6.12}$$

that is, at the Fermi level the probability of occupation is $\frac{1}{2}$.

As may be seen in Fig. 6.1, $F(E)$ looks very different from the classical

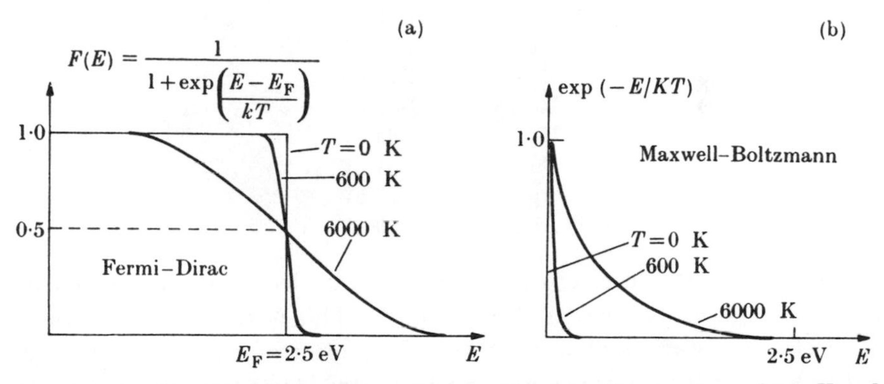

FIG. 6.1. (a) The Fermi–Dirac distribution function for a Fermi energy of 2·5 eV and for temperatures of 0 K, 600 K, and 6000 K. (b) The classical Maxwell–Boltzmann distribution function of energies for the same temperatures.

distribution $\exp(-E/kT)$. Let us analyse its properties in the following cases:

(1) At $T = 0$,

$$\begin{aligned} F(E) &= 1 \\ F(E) &= 0 \end{aligned} \quad \text{for} \quad \begin{aligned} E &< E_{\mathrm{F}} \\ E &> E_{\mathrm{F}}. \end{aligned} \tag{6.13}$$

Thus at absolute zero temperature all the available states are occupied up to E_{F} and all the states above E_{F} are empty. But, remember, $Z(E)\,\mathrm{d}E$ is the number of states between E and $E+\mathrm{d}E$. Thus the total number of states is

$$\int_0^{E_{\mathrm{F}}} Z(E)\,\mathrm{d}E, \tag{6.14}$$

which must equal the total number of electrons NL^3 where N is the number of electrons per unit volume. Thus substituting eqn. (6.10) into eqn. (6.14) the following equation must be satisfied

$$(4\pi L^3(2m)^{\frac{3}{2}}/h^3)\int_0^{E_F} E^{\frac{1}{2}}\,dE = NL^3. \tag{6.15}$$

Integrating and solving for E_F we get

$$E_F = \frac{h^2}{2m}\left(\frac{3N}{8\pi}\right)^{\frac{2}{3}}. \tag{6.16}$$

E_F, calculated from the number of free electrons, is shown in Table 6.1.

TABLE 6.1. *Fermi levels calculated from eqn. (6.16)*

Li	4·72 eV
Na	3·12 eV
K	2·14 eV
Rb	1·82 eV
Cs	1·53 eV
Cu	7·04 eV
Ag	5·51 eV
Al	11·70 eV

Thus, even at absolute zero temperature, the electrons' energies cover a wide range up to several electron volts. This is strongly in contrast with classical statistics where at $T = 0$ all electrons have zero energy.

(2) For electron energies above the Fermi level, so that

$$E - E_F \gg kT, \tag{6.17}$$

the term unity in eqn. (6.11) may be neglected, leading to

$$F(E) \cong \exp\{-(E-E_F)/kT\}, \tag{6.18}$$

which you may recognize as the classical Maxwell–Boltzmann distribution. That is, for sufficiently large energies the Fermi–Dirac distribution is reduced to the Maxwell–Boltzmann distribution, generally referred to as the 'Boltzmann tail'.

(3) For electron energies below the Fermi level, so that

$$E_F - E \gg kT, \tag{6.19}$$

eqn. (6.11) may be approximated by

$$F(E) \cong 1 - \exp(E-E_F)/kT, \tag{6.20}$$

that is, the probability of a state being occupied is very close to unity.

It is sometimes useful to talk about the probability of a state *not* being occupied and use the function $1-F(E)$. We may say then for the present case that the probability of non-occupation varies exponentially.

(4) In the range $E \approx E_F$ the distribution function changes rather abruptly from nearly unity to nearly zero. The rate of change depends on kT. For absolute zero temperature the change is infinitely fast, for higher temperatures (as can be seen in Fig. 6.1) it is more gradual. We may take this central region (quite arbitrarily) as between $F(E) = 0{\cdot}9$ and $F(E) = 0{\cdot}1$. The width of the region comes out then (by solving eqn. (6.11)) to about $4{\cdot}4kT$.

Summarizing, we may distinguish three regions for finite temperatures: from $E = 0$ to $E = E_F - 2{\cdot}2kT$, where the probability of occupation is close to unity, and the probability of non-occupation varies exponentially; from $E = E_F - 2{\cdot}2kT$ to $E = E_F + 2{\cdot}2kT$, where the distribution function changes over from nearly unity to nearly zero; and from $E = E_F + 2{\cdot}2kT$ to $E = \infty$, where the probability of occupation varies exponentially.

6.3. The specific heat of electrons

Classical theory, as I have mentioned before, failed to predict the specific heat of electrons. Now we can see the reason. The real culprit is not the wave nature of the electron nor Schrödinger's equation but Pauli's principle. Since only one electron can occupy a state, electrons of lower energy are not in a position to accept the small amount of energy offered to them occasionally. The states above them are occupied, so they stay where they are. Only the electrons in the vicinity of the Fermi level have any reasonable chance of getting into states of higher energy; so they are the only ones capable of contributing to the specific heat.

The specific heat at constant volume per electron is defined as

$$c_V = \frac{d\langle E\rangle}{dT} \tag{6.21}$$

where $\langle E\rangle$ is the average energy of electrons.

A classical electron would have an average energy $\frac{3}{2}kT$. On the basis of the Fermi–Dirac distribution we may make a very rough estimate, saying that only the electrons in the region $E = E_F - 2{\cdot}2kT$ to $E = E_F$ get promoted, that is, only a portion $2{\cdot}2kT/E_F$ of the electrons behave as classical electrons. Hence the average energy is

$$\langle E\rangle = \tfrac{3}{2}kT\frac{2{\cdot}2kT}{E_F}, \tag{6.22}$$

which gives for the specific heat

$$c_V = 6{\cdot}6\frac{k^2}{E_F}T. \tag{6.23}$$

A proper derivation of the specific heat would run into mathematical difficulties but it is very simple in principle. The average energy of an electron following a distribution $F(E)$ is given by

$$\langle E\rangle = \frac{1}{N}\int_0^\infty F(E)Z(E)E\,\mathrm{d}E, \tag{6.24}$$

which should be evaluated as a function of temperature† and differentiated. The result is

$$c_V = \frac{\pi^2}{2}\frac{k^2}{E_F}T, \tag{6.25}$$

which agrees reasonably well with eqn. (6.23) obtained by heuristic arguments. This electronic specific heat is vastly lower than the classical value $(3/2)k$ for any temperature at which a material can remain solid.

6.4. The work function

If the metal is heated or light waves are incident upon it then electrons may leave the metal. A more detailed experimental study would reveal that there is a certain threshold energy the electrons should possess in order to be able to escape. We call this energy (for historical reasons) the *work function* and denote it by ϕ. Thus our model is as shown in Fig. 6.2. At absolute zero temperature all the states are filled up to E_F and there is an external potential barrier ϕ.

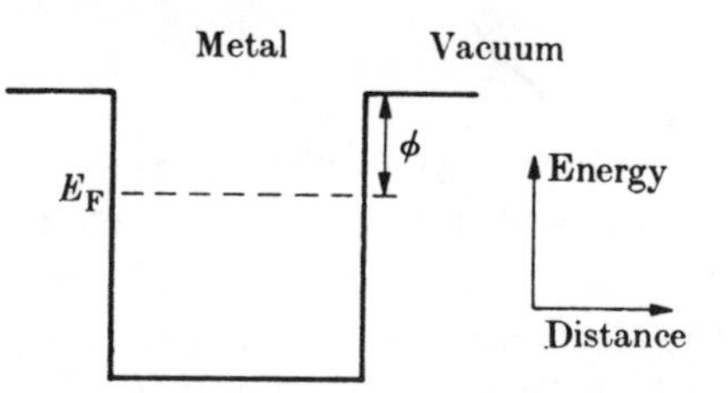

FIG. 6.2. Model for thermionic emission calculation. The potential barrier that keeps the electrons in the metal is above the Fermi energy level by an energy **T**.

† See, for example, F. Seitz, *Modern Theory of Solids*, McGraw-Hill, New York, 1940, p. 146.

It must be admitted that our new model is somewhat at variance with the old one. Not long ago, we calculated the energy levels of the electrons on the assumption that the external potential barrier is infinitely large, and now I go back on my word and say that the potential barrier is finite after all. Is this permissible? Strictly speaking, no. To be consistent we should solve Schrödinger's equation subject to the boundary conditions for a finite potential well. But since the potential well is deep enough and the number of electrons escaping is relatively small there is no need to recalculate the energy levels. So I am cheating, but not excessively.

6.5. Thermionic emission

In this section we shall be concerned with the emission of electrons at high temperatures (hence the adjective *thermionic*). As we agreed before, the electron needs at least $E_F+\phi$ energy in order to escape from the metal, and all this, of course, should be available in the form of kinetic energy. Luckily, in the free-electron model, all energy is kinetic energy (since the potential energy is zero and the electrons do not interact); so the relationship between energy and momentum is simply

$$E = \frac{1}{2m}(p_x^2+p_y^2+p_z^2). \tag{6.26}$$

A further condition is that the electron, besides having the right amount of energy, must go in the right direction. Taking x as the coordinate perpendicular to the surface of the metal, the electron momentum must satisfy the inequality

$$\frac{p_x^2}{2m} > \frac{p_{x0}^2}{2m} = E_F+\phi. \tag{6.27}$$

However, this is still not enough. An electron may not be able to scale the barrier even if it has the right energy in the right direction. According to the rules of quantum mechanics it may still suffer reflection. Thus the probability of escape is $1-r(p_x)$, where $r(p_x)$ is the reflection coefficient. If the number of electrons having a momentum between p_x and $p_x+\mathrm{d}p_x$ is $N(p_x)\,\mathrm{d}p_x$, then the number of electrons arriving at the surface per second per unit area is

$$\frac{p_x}{m}N(p_x)\,\mathrm{d}p_x \tag{6.28}$$

and the number of those escaping is

$$\{1-r(p_x)\}\frac{p_x}{m}N(p_x)\,\mathrm{d}p_x. \tag{6.29}$$

Adding the contributions from all electrons that have momenta in excess of p_{x_0}, we can write for the emission current density

$$J = \frac{e}{m}\int_{p_{x_0}}^{\infty}\{1-r(p_x)\}p_x N(p_x)\,\mathrm{d}p_x. \tag{6.30}$$

We may obtain the number of electrons in an infinitesimal momentum range in the same way as for the infinitesimal energy range. First of all it consists of two factors, the density of states multiplied by the probability of occupation. The density of states, $Z(p_x)$, can be easily obtained by noting from eqns. (6.2) and (6.26) that

$$p_x = \frac{h}{2}n_x,\quad p_y = \frac{h}{2}n_y,\quad p_z = \frac{h}{2}n_z. \tag{6.31}$$

The number of states in a cube of side one is exactly one. Therefore the number of states in a volume of sides $\mathrm{d}n_x$, $\mathrm{d}n_y$, $\mathrm{d}n_z$ is equal to $\mathrm{d}n_x\,\mathrm{d}n_y\,\mathrm{d}n_z$, which with the aid of eqn. (6.31) can be expressed as

$$\left(\frac{2}{h}\right)^3 \mathrm{d}p_x\,\mathrm{d}p_y\,\mathrm{d}p_z. \tag{6.32}$$

Dividing again by 8 (because only positive integers matter) and multiplying by two (because of spin) we get

$$Z(p_x, p_y, p_z) = \frac{2}{h^3}. \tag{6.33}$$

Hence the number of electrons in the momentum range p_x, $p_x+\mathrm{d}p_x$; p_y, $p_y+\mathrm{d}p_y$; p_z, $p_z+\mathrm{d}p_z$ is

$$N(p_x, p_y, p_z)\,\mathrm{d}p_x\,\mathrm{d}p_y\,\mathrm{d}p_z = \frac{2}{h^3}\frac{\mathrm{d}p_x\,\mathrm{d}p_y\,\mathrm{d}p_z}{\exp\left[\left\{\frac{1}{2m}(p_x^2+p_y^2+p_z^2)-E_\mathrm{F}\right\}\Big/kT\right]+1}. \tag{6.34}$$

To get the number of electrons in the momentum range p_x, $p_x+\mathrm{d}p_x$, the above equation needs to be integrated for all values of p_y and p_z

$$N(p_x)\,\mathrm{d}p_x = \frac{2}{h^3}\,\mathrm{d}p_x\int_{-\infty}^{\infty}\int_{-\infty}^{\infty}\frac{\mathrm{d}p_y\,\mathrm{d}p_z}{\exp\left[\left\{\frac{1}{2m}(p_x^2+p_y^2+p_z^2)-E_\mathrm{F}\right\}\Big/kT\right]+1}. \tag{6.35}$$

This integral looks rather complicated but since we are interested only in those electrons exceeding the threshold ϕ ($\gg kT$) we may neglect the unity term in the denominator. We are left then with some Gaussian functions whose integrals between $\pm\infty$ can be found in the better integral tables (you can derive them for yourself if you are fond of doing integrals). This leads us to

$$N(p_x)\,\mathrm{d}p_x = \frac{4\pi mkT}{h^3}\mathrm{e}^{E_F/kT}\mathrm{e}^{-p_x^2/2mkT}\,\mathrm{d}p_x. \qquad (6.36)$$

Substituting eqn. (6.36) into (6.30) and assuming that $r(p_x) = r$ is independent of p_x (which is not true but gives a good enough approximation), the integration can be easily performed, leading to

$$J = A_0(1-r)T^2\mathrm{e}^{-\phi/kT}, \qquad (6.37)$$

where

$$A_0 = 4\pi emk^2/h^3 = 1{\cdot}2\times10^6\ \mathrm{A\ m^{-2}\ K^{-2}}. \qquad (6.38)$$

The most important factor in eqn. (6.37) is $\exp(-\phi/kT)$, which is strongly dependent both on temperature and on the actual value of the work function. Take, for example, tungsten (the work functions for a number of metals are given in Table 6.2), for which $\phi \simeq 4{\cdot}5$ eV and

TABLE 6.2. *Work functions*

Na	2·3 eV
K	2·2 eV
Ca	3·2 eV
Cs	1·8 eV
Ba	2·5 eV
Pt	5·3 eV
Ta	4·2 eV
W	4·5 eV

take $T = 2500$ K. Then a 10 per cent change in the work function or temperature changes the emission by a factor of 8.

The main merit of eqn. (6.37) is to show the exponential dependence on temperature that is well borne out by experimental results. The actual numerical values are usually below those predicted by the equation but this is not very surprising in view of the many simplifications we had to introduce. In a real crystal ϕ is a function of temperature, of the surface conditions, and of the directions of the crystallographic axes, which our simple model did not take into account.

There is one more thing I would like to discuss, which is really so trivial that most textbooks don't even bother to mention it. Our analysis was done for a piece of metal in isolation. The electron current obtained in eqn. (6.37) is the current that would start to flow if the sample were suddenly heated to a temperature T. But this current would not flow for long because, as electrons leave the metal, it becomes positively charged, making it more difficult for further electrons to leave. Thus our formulae are valid only if we have some means of replenishing the electrons lost by emission. That is, we need an electric circuit like the one in Fig. 6.3a. As soon as an electron is emitted from

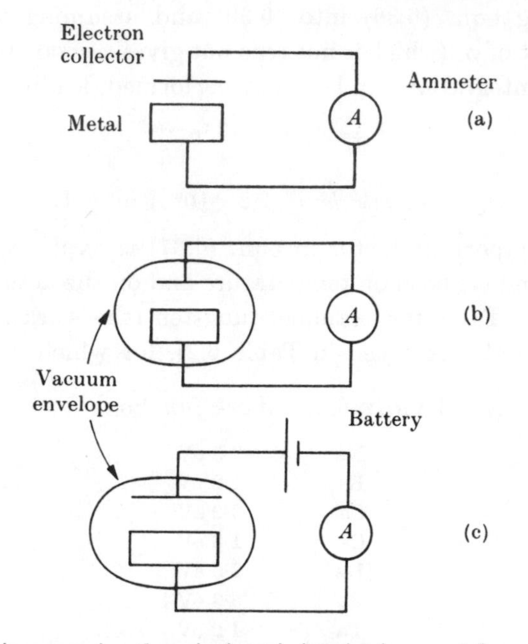

FIG. 6.3. Stages in measuring thermionic emission. (a) A current flows but it is impeded by air molecules. (b) A current flows in a vacuum until it builds up a charge cloud that repels further electrons. The steady-state ammeter reading is much less than the total emission current. (c) By employing a battery, all the emission current is measured.

our piece of metal another electron will enter from the circuit. The current flowing can be measured by an ammeter.

A disadvantage of this scheme is that the electrons travelling to the electrode will be scattered by air molecules; we should really evacuate the place between the emitter and the receiving electrode, making up the usual cathode–anode configuration of a vacuum tube. This is denoted in Fig. 6.3b by the envelope shown. The electrons are now free to reach the anode but also free to accumulate in the vicinity of the cathode. This is bad again, because by their negative charge they will compel many of their fellow electrons to interrupt their planned journey to the anode and return instead to the emitter. So again we do not measure the 'natural' current.

In order to prevent the accumulation of electrons in front of the cathode, a d.c. voltage may be applied to the anode (Fig. 6.3c); this will sweep out most of the unwanted electrons from the cathode–anode region. This is the arrangement used for measuring thermionic current.

The requirements to be fulfilled by cathode materials vary considerably according to the particular application. The cathodes must have a large emission current for high-power applications, low temperature for low-noise amplification, and long life for amplification in submarine repeaters. All these various requirements are admirably met by industry, though the feat should not be attributed to the powers of science. To make a good cathode is still an art, and a black art at that.

6.6. The Schottky effect

We are now going to refine our model for thermionic emission further by including (a) image force and (b) electric field.

It is a simple and rather picturesque consequence of the laws of electrostatics that the forces on an electron in front of an infinitely conducting sheet are correctly given by replacing the sheet by the 'mirror' charge (a positively charged particle the same distance behind the sheet as shown in Fig. 6.4). The force between these two charges is

$$F = \frac{e^2}{4\pi\varepsilon_0}\frac{1}{(2x)^2} \tag{6.39}$$

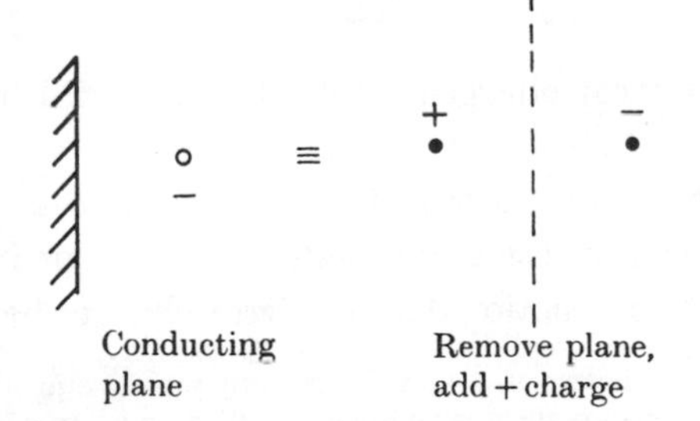

FIG. 6.4. The 'image charge' theorem. The effect of a plane conductor on the static field due to a charged particle is equivalent to a second, oppositely charged, particle in the mirror image position.

and the potential energy is the integral of this force from the point x to infinity:

$$V(x) = \int_x^\infty F(y)\,dy = -\frac{e^2}{16\pi\varepsilon_0 x}. \tag{6.40}$$

In the above calculation we took the potential energy as zero at $x = \infty$ to agree with the usual conventions of electrostatics, but remember that our zero a short while ago was that of a valence electron at rest. Hence, to be consistent, we must redraw the energy diagram inside and outside the metal as shown in Fig. 6.5a. If we include now

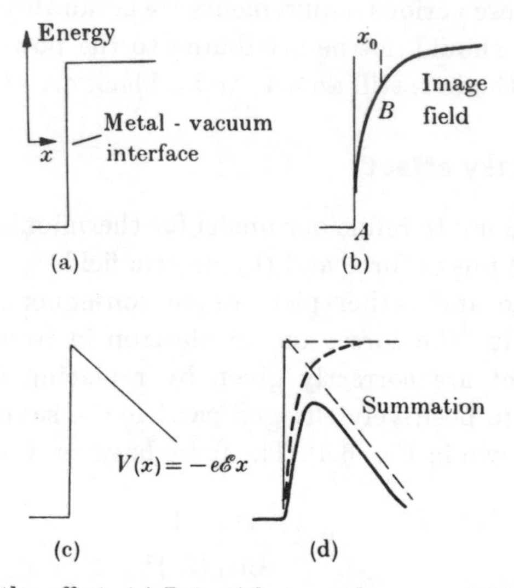

FIG. 6.5. The Schottky effect. (a) Potential at metal-vacuum interface. (b) Potential changed by image charge field. (c) Potential due to applied anode voltage in vacuum region. (d) Total potential field showing reduction in height of the potential barrier compared with (a).

the effect of the mirror charges† then the potential barrier modifies to that shown in Fig. 6.5b.

In the absence of an electric field this change in the shape of the potential barrier has practically no effect at all. But if we do have electric fields, the small correction due to mirror charges becomes significant.

† Note that the curve between A and B does not satisfy eqn. (6.40). This is because the concept of a homogeneous sheet is no longer applicable when x is comparable with the interatomic distance. The energy is, however, given for $x = 0$ (an electron resting on the surface must have the same energy as an electron at rest inside the metal); so we simply assume that eqn. (6.40) is valid for $x > x_0$ and connect the points A and B by a smooth line.

For simplicity let us investigate the case when the electric field is constant. Then

$$V(x) = -e\mathscr{E}x \tag{6.41}$$

as shown in Fig. 6.5c. If both an electric field is present and the mirror charges are taken into account then the potentials should be added, leading to the potential barrier shown in Fig. 6.5d. Clearly, there is a maximum that can be calculated from the condition

$$\frac{\mathrm{d}}{\mathrm{d}x}\left(-\frac{e^2}{16\pi\varepsilon_0 x}-e\mathscr{E}x\right) = 0, \tag{6.42}$$

leading to

$$V_{\max} = -e\left(\frac{e\mathscr{E}}{4\pi\varepsilon_0}\right)^{\frac{1}{2}}, \tag{6.43}$$

that is, the energy needed to escape from the metal is reduced by $-V_{\max}$. The effective work function is thus reduced from ϕ to

$$\phi_{\text{eff}} = \phi-e(e\mathscr{E}/4\pi\varepsilon_0)^{\frac{1}{2}}. \tag{6.44}$$

Substituting this into eqn. (6.37) we get the new formula for thermionic emission

$$J = A_0(1-r)T^2 \exp[-\{\phi-e\surd(e\mathscr{E}/4\pi\varepsilon_0)\}/kT]. \tag{6.45}$$

The reduction in the effective value of the work function is known as the *Schottky effect*, and plotting $\log J$ against $\mathscr{E}^{\frac{1}{2}}$ we get the so-called *Schottky line*. A comparison with experimental results in Fig. 6.6 shows

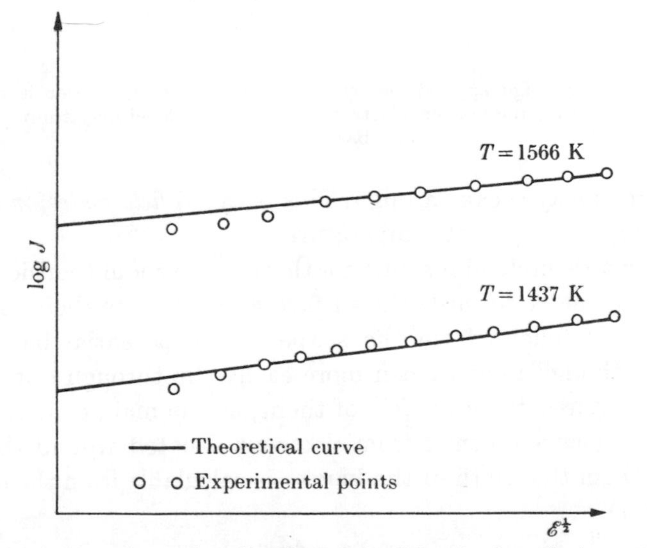

FIG. 6.6. Experimental verification of the Schottky formula (eqn. (6.45)).

that above a certain value of the electric field the relationship is quite accurate. Do not be too much impressed, though; in graphs of this sort the constants are generally fiddled to get the theoretical and experimental curves on top of each other. But it certainly follows from Fig. 6.6 that the functional relationship between J and $\mathscr{E}^{\frac{1}{2}}$ is correct.

6.7. Field emission

As we have seen in the previous section, the presence of an electric field increases the emission current because more electrons can escape over the reduced barrier. If we increase the electric field further (towards 10^9 V/m) then a new escape route opens up. Instead of going *over* the potential barrier, the electrons *tunnel across* it. It may be seen in Fig. 6.7 that for high-enough electric fields the barrier is thin and

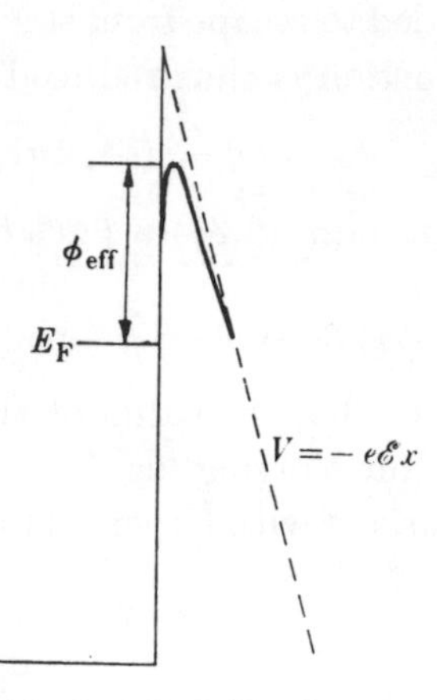

FIG. 6.7. With very high applied electric fields the potential barrier is thin, thus, instead of moving over the barrier, electrons at the Fermi level may tunnel across the barrier.

thus electrons may sneak through. This is called *field emission* and it is practically independent of temperature.

To derive a theoretical formula for this case we should consider all the electrons that move towards the surface and calculate their tunnelling probability. It follows from the shape of the potential barrier that electrons with higher energy can more easily slip through but (at ordinary temperatures) there are few of them; so the main contribution to the tunnelling current comes from electrons situated around the Fermi level. For them the width of the barrier is calculable from the equation (see Fig. 6.7)

$$-\phi = -e\mathscr{E}x_F \qquad (6.46)$$

and the height of the potential barrier they face is ϕ_{eff}. Hence, very approximately, we may represent the situation by the potential profile of Fig. 6.8. It may be shown (see Example 3.6) that the tunnelling

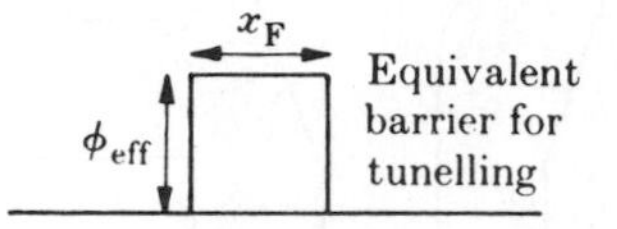

FIG. 6.8. Equivalent barrier, for simplifying the calculation of tunnelling current in Fig. 6.7.

current varies approximately exponentially with barrier width,

$$J \sim \exp\left(-\frac{(2m\phi_{\text{eff}})^{\frac{1}{2}}}{\hbar}x_F\right), \tag{6.47}$$

which, with the aid of eqn. (6.46) reduces to

$$J \sim \exp\left(-\frac{(2m)^{\frac{1}{2}}}{\hbar e}\frac{\phi_{\text{eff}}^{\frac{1}{2}}\phi}{\mathscr{E}}\right). \tag{6.48}$$

The exponential factor in eqn. (6.48) represents quite a good approximation to the exact formula, which is unfortunately too long to quote. It may be noted that the role of temperature in eqns. (6.37) and (6.45) is taken over here by the electric field.

The theory has been fairly well confirmed by experiments. The major difficulty in the comparison is to take account of surface irregularities. The presence of any protuberances considerably alters the situation because the electric field is higher at those places. This is a disadvantage as far as the interpretation of the measurements is concerned but the existence of the effect made possible the invention of an ingenious device called the field-emission microscope.

6.8. The field-emission microscope

The essential part of a field–emission microscope is a very sharp tip ($\approx$100 nm in diameter), which is placed in an evacuated chamber (Fig. 6.9). A potential of a few thousand volts is applied between the tip (made usually of tungsten) and the anode, which creates at the tip an electric field high enough to draw out electrons. The emitted electrons follow the lines of force and produce a magnified picture (magnification $= r_2/r_1$ where r_2 = radius of the screen and r_1 = radius of the tip) on the fluorescent screen. Since the magnification may be as large as 10^6

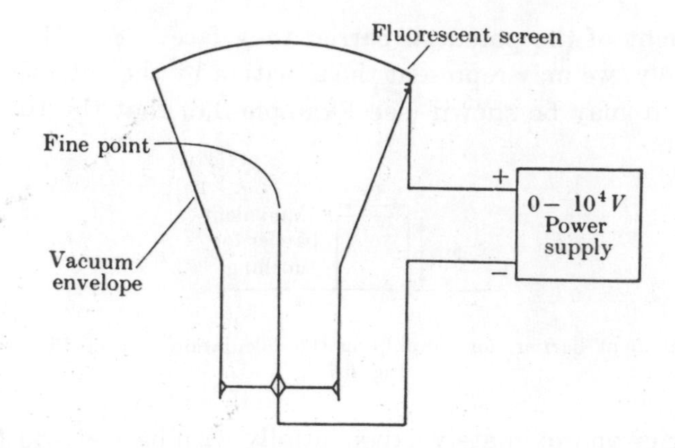

Fig. 6.9. Sketch showing the principle of field-emission microscope.

we could expect to see the periodic variation in the electron emission caused by the atomic structure. The failure to observe this is explained by two reasons: quantum-mechanical diffraction, and deviation from the 'theoretical' course owing to a random transverse component in the electron velocity when leaving the metal.

The limitations we have just mentioned can be overcome by introducing helium into the chamber and reversing the polarity of the applied potential. The helium atoms that happen to be in the immediate vicinity of the tungsten tip become ionized owing to the large electric field, thus acquiring a positive charge, and move to the screen. Both the quantum-mechanical diffraction (remember, the de Broglie wavelength is inversely proportional to mass) and the random thermal velocities are now smaller, so that the resolution is higher and individual atoms can indeed be distinguished as shown in Plate 1. This device, called the *field–ion microscope*, was the first in the history of science to make individual atoms visible. Thus just about two and a half millennia after introducing the concept of atoms it proved possible to see them.

6.9. The photoelectric effect

Emission of electrons owing to the incidence of electromagnetic waves is called the *photoelectric effect.* The word *photo* (*light* in Greek) came into the description because the effect was first found in the visible range. Interestingly, it is one of the earliest phenomena that cast serious doubts on the validity of classical physics and was instrumental in the birth of quantum physics.

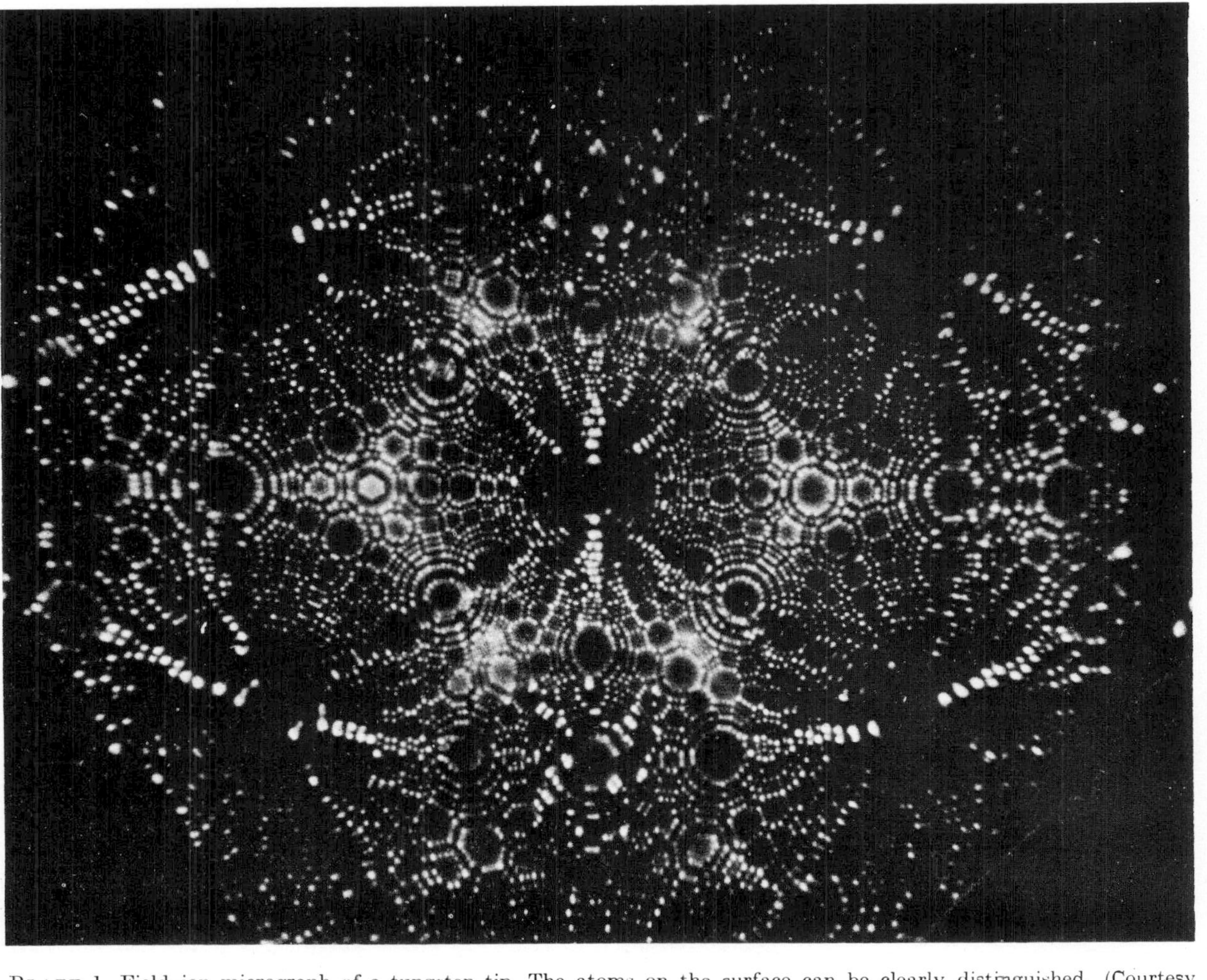

PLATE 1. Field–ion micrograph of a tungsten tip. The atoms on the surface can be clearly distinguished. (Courtesy of E. W. Muller)

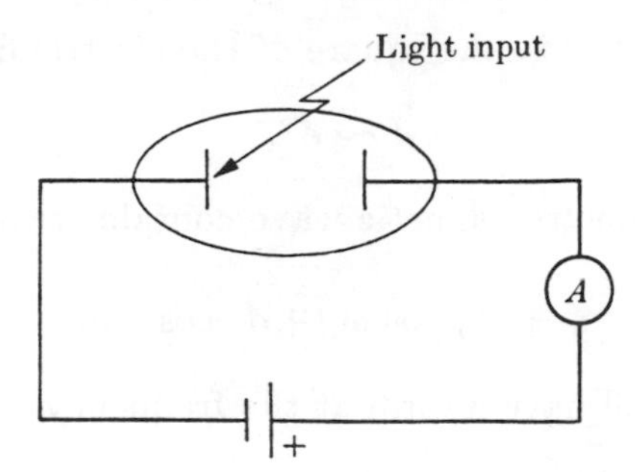

FIG. 6.10. An experiment showing the photoelectric effect. If the frequency of the light is above a certain threshold value, the incident photons knock out electrons from the cathode. These cause a current in the external circuit by moving to the anode.

The basic experimental set-up may be seen in Fig. 6.10. When an electromagnetic wave is incident, an electric current starts to flow between the electrodes. The magnitude of the current is proportional to the input electromagnetic power but there is no current unless the frequency is high enough to make

$$\hbar\omega > \phi. \tag{6.49}$$

The threshold occurs when the energy of the incident photon is just large enough to kick out an electron.

A detailed calculation of the current is not easy because a photon is under no obligation to give its energy to an electron. One must calculate transition probabilities, which are different at the surface and in the bulk of the material. The problem is rather complex; we shall not go more deeply into the theory. It might be some consolation for you that the first engineers who used and designed photocells (the commercially available device based on the photoelectric effect) knew much less about its functioning than you do.

While discussing applications I should like to mention a recent one made possible by the advent of the laser. Lasers produce light waves of a single frequency (well, nearly a single frequency). Assume now that there is an incident light wave that carries some information: telephone conversations, say, or a couple of hundred television channels,† and we want to demodulate it. How should we do it? Rather as in a radio receiver, we can have a local oscillator and then with the aid of a non-linear element we can produce the difference frequency. The photoelectric effect is certainly nonlinear: the electric current is proportional

† There is no demand at the present to take hundreds of television channels from point A to point B, but if the demand ever arises lasers would be capable of doing it.

to input power, that is to the square of the electric field intensity

$$J \sim \mathscr{E}^2. \tag{6.50}$$

Thus, if the input electromagnetic wave contains two distinct frequencies,

$$\mathscr{E} = \mathscr{E}_1 \cos \omega_1 t + \mathscr{E}_2 \cos \omega_2 t, \tag{6.51}$$

then the current will have a term at the frequency $\omega_1 - \omega_2$.

6.10. The junction between two metals

If two metals of different work functions are brought into contact (Fig. 6.11) the situation is clearly unstable. Electrons will cross from

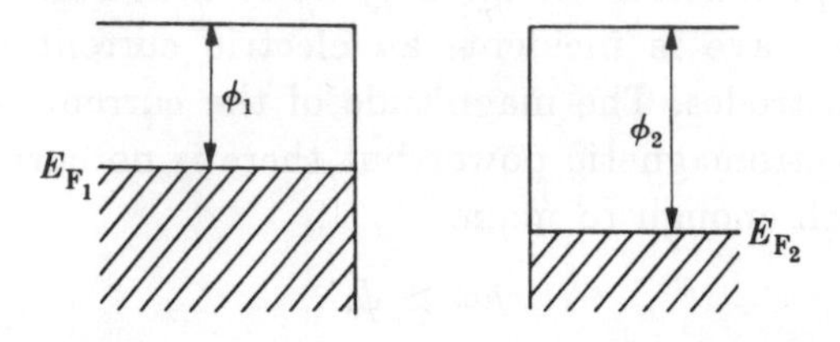

FIG. 6.11. The Fermi levels and work functions of two metals to be brought into contact.

left to right to occupy the lower energy states available. However, as electrons cross over there will be an excess of positive charge on the left-hand side and an excess of negative charge on the right-hand side. Consequently an electric field is set up with a polarity that hinders the flow of electrons from left to right and encourages the flow of electrons from right to left. A dynamic equilibrium is established when equal numbers of electrons cross in both directions. At what potential difference will this occur? An exact solution of this problem belongs to the domain of statistical thermodynamics. The solution is fairly lengthy but the answer (as so often in thermodynamics) could hardly be simpler.

The potential difference between the two metals (called the *contact potential*) is equal to the difference between the two work functions; or, in more general terms, the potential difference may be obtained by equating the Fermi levels of the two media in contact. This is a general law valid for any number of materials in equilibrium at any temperature.

The resulting energy diagram is shown in Fig. 6.12. The potential difference appearing between the two metals is a real one. If we could put an extra electron in the contact region, it would feel a force towards

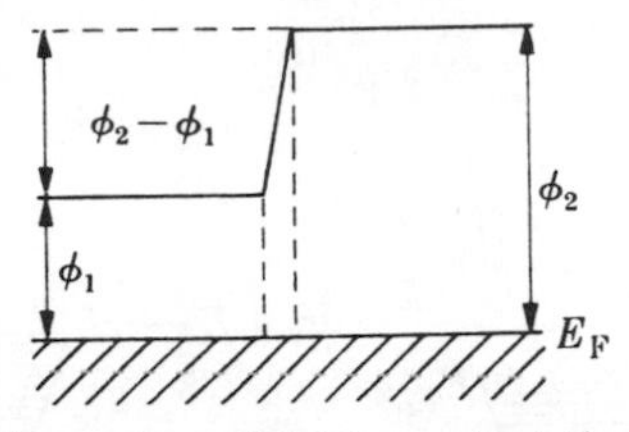

FIG. 6.12. When the two metals are brought into contact there is a potential difference $\phi_2-\phi_1$ between them.

the left. The potential difference is real but, alas, it cannot perform the function of a battery. Why? Because extracting power from an equilibrium state is against the second law of thermodynamics.

Examples

1. Evaluate the Fermi function for an energy kT above the Fermi energy.
   Find the temperature at which there is a 1 per cent probability that a state, with an energy 0·5 eV above the Fermi energy, will be occupied by an electron.
2. Indicate the main steps in the derivation of the Fermi level and calculate its value for sodium from the data given in example 1.4.
3. Show that the average kinetic energy of free electrons following Fermi–Dirac statistics is $(\frac{3}{5})E_F$ at $T = 0$ K.
4. The work function of tungsten is 4·5 eV. Calculate the thermionic emission of a filament 0·05 m long and 10^{-4} m diameter that is at a temperature of (a) 2400 K, (b) 2800 K. How much would the currents increase if a field of 10^7 V/m is applied to the surface?
5. Light from a He–Ne laser (wavelength 632·8 nm) falls on a photo-emissive cell with a quantum efficiency of 10^{-4} (the number of electrons emitted per incident photon. If the laser power is 2 mW, and all liberated electrons reach the anode, how large is the current? Could you estimate the work function of the cathode material by varying the anode voltage of the photocell?
6. Work out the Fermi level for conduction electrons in copper. Estimate its specific heat at room temperature; what fraction of it is contributed by the electrons? Check whether your simple calculation agrees with data on specific heat given in a reference book.
   Assume one conduction electron per atom. The atomic weight of copper is 63·5 and its density 9·4 10^3 kg/m^3.
7. Fig. 6.13a shows the energy diagram for a metal-insulator-metal sandwich at thermodynamic equilibrium. Take the insulator as representing a high potential barrier. The temperature is sufficiently low for all states above the Fermi level to be regarded as unoccupied. When a voltage U is applied (Fig. 6.13b) electrons may tunnel through the insulator from left to right. Assume that the tunnelling current in each energy range dE is proportional to the number of filled states from which tunnelling is possible and to the number of empty states on the other side into which electrons can tunnel.

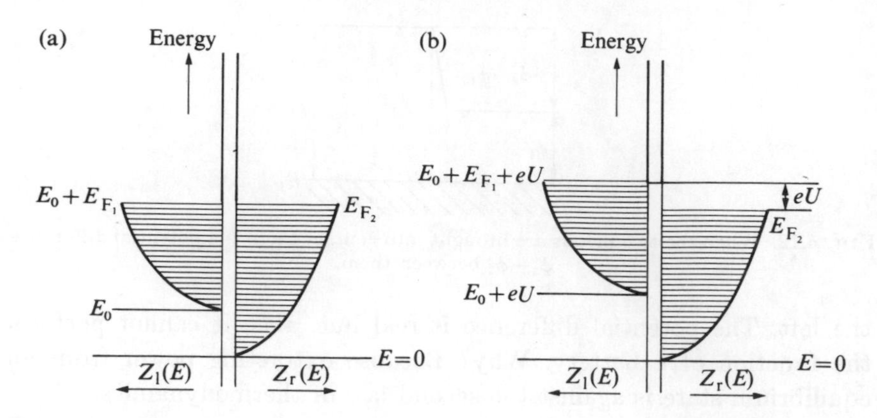

FIG. 6.13. Energy against density of states for a metal-insulator-metal tunnel junction.

In the co-ordinate system of Fig. 6.13a the density of states as a function of energy may be written as

$$Z_l(E) = C_l(E-E_0)^{\frac{1}{2}} \quad \text{for } E > E_0$$

and

$$Z_r(E) = C_r E^{\frac{1}{2}} \quad \text{for } E > 0,$$

where C_l, C_r, and E_0 are constants.

Show (i) that the tunnelling current takes the form

$$I \sim \int_{E_{F_2}}^{E_{F_2}+eU} (E-eU-E_0)^{\frac{1}{2}} E^{\frac{1}{2}}\,\mathrm{d}E$$

and (ii) that Ohm's law is satisfied for small voltages.

7. The band theory of solids

Band of all evils, cradle of causeless care.

SIR PHILIP SIDNEY

7.1. Introduction

MOST properties of metals can be well explained with the aid of the free-electron model but when we come to insulators and semiconductors the theory fails. This is not very surprising because the term 'free electron', by definition, means an electron free to roam around and conduct electricity; and we know that the main job of an insulator is to insulate, that is, *not* to conduct electricity. It is not particularly difficult to find a model explaining the absence of electrical conductivity. We only need to imagine that the valence electrons cling desperately to their respective lattice ions and are unwilling to move away. So we are all right at the two extremes; free electrons mean high conductivity, tightly bound electrons mean no conductivity. Now what about semiconductors? They are neither good conductors nor insulators; so neither model is applicable. What can we do? Well, we have touched upon this problem before. Silicon and germanium are semiconductors in spite of the covalent bonds between the atoms. The bonding process uses up all the available electrons, so at absolute zero temperature there are no electrons available for conduction. At finite temperatures however, some of the electrons may escape. The lattice atoms vibrate randomly, having *occasionally* much more than the average thermal energy. Thus at certain instants at certain atoms there is enough energy to break the covalent bond and liberate an electron. The higher the temperature the more likely it is that some electrons escape. This is a possible description of the electrical properties of semiconductors and, physically, it seems quite plausible. It involves no more than developing our physical picture of the covalent bond a little further by taking account of thermal vibrations as well. All we need to do is to put these arguments into some quantitative form and we shall have a theory of semiconductors. It can be done, but somehow the ensuing theory never caught the engineers' imagination.

The theory that did gain wide popularity is the one based on the concept of energy bands. This theory is more difficult to comprehend initially but once digested and understood it can provide a solid foundation for the engineers' flights of fancy.

The job of engineers is to invent. Physicists discover the laws of nature, and engineers exploit the phenomena for some useful (sometimes not so useful) end. But in order to exploit them the engineer needs to combine the phenomena, to regroup them, to modify them, to interfere with them; that is, to create something new from existing components. Invention has never been an easy task, but at least in the good old days the basic mechanism was simple to understand. It wasn't very difficult to be wise after the event. It was, for example, an early triumph of engineering to turn the energy of steam into a steam engine but, having accomplished the feat, most people could comprehend that the expanding steam moved a piston, which was connected to a wheel, etc. etc. It needed perhaps a little more abstract thought to appreciate Watt's invention of the separate condenser, but even then any intelligent man willing to devote half an hour of his time to the problem of heat exchange could realize the advantages. Alas, these times have gone. No longer can a layman hope to understand the working principles of complex mechanisms, and this is particularly true in electronics. And most unfortunately it is true not only for the layman. Even electronic engineers find it hard nowadays to follow the phenomena in an electronic device. Engineers may nowadays be expected to reach for the keyboard of a computer at the slightest provocation, but the fundamental equations are still far too complicated for a direct numerical attack. We need models. The models need not be simple ones but they should be comprehensive and valid under a wide range of conditions. They have to serve as a basis for intuition. Such a model and the concurrent physical picture are provided by the band theory of solids. It may be said without undue exaggeration that the spectacular advance in solid-state electronic devices in the last twenty years owes its existence to the power and simplicity of the band theory of solids.

Well, after this rather lengthy introduction, let us see what this theory is about. There are several elementary derivations, each one giving a slightly different physical picture. Since our aim is a thorough understanding of the basic ideas involved it is probably the best to show you all the three approaches I know.

7.2. The Kronig–Penney model

This model is historically the first (1930), and is concerned with the solution of Schrödinger's equation, assuming a certain potential distribution inside the solid. According to the free-electron model the potential inside the solid is uniform; the Kronig–Penney model goes one step further by taking into account the variation of potential due to the presence of immobile lattice ions.

Considering a one-dimensional case for simplicity, the potential energy of an electron is shown in Fig. 7.1. The highest potential is half-

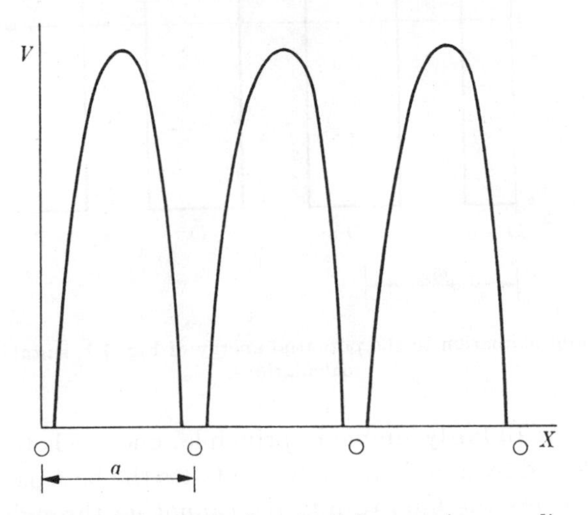

FIG. 7.1. The variation of the electron's potential energy in a one-dimensional crystal.

way between the ions and the potential tends to minus infinity as the position of the ions is approached. This potential distribution is still very complicated for a straightforward mathematical solution. We shall therefore replace it by a simpler one that still displays the essential features of the function, namely (i) it has the same period as the lattice; (ii) the potential is lower in the vicinity of the lattice ion and higher between the ions.

The potential distribution chosen is shown in Fig. 7.2. The ions are located at $x = 0$, a, $2a$, ... etc. The potential wells are separated from each other by potential barriers of height V_0 and width w.

The solution of the time-independent Schrödinger equation

$$\frac{\hbar^2}{2m}\frac{\mathrm{d}^2\psi}{\mathrm{d}x^2}+\{E-V(x)\}\psi = 0 \tag{7.1}$$

for the above chosen potential distribution is not too difficult. We can solve it for the $V(x) = V_0/2$ and $V(x) = -V_0/2$ regions separately, match the solutions at the boundaries, and take good care that the solution

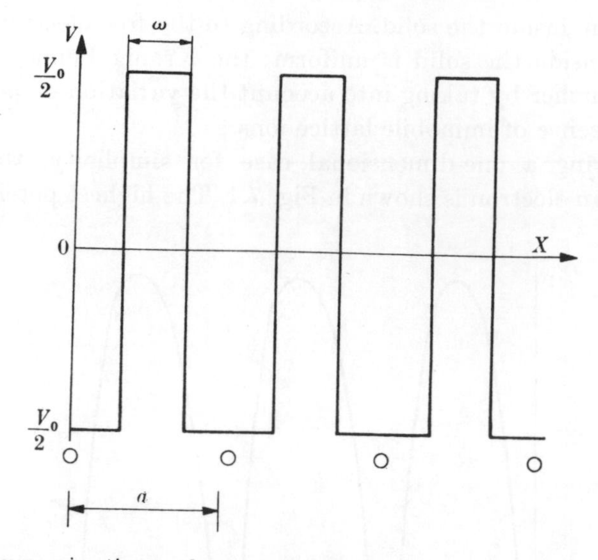

FIG. 7.2. An approximation to the potential energy of Fig. 7.1, suitable for analytical calculations.

is periodic. It is all fairly simple in principle; one needs to prove a new theorem followed by a derivation, which takes the best part of an hour, and then one gets the final result. We cannot go through the lot so I shall just say that the wave functions are assumed to be of the form

$$u_k(x)\mathrm{e}^{ikx}, \tag{7.2}$$

where $u_k(x)$ is a periodic function and a solution exists if k is related to the energy E by the following equation†

$$\cos ka = P\frac{\sin \alpha a}{\alpha a} + \cos \alpha a, \tag{7.3}$$

where

$$P = \frac{ma}{\hbar^2}V_0 w \tag{7.4}$$

and

$$\alpha = \frac{1}{\hbar}\sqrt{(2mE)}. \tag{7.5}$$

Remember, for a free electron we had the relationship

$$E = \frac{\hbar^2 k^2}{2m}. \tag{7.6}$$

† There is actually one more mathematical simplification introduced in arriving at eqn. (7.3), namely w and V_0 are assumed to tend to zero and infinity respectively, with their product $V_0 w$ kept constant.

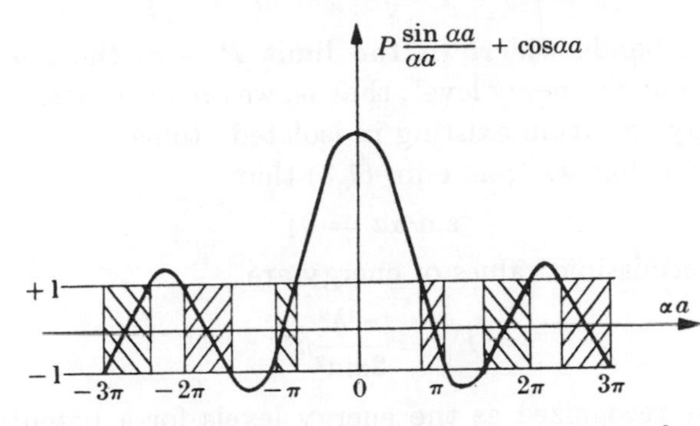

FIG. 7.3. The right-hand side of eqn. (7.3) for $P = 3\pi/2$ as a function of αa.

The relationship is now different, implying that the electron is no longer free.

In order to find the E–k curve we plot the right-hand side of eqn. (7.3) in Fig. 7.3 as a function of αa. Since the left-hand side of eqn. (7.3) must always be between +1 and −1, a solution exists only for those values of E for which the right-hand side is between the same limits; that is, there is a solution for the shaded region and no solution outside the shaded region. Since α is related to E this means that the electron may possess energies within certain bands but not outside them. Expressed in another way: *there are allowed and forbidden bands of energy.* This is our basic conclusion but we can draw some other interesting conclusions from eqn. (7.3).

(i) If $V_0 w$ is large, that is, if P is large, the function described by the right-hand side of eqn. (7.3) crosses the +1, −1 region at a steeper angle, as shown in Fig. 7.4. Thus the allowed bands are narrower and

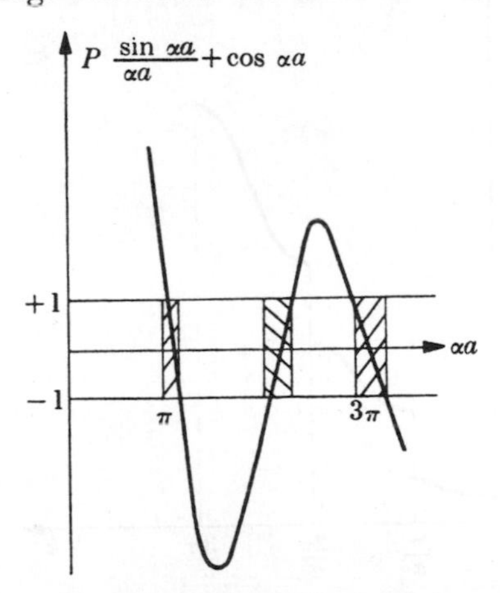

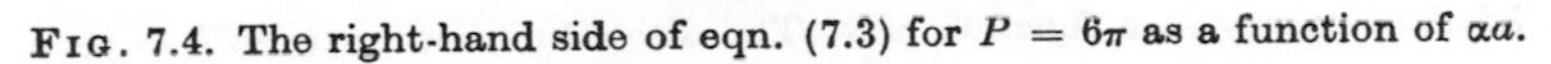
FIG. 7.4. The right-hand side of eqn. (7.3) for $P = 6\pi$ as a function of αa.

the forbidden bands wider. In the limit $P \to \infty$ the allowed band reduces to one single energy level; that is, we are back to the case of the discrete energy spectrum existing in isolated atoms.

For $P \to \infty$ it follows from eqn. (7.3) that

$$\sin \alpha a = 0; \tag{7.7}$$

that is, the permissible values of energy are

$$E_n = \frac{\pi^2 \hbar^2}{2ma^2} n^2, \tag{7.8}$$

which may be recognized as the energy levels for a potential well of width a. Accordingly all electrons are independent of each other and each one is confined to one atom by an infinite potential barrier.

(ii) In the limit $P \to 0$, we get

$$\cos \alpha a = \cos ka; \tag{7.9}$$

that is,

$$E = \frac{\hbar^2 k^2}{2m}, \tag{7.10}$$

as for the free electron. Thus by varying P from zero to infinity we cover the whole range from the completely free electron to the completely bound electron.

(iii) At the boundary of an allowed band $\cos ka = \pm 1$; that is,

$$k = \frac{n\pi}{a}, \qquad n = 1, 2, 3 \ldots \tag{7.11}$$

Looking at a typical energy versus k plot (Fig. 7.5) we can see that

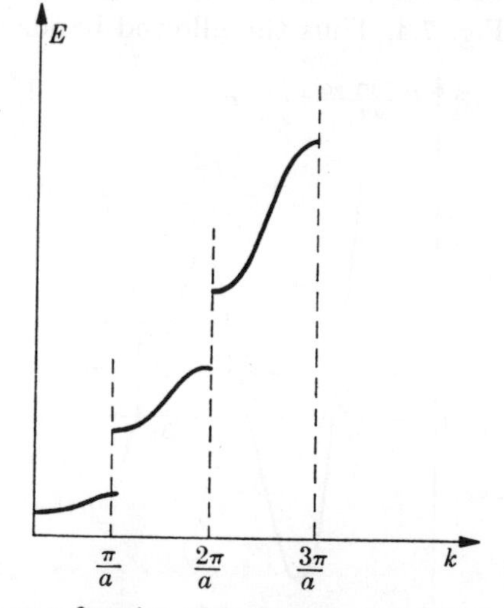

FIG. 7.5. The energy as a function of k. The discontinuities occur at $k = n\pi/a$, $n = 1, 2, 3 \ldots$.

the discontinuities in energy occur at the values of k specified above. We shall say more about this curve, mainly about the discontinuities in energy, but let us see first what the other models can tell us.

7.3. The Ziman model

This derivation relies somewhat less on mathematics and more on physical intuition. We may start again with the assertion that the presence of lattice ions will make the free electron model untenable—at least under certain circumstances.

Let us concentrate now on the wave aspect of the electron and look upon a free electron as a propagating plane wave. Its wave function is then

$$\psi_k = e^{ikx}. \tag{7.12}$$

We know that waves (whether X-rays or electron waves) can easily move across a crystal lattice; so plane waves (that is free electrons) may after all represent the truth. Waves *may* move across a crystal lattice, but not always. There is strong disturbance when the individual reflections add in phase; that is, when

$$n\lambda = 2\,a\cos\theta, \qquad n = 1, 2, 3\ldots, \tag{7.13}$$

as follows from the sketch in Fig. 7.6. This is a well-known relationship (called *Bragg reflection*) for X-rays and, of course, it is equally applicable to electron waves. So we may argue that the propagation of electrons is

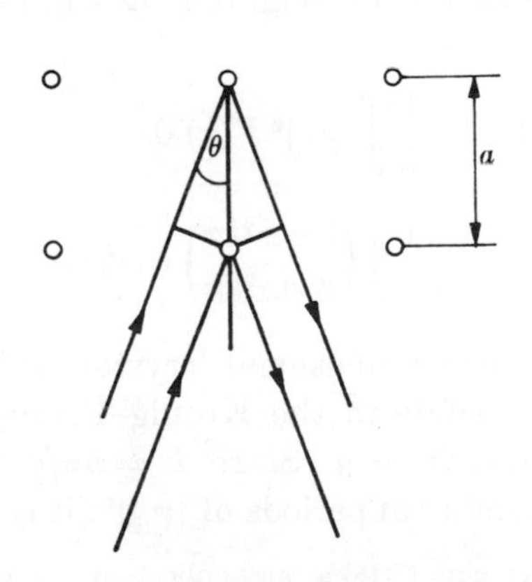

FIG. 7.6. Geometry of reflection from atomic planes.

strongly disturbed whenever eqn. (7.13) is satisfied. In one dimension the condition reduces to

$$n\lambda = 2a. \tag{7.14}$$

Using the relationship between wavelength and wave number the above equation may be rewritten as

$$k = \frac{n\pi}{a}. \tag{7.15}$$

Thus we may conclude that our free-electron model is not valid when eqn. (7.15) applies. The wave is reflected, so the wave function should also contain a term representing a wave in the opposite direction

$$\psi_{-k} = \mathrm{e}^{-ikx}. \tag{7.16}$$

Since waves of that particular wavelength are reflected to and fro we may expect the forward- and backward-travelling waves to be present in the same proportion; that is, we shall assume wave functions in the form

$$\psi_{\pm} = \frac{1}{\sqrt{2}}(\mathrm{e}^{ikx} \pm \mathrm{e}^{-ikx}) = \sqrt{2}\begin{pmatrix}\cos kx \\ \mathrm{i}\sin kx\end{pmatrix}, \tag{7.17}$$

where the constant is chosen for correct normalization.

Let us now calculate the potential energies of the electrons in both cases. Be careful; we are not here considering potential energy in general but the potential energy of the electrons that happen to have the wave functions $\psi_{\pm}$. These electrons have definite probabilities of turning up at various places, so their potential energy† may be obtained by averaging the *actual* potential $V(x)$ weighted by the probability function $|\psi_{\pm}|^2$. Hence

$$V_{\pm} = \frac{1}{L}\int |\psi_{\pm}|^2\, V(x)\,\mathrm{d}x$$

$$= \frac{1}{L}\int \begin{pmatrix}2\cos^2 kx \\ 2\sin^2 kx\end{pmatrix} V(x)\,\mathrm{d}x. \tag{7.18}$$

L is the length of the one-dimensional 'crystal' and $V(x)$ is the same function that we met before in the Kronig–Penney model but now, for simplicity, we take $2w = a$. Since $k = n\pi/a$ the function $V(x)$ contains an integral number of periods of $|\psi_{\pm}|^2$; it is therefore sufficient

† You may also look upon eqn. (7.18) as an application of the general formula given by eqn. (3.45).

to average over one period. Hence

$$V_{\pm} = \frac{1}{a}\int_0^a \begin{pmatrix} 2\cos^2 kx \\ 2\sin^2 kx \end{pmatrix} V(x)\,\mathrm{d}x$$

$$= \frac{1}{a}\int_0^a \begin{pmatrix} 1+\cos 2kx \\ 1-\cos 2kx \end{pmatrix} V(x)\,\mathrm{d}x$$

$$= \pm\frac{1}{a}\int_0^a \cos 2kx\, V(x)\,\mathrm{d}x, \tag{7.19}$$

since $V(x)$ integrates to zero. Therefore

$$V_{\pm} = \pm V_n. \tag{7.20}$$

The integration in eqn. (7.19) can be easily performed but we are not really interested in the actual numerical values. The important thing is that $V_n \neq 0$, and has opposite signs for the wave functions $\psi_{\pm}$.

Let us go through the argument again. If the electron waves have certain wave numbers (satisfying eqn. (7.15)) they are reflected by the lattice. For each value of k two distinct wave functions ψ_+ and ψ_- can be constructed, and the corresponding potential energies turn out to be $+V_n$ and $-V_n$.

The kinetic energies are the same for both wave functions, namely

$$E = \frac{\hbar^2 k^2}{2m}. \tag{7.21}$$

Thus the total energies are

$$E_{\pm} = \frac{\hbar^2 k^2}{2m} \pm V_n. \tag{7.22}$$

This is shown in Fig. 7.7 for $k = \pi/a$. The energy of the electron may be

$$\frac{\hbar^2 k^2}{2m} - V_1 \quad \text{or} \quad \frac{\hbar^2 k^2}{2m} + V_1 \tag{7.23}$$

but *cannot be any value in between*. There is an energy gap.

What will happen when $k \neq n\pi/a$. The same argument can be developed further and a general form may be obtained for the energy.†

† J. M. Ziman, *Electrons in metals, a short guide to the Fermi surface*, Taylor and Francis Ltd.

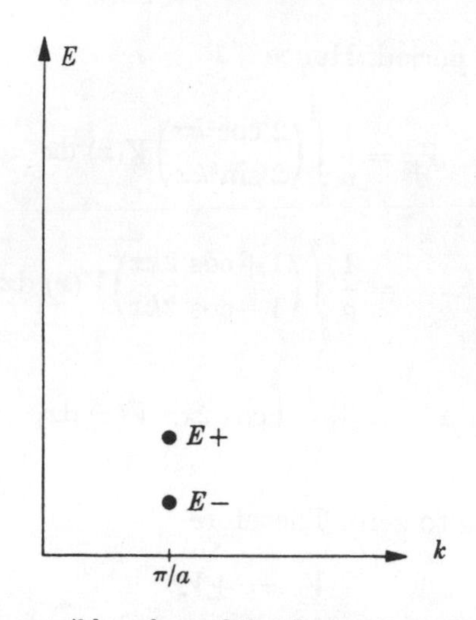

FIG. 7.7. The two possible values of the electron's total energy at $k = \pi/a$.

It is, however, not unreasonable to assume that apart from the discontinuities already mentioned the E–k curve will proceed smoothly. So we could construct it in the following manner. Draw the free electron parabola (dotted lines in Fig. 7.8) add and subtract V_n at the points $k = n\pi/a$, and connect the end points with a smooth curve keeping close to the parabola. Not unexpectedly, Fig. 7.8 looks like Fig. 7.5 obtained from the Kronig–Penney model.

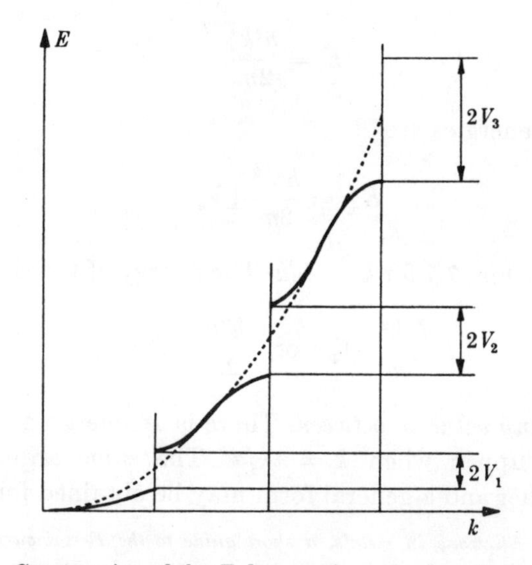

FIG. 7.8. Construction of the E–k curve from the free electron parabola.

7.4. The Feynman model

This is the one I like best because it combines mathematical simplicity with an eloquent physical picture. It is essentially a generalization of the model we used before to understand the covalent bond; another use of the coupled mode approach.

Remember, the energy levels of two interacting atoms are split; one is slightly higher, the other slightly below the original (uncoupled) energy. What happens when n atoms are brought close together? It is not unreasonable to expect that there will be an n-fold split in energy. So if the n atoms are far away from each other, each one has its original energy levels (denoted by E_1 and E_2 in Fig. 7.9a), but when

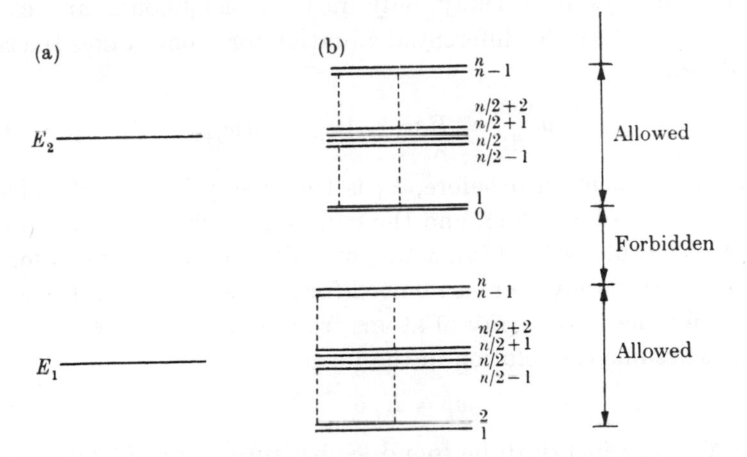

FIG. 7.9. There is an n-fold split in energy when n atoms are brought close to each other, resulting in a band of allowed energies, when n is large.

there is interaction they split into n separate energy levels. Now looking at this cluster of energy levels displayed in Fig. 7.9b we are perfectly entitled to refer to allowed energy bands and to forbidden gaps between them.

To make the relationship a little more quantitative let us consider the one-dimensional array of atoms shown in Fig. 7.10. We shall now

$j-2$ $\quad$ $j-1$ $\quad$ j $\quad$ $j+1$ $\quad$ $j+2$

a

FIG. 7.10. A one-dimensional array of atoms.

put a single electron on atom j into an energy level E_1 and define by this the state (j). Just as we discussed before in connection with the hydrogen molecular ion, there is a finite probability that the electron will jump from atom j to atom $j+1$, that is from state (j) into state $(j+1)$. There is of course no reason why the electron should jump only in one direction; it has a chance of jumping the other way too. So the transition from state (j) into state $(j-1)$ must have equal probability. It is quite obvious that a direct jump to an atom farther away is also possible but much less likely; we shall therefore disregard that possibility.

We have now a large number of states so we should turn to eqn. (5.24), which looks formidable with j running through all the atoms of the one-dimensional crystal. Luckily only nearest neighbours are coupled (or so we claim), so the differential equation for atom j takes the rather simple form

$$\mathrm{i}\hbar\frac{\mathrm{d}w_j}{\mathrm{d}t} = E_1 w_j - Aw_{j-1} - Aw_{j+1}, \tag{7.24}$$

where, as we mentioned before, E_1 is the energy level of the electron in the absence of coupling, and the coupling coefficient is taken again as $-A$. We could write down analogous differential equations for each atom but fortunately there is no need for it. We can obtain the general solution for the whole array of atoms from eqn. (7.24).

Let us assume the solution in the form

$$w_j = K_j\,\mathrm{e}^{-\mathrm{i}Et/\hbar}, \tag{7.25}$$

where E is the energy to be found. Substituting eqn. (7.25) into eqn. (7.24) we get

$$EK_j = E_1 K_j - A(K_{j-1} + K_{j+1}). \tag{7.26}$$

Note now that atom j is located at x_j and its neighbours at $x_j \pm a$ respectively. We may therefore look upon the amplitudes K_j, K_{j+1}, and K_{j-1} as functions of the x-coordinate. Rewriting eqn. (7.26) in this new form we get

$$EK(x_j) = E_1 K(x_j) - A\{K(x_j+a) + K(x_j-a)\}. \tag{7.27}$$

This is called a *difference equation* and may be solved by the same method as a differential equation. We can assume the trial solution

$$K(x_j) = \mathrm{e}^{\mathrm{i}kx_j}, \tag{7.28}$$

which, substituted into eqn. (7.27) gives

$$E\mathrm{e}^{\mathrm{i}kx_j} = E_1\mathrm{e}^{\mathrm{i}kx_j} - A\{\mathrm{e}^{\mathrm{i}k(x_j+a)} + \mathrm{e}^{\mathrm{i}k(x_j-a)}\}. \tag{7.29}$$

Dividing by $\exp(ikx_j)$ the eqn. (7.29) reduces to the final form

$$E = E_1 - 2A\cos ka, \tag{7.30}$$

which is plotted in Fig. 7.11. Thus, once more, we get the result that energies within a band (between E_1-2A and E_1+2A) are allowed and outside that range are forbidden.

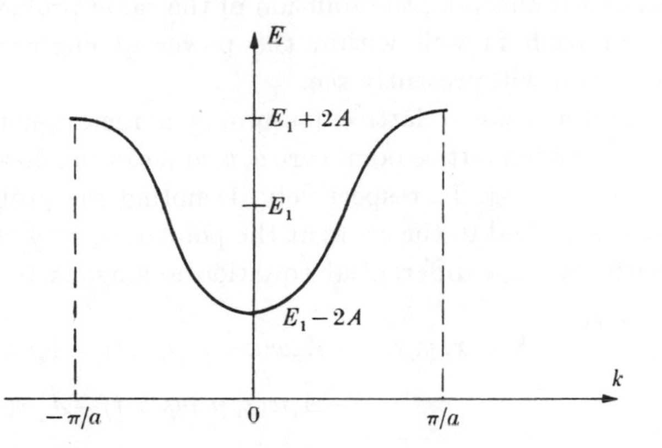

FIG. 7.11. Energy as a function of k obtained from the Feynman model.

It is a great merit of the Feynman model that we have obtained a very simple mathematical relationship for the E–k curve within a given energy band. But what about other energy bands that have automatically come out from the other models? We could obtain the next energy band from the Feynman model by planting our electron into the next higher energy level of the isolated atom, E_2, and following the same procedure as before. We could then obtain for the next band

$$E = E_2 - 2B\cos ka, \tag{7.31}$$

where B again is the coupling between nearest neighbours.

Another great advantage of the Feynman model is that it is by no means restricted to electrons. It could apply to any other particles. But, you may ask, what other particles can be there? Well, we have talked about positively charged particles called holes. They may be represented as deficiency of electrons and so they too can jump from atom to atom. But there are even more interesting possibilities. Consider, for example, an atom that somehow gets into an excited state (meaning that one of its electrons is in a state of higher energy) and can with a certain probability transfer its energy to the next atom down the line. The concepts are all familiar, and so we may describe this process in terms of a particle moving across the lattice. It is called an *exciton* and it is believed to play an important role as carrier of energy in certain biological processes.

Yet another advantage of this model is its easy applicability to three-dimensional problems. Whereas the three-dimensional solution of the Kronig–Penney model would send shudders down the spines of trained numerical analysts, the solution of the same problem using the Feynman approach is well within the power of engineering undergraduates, as you will presently see.

In a three-dimensional lattice (assuming a rectangular structure) the distances between lattice points are a, b, and c in the directions of the coordinate axes x, y, and z respectively. Denoting the probability that an electron is attached to the atom at the point x, y, z by $|w(x, y, z, t)|^2$, we may write down a differential equation analogous to eqn. (7.24):

$$\begin{aligned} i\hbar\frac{\partial w(x, y, z, t)}{\partial t} = E_1 w(x, y, z, t) - A_x w(x+a, y, z, t) - A_x w(x-a, y, z, t) \\ - A_y w(x, y+b, z, t) - A_y w(x, y-b, z, t) \\ - A_z w(x, y, z+c, t) - A_z w(x, y, z-c, t), \end{aligned} \tag{7.32}$$

where A_x, A_y, A_z are the coupling coefficients between nearest neighbours in the x, y, z directions respectively.

The solution of the above differential equation can be easily guessed by analogy with the one-dimensional solution in the form

$$w(x, y, z, t) = \exp(-iEt/\hbar)\exp\{i(k_x x + k_y y + k_z z)\}, \tag{7.33}$$

which, substituted in eqn. (7.32), gives in a few easy steps

$$E = E_1 - 2A_x \cos k_x a - 2A_y \cos k_y b - 2A_z \cos k_z c. \tag{7.34}$$

Thus in the three-dimensional case the energy band extends from the minimum energy

$$E_{\min} = E_1 - 2(A_x + A_y + A_z) \tag{7.35}$$

to the maximum energy

$$E_{\max} = E_1 + 2(A_x + A_y + A_z). \tag{7.36}$$

7.5. The effective mass

It has been known for a long time that an electron has a well-defined mass, and when accelerated by an electric field it obeys Newtonian mechanics. What happens when the electron to be accelerated happens to be inside a crystal? How will it react to an electric field? We have already given away the secret when talking about cyclotron resonance. The mass of an electron in a crystal appears, in general, different from the free electron mass, and is usually referred to as the *effective mass.*

We shall obtain the answer by using a semi-classical picture, which, as the name implies, is fifty per cent classical and fifty per cent quantum-mechanical. The quantum-mechanical part describes the velocity of the electron in a one-dimensional lattice by its group velocity

$$v_g = \frac{1}{\hbar}\frac{\partial E}{\partial k}, \tag{7.37}$$

which depends on the actual E–k curve. The classical part expresses $\mathrm{d}E$ as the work done by a classical particle travelling a distance $v_g\,\mathrm{d}t$ under the influence of a force $e\mathscr{E}$ yielding

$$\begin{aligned} \mathrm{d}E &= e\mathscr{E}v_g\,\mathrm{d}t \\ &= e\mathscr{E}\frac{1}{\hbar}\frac{\partial E}{\partial k}\,\mathrm{d}t. \end{aligned} \tag{7.38}$$

We may obtain the acceleration by differentiating eqn. (7.37) as follows

$$\frac{\mathrm{d}v_g}{\mathrm{d}t} = \frac{1}{\hbar}\frac{\mathrm{d}}{\mathrm{d}t}\frac{\partial E}{\partial k} = \frac{1}{\hbar}\frac{\partial^2 E}{\partial k^2}\frac{\mathrm{d}k}{\mathrm{d}t}. \tag{7.39}$$

Expressing now $\mathrm{d}k/\mathrm{d}t$ from eqn. (7.38) and substituting it into eqn. (7.39) we get

$$\frac{\mathrm{d}v_g}{\mathrm{d}t} = \frac{1}{\hbar^2}\frac{\partial^2 E}{\partial k^2}e\mathscr{E}. \tag{7.40}$$

Comparing this formula with that for a free, classical particle

$$m\frac{\mathrm{d}v}{\mathrm{d}t} = e\mathscr{E}, \tag{7.41}$$

we may define

$$m^* = \hbar^2\left(\frac{\partial^2 E}{\partial k^2}\right)^{-1} \tag{7.42}$$

as the effective mass of an electron. Thus the answer to the original question is that an electron in a crystal lattice *does* react to an electric field but its mass is given by eqn. (7.42) in contrast to the mass of a free electron. Let us just check whether we run into any contradiction with our E–k curve for a free electron. Then

$$E = \frac{\hbar^2 k^2}{2m},$$

and thus

$$\frac{\partial^2 E}{\partial k^2} = \frac{\hbar^2}{m},$$

which substituted into eqn. (7.42) gives

$$m^* = m.$$

So everything is all right.

For an electron in a one-dimensional lattice we may take E in the form of eqn. (7.30), giving

$$m^* = \frac{\hbar^2}{2Aa^2} \sec ka. \tag{7.43}$$

The graphs of energy, group velocity, and effective mass are plotted for this case in Fig. 7.12 as a function of k between $-\pi/a$ and π/a.

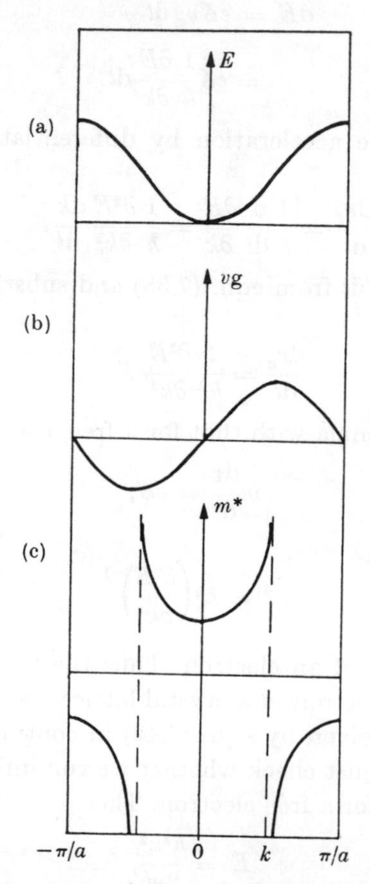

FIG. 7.12. Energy, group velocity and effective mass as a function of k.

Oddly enough, m^* may go to infinity and may take on negative values as well.

If an electron, initially at rest at $k = 0$, is accelerated by an electric field, it will move to higher values of k, and will become heavier and heavier, reaching infinity at $k = \pi/2a$. For even higher values of k the effective mass becomes negative, heralding the advent of a new particle, the hole, which we have casually met from time to time and shall often meet in the rest of this course.

The definition of effective mass as given in eqn. (7.42) is for a one-dimensional crystal but it can be easily generalized for three dimensions. If the energy is given in terms of k_x, k_y, and k_z, as for example in eqn. (7.34), then the effective mass in the x-direction is

$$\begin{aligned} m_x^* &= \hbar^2\left(\frac{\partial^2 E}{\partial k_x^2}\right)^{-1} \\ &= \frac{\hbar^2}{2A_x a^2}\sec k_x a. \end{aligned} \tag{7.44}$$

In the y-direction it is

$$\begin{aligned} m_y^* &= \hbar^2\left(\frac{\partial^2 E}{\partial k_y^2}\right)^{-1} \\ &= \frac{\hbar^2}{2A_y b^2}\sec k_y b. \end{aligned} \tag{7.45}$$

A similar formula applies in the z-direction. So, oddly enough, the effective mass may be quite different in different directions. Physically this means that the same electric field applied in different directions will cause varying amounts of acceleration. This is bad enough, but something even worse may happen. There can be a term like

$$m_{xy}^* = \hbar^2\left(\frac{\partial^2 E}{\partial k_x \partial k_y}\right)^{-1} \tag{7.46}$$

With our simple model m_{xy}^* turns out to be infinitely large but it is worth noting that in general an electric field applied in the x-direction may accelerate an electron in the y-direction. As far as I know there are no electronic devices making use of this effect; if you want to invent something quickly, bear this possibility in mind.

If you are fond of mathematics you may think of the effective mass (or rather of its reciprocal) as a tensor quantity, but if you dislike tensors just regard the electron in a crystal as an extremely whimsical particle which, in response to an electric field in the (say) z-direction, may move in a different direction.

7.6. The effective number of free electrons

Let us now leave the fanciful world of three dimensions and return to the mathematically simpler one-dimensional case. In a manner rather similar to the derivation of effective mass we can derive a formula for the number of electrons available for conduction. According to eqn. (7.40),

$$\frac{\mathrm{d}v_g}{\mathrm{d}t} = \frac{1}{\hbar^2}\frac{\partial^2 E}{\partial k^2}e\mathscr{E}. \tag{7.47}$$

We have here the formula for the acceleration of an electron. But we have not only one electron, we have lots of electrons. Every available state may be filled by an electron; so the total effect of accelerating all the electrons may be obtained by a summation over all the occupied states. We wish to sum $\mathrm{d}v_g/\mathrm{d}t$ for all electrons. Multiplying by the electronic charge,

$$\sum \frac{\mathrm{d}}{\mathrm{d}t}(ev_g)$$

is nothing else but the rate of change of electric current that flows initially when an electric field is applied.† Thus,

$$\begin{aligned}\frac{\mathrm{d}I}{\mathrm{d}t} &= \sum \frac{\mathrm{d}}{\mathrm{d}t}(ev_g)\\ &= \frac{1}{\hbar^2}e^2\mathscr{E}\sum\frac{\partial^2 E}{\partial k^2},\end{aligned} \tag{7.48}$$

or, going over to integration,

$$\frac{\mathrm{d}I}{\mathrm{d}t} = \frac{1}{\hbar^2}e^2\mathscr{E}\frac{1}{\pi}\int\frac{\mathrm{d}^2E}{\mathrm{d}k^2}\,\mathrm{d}k, \tag{7.49}$$

where the density of states in the range $\mathrm{d}k$ is $\mathrm{d}k/\pi$.‡

† Do not mistake this for the rate of change of electric current under stationary conditions. For the steady state to apply one must take collisions into account as well.

‡ We have already done it twice before, but since the density of states is a rather difficult concept (making something continuous having previously stressed that it must be discrete), and since this is a slightly different situation, we shall do the derivation again. Remember, we are in one dimension and we are interested in the number of states in momentum space in an interval $\mathrm{d}p_x$. According to eqn. (6.31)

$$p_x = \frac{h}{2}n_x,$$

where n_x is an integer. So for unit length there is exactly one state and for a length x

If there were N non-interacting free electrons we should obtain

$$\frac{\mathrm{d}I}{\mathrm{d}t} = \frac{e^2\mathscr{E}}{m}N. \tag{7.50}$$

For free electrons eqn. (7.50) applies; for electrons in a crystal eqn. (7.49) is true. Hence, if we wish to create a mental picture in which the electrons in the crystal are replaced by 'effective' electrons, we may define the number of effective electrons by equating eqn. (7.49) with eqn. (7.50). Hence

$$N_{\text{eff}} = \frac{1}{\pi}\frac{m}{\hbar^2}\int\frac{\mathrm{d}^2E}{\mathrm{d}k^2}\,\mathrm{d}k. \tag{7.51}$$

This, as you may have already guessed, applies only at absolute zero because we did not include the probability of occupation. In this case all the states are occupied up to an energy $E = E_a$, and all the states above E_a are empty.

If E_a happens to be somewhere inside an energy band (as shown in Fig. 7.13) then the integration goes from $k = -k_a$ to $k = k_a$. Performing the integration:

$$\begin{aligned} N_{\text{eff}} &= \frac{1}{\pi}\frac{m}{\hbar^2}\left\{\left(\frac{\mathrm{d}E}{\mathrm{d}k}\right)_{k=k_a} - \left(\frac{\mathrm{d}E}{\mathrm{d}k}\right)_{k=-k_a}\right\} \\ &= \frac{2}{\pi}\frac{m}{\hbar^2}\left(\frac{\mathrm{d}E}{\mathrm{d}k}\right)_{k=k_a} \end{aligned} \tag{7.52}$$

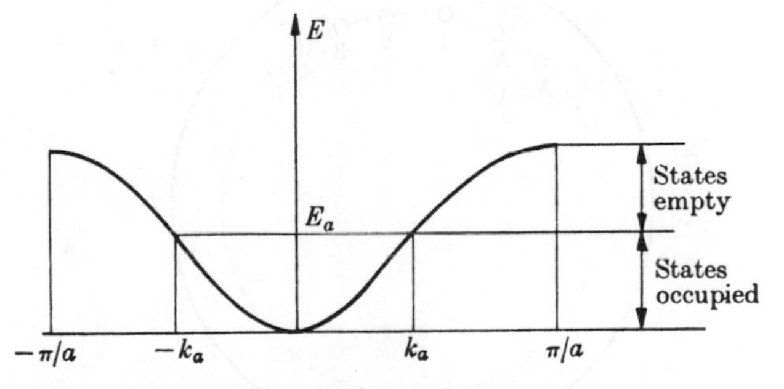

FIG. 7.13. One-dimensional energy band filled up to k_a at $T = 0$ K.

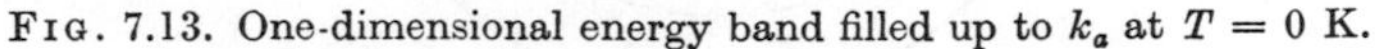

the number of states, $\mathrm{d}n_x$, is $(2/h)\,\mathrm{d}p_x$, which is equal to $(1/\pi)\,\mathrm{d}k_x$. We have to divide by two because only positive values of n are permitted, and have to multiply by two because of the two possible values of spin. Thus the number of states in a $\mathrm{d}k_x$ interval remains

$$\frac{1}{\pi}\,\mathrm{d}k_x.$$

This is a very important result. It says that the effective number of electrons capable of contributing to electrical conduction depends on the slope of the E–k curve at the highest occupied energy level.

At the highest energy in the band $\mathrm{d}E/\mathrm{d}k$ vanishes. We thus come to the conclusion that the number of effective electrons for a full band is zero. If the energy band is filled there is no electrical conduction.

7.7. The number of possible states per band

In order to find the number of states we must introduce boundary conditions. The simplest one (though physically the least defensible) is the so-called 'periodic boundary condition'.† It is based on the argument that a macroscopic crystal is so large in comparison with atomic dimensions that the detailed nature of the boundary conditions does not matter and we should choose them for mathematical convenience.

In the case of the periodic boundary condition we may simply imagine the one-dimensional crystal biting its own tail. This is shown in Fig. 7-14, where the last atom is brought into contact with the first atom. For this particular configuration it must be valid that

$$\psi(x+L) = \psi(x). \tag{7.53}$$

Then with the aid of eqn. (7.2) it follows that

$$\mathrm{e}^{\mathrm{i}k(x+L)}u_k(x+L) = \mathrm{e}^{\mathrm{i}kx}u_k(x). \tag{7.54}$$

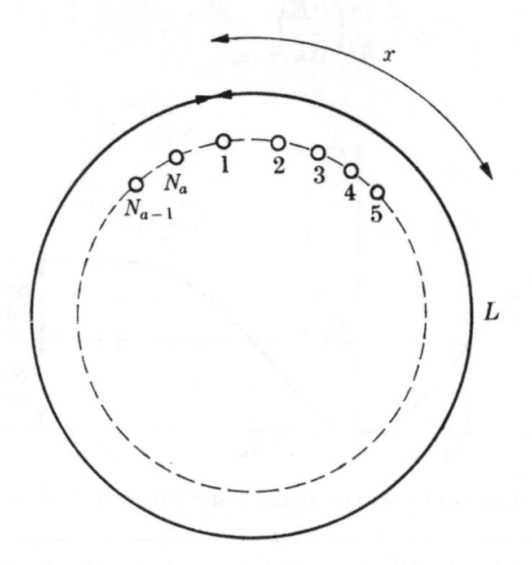

FIG. 7.14. Illustration of periodic boundary condition for a one-dimensional crystal.

† It would be more logical to demand that the wavefunction should disappear at the boundary but that would involve us only in more mathematics without changing any of the conclusions. So I must ask you to accept the rather artificial boundary condition expressed by eqn. (7.53).

Since u_k is a periodic function repeating itself from atom to atom,

$$u_k(x+L) = u_k(x) \tag{7.55}$$

and, therefore, to satisfy eqn. (7.54), we must have

$$kL = 2\pi r, \tag{7.56}$$

where r is a positive or negative integer.

It follows from the Kronig–Penney and from the Ziman models that in an energy band (that is in a region without discontinuity in energy) k varies from $n\pi/a$ to $(n+1)\pi/a$.† Hence

$$k_{\max} \equiv (n+1)\pi/a = 2\pi r_{\max}/L \tag{7.57}$$

and

$$k_{\min} \equiv n(\pi/a) = 2\pi r_{\min}/L. \tag{7.58}$$

Rearranging, we have

$$\begin{aligned} r_{\max} - r_{\min} &= L/2a \\ &= N_a L/2, \end{aligned} \tag{7.59}$$

where N_a is the number of atoms per unit length. Since r may have negative values as well, the total number of permissible values of k is

$$2(r_{\max} - r_{\min}) = N_a L. \tag{7.60}$$

Now to each value of k belongs a wavefunction; so the total number of wavefunctions is $N_a L$, and thus, including spin, the total number of available states is $2N_a L$.

7.8. Metals and insulators

At absolute zero some materials conduct well, some others are insulators. Why? The answer can be obtained from the formulae we have derived.

If each atom in our one-dimensional crystal contains one electron, then the total number of electrons is $N_a L$ and the band is half-filled. Since dE/dk is large in the middle of the band this means high effective number of electrons; that is, high conductivity.

If each atom contains two electrons, the total number of electrons is $2N_a L$; that is, each available state is filled. There is no conductivity: the solid is an insulator.

† The Feynman model gives only one energy band at a time but it shows clearly that the energy is a periodic function of ka, that is the same energy may be described by many values of k. Hence it would have been equally justified (as some people prefer) to choose the interval from $k = 0$ to $k = \pi/a$.

If each atom contains three electrons, the total number of electrons is $3N_aL$; that is, the first band is filled and the second band is half-filled. The value of dE/dk is large in the middle of the second band; therefore a solid containing atoms with three electrons each (it happens to be lithium) is a good conductor.

It is not difficult to see the general trend. Atoms with even numbers of electrons make up the insulators, whereas atoms with odd numbers of electrons turn out to be metals. Thus all we need to know is the number of electrons, even or odd, and the electric behaviour of the solid is determined. Diamond, with six electrons, must be an insulator and aluminium, with thirteen electrons, must be a metal. Simple, isn't it?

It is a genuine triumph of the one-dimensional model that the electric properties of a large number of elements may be promptly predicted. Unfortunately, it doesn't work always. Beryllium with four electrons and magnesium with twelve electrons should be insulators. They are not. They are metals; though metals of an unusual type in which electric conduction (evidenced by Hall-effect measurements) takes place both by holes and electrons. What is the mechanism responsible? For that we need a more rigorous definition of holes.

7.9. Holes

We first met holes as positively charged particles that enjoy a gay existence quite separately from electrons. The truth is that they are not separate entities but merely by-products of the electrons' motion in a periodic potential. There is no such thing as a free hole that can be fired from a hole gun. Holes are artifices but quite lively ones. The justification for their existence is as follows:

Using our definition of effective mass (eqn. (7.42)) we may rewrite eqn. (7.48) in the following manner:

$$\frac{dI}{dt} = e^2\mathscr{E}\sum_i \frac{1}{m_i^*}, \tag{7.61}$$

where the summation is over the occupied states.

If there is only one electron in the band, then

$$\frac{dI_e}{dt} = \frac{e^2\mathscr{E}}{m^*}. \tag{7.62}$$

If the band is full, then according to eqn. (7.52) the effective number of electrons is zero; that is,

$$\frac{dI}{dt} = e^2\mathscr{E}\sum_i \frac{1}{m_i^*} = 0. \tag{7.63}$$

Assume now that somewhere towards the top of the band an electron (denoted by j) is missing. Then the summation in eqn. (7.61) must omit the state j, which we may write as

$$\frac{\mathrm{d}I_{\mathrm{h}}}{\mathrm{d}t} = e^2 \sum_{\substack{i \\ i \neq j}} \frac{1}{m_i^*}. \tag{7.64}$$

But from eqn. (7.63)

$$e^2\mathscr{E}\left(\frac{1}{m_j} + \sum_{\substack{i \\ i \neq j}} \frac{1}{m_i^*}\right) = 0. \tag{7.65}$$

Eqn. (7.64) therefore reduces to

$$\frac{\mathrm{d}I_{\mathrm{h}}}{\mathrm{d}t} = -e^2\mathscr{E}\frac{1}{m_j^*}. \tag{7.66}$$

In the upper part of the band, however, the effective mass is negative; therefore

$$\frac{\mathrm{d}I_{\mathrm{h}}}{\mathrm{d}t} = \frac{e^2\mathscr{E}}{|m_j^*|}. \tag{7.67}$$

Hence, an electron missing from the top of the band leads to exactly the same formula as an electron present at the bottom of the band.

Now there is no reason why we should not always refer to this phenomenon as a current due to a missing electron that has a negative mass. But it is a lot shorter, and a lot more convenient, to say that the current is caused by a positive particle, called a *hole*. We can also explain the reason why the signs of eqn. (7.62) and of eqn. (7.67) are the same. In response to an electric field holes move in an *opposite* direction carrying an *opposite* charge; their contribution to electric current is therefore the same as that of electrons.

7.10. Divalent metals

We may now return to the case of beryllium and magnesium and to their colleagues, generally referred to as divalent metals. One-dimensional theory is unable to explain their electric properties; let us try two dimensions.

The E–k_x, k_y surface may be obtained from eqn. (7.34) as follows:

$$E = E_1 - 2A_x \cos k_x a - 2A_y \cos k_y b. \tag{7.68}$$

Let us plot now the constant energy curves in the k_x–k_y plane for the simple case when

$$E_1 = 1, \qquad A_x = A_y = \tfrac{1}{4}, \qquad a = b. \tag{7.69}$$

It may be seen in Fig. 7.15 that the minimum energy $E = 0$ is at

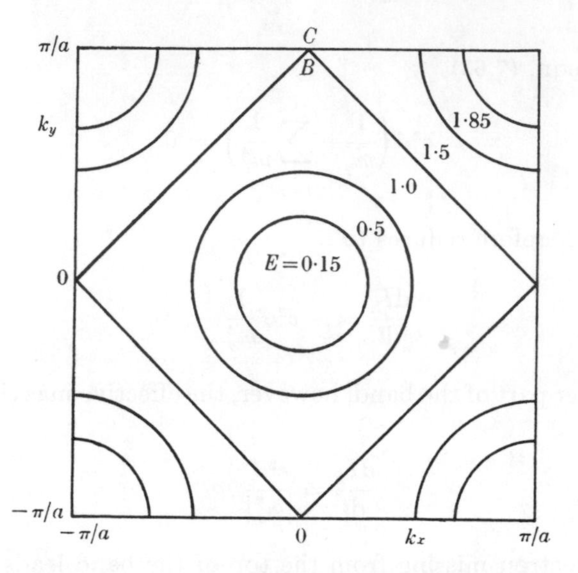

FIG. 7.15. Constant energy contours for a two-dimensional crystal in the k_x–k_y plane on the basis of the Feynman model.

the origin and for higher values of k_x and k_y the energy increases. Note well that the boundaries $k_x = \pm\pi/a$ and $k_y = \pm\pi/a$ represent a discontinuity in energy. (This is something we have proved only for the one-dimensional case but the generalization to two dimensions is fairly obvious.) There is an energy gap there. If the wave vector changes from point B (just inside the rectangle) to point C (just outside the rectangle), the corresponding energy may jump from one unit to (say) 1·5 units.

Let us now follow what happens at $T = 0$ as we fill up the available states with electrons. There is nothing particularly interesting until all the states up to $E = 1$ are filled, as shown in Fig. 7.16a. The next electron coming has an itch to leave the rectangle; it looks out, sees that the energy outside is 1·5 units, and therefore stays inside. This will go on until all the states are filled up to an energy $E = 1{\cdot}5$, as shown in Fig. 7.16b. The remaining states inside our rectangle have energies in excess of 1·5, and so the next electron in its search for lowest energy will go outside. It will go into a higher band because there are lower

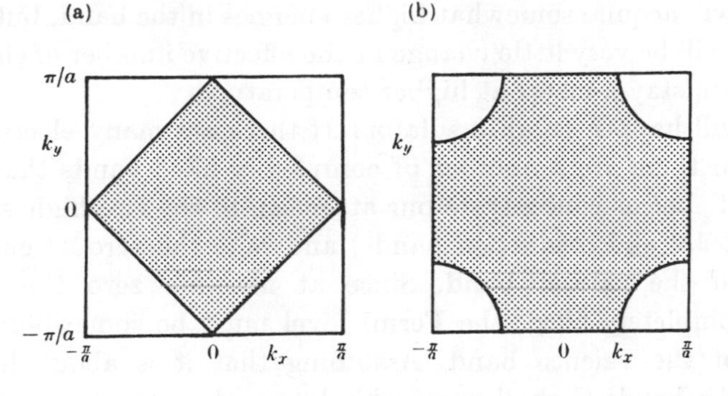

FIG. 7.16. (a) All energy levels filled up to $E = 1$. (b) All energy levels filled up to $E = 1{\cdot}5$.

energy states in that higher band (in spite of the energy gap) than inside the rectangle.

The detailed continuation of the story depends on the variation of energy with k in the higher band, but one thing is certain: the higher band will not be empty.

For an atom with two electrons the number of available states is equal to the number of electrons. If the energy gap is large (say two units instead of half a unit), then all the states in the rectangle would be filled and the material would be an insulator. If the energy gap is small (half a unit in our example), then some states will remain unfilled in the rectangle, and some states will be filled in the higher band. This means that both bands will contribute to electrical conduction. There will be holes coming from the rectangle and electrons from the higher band. This is how it happens that in some metals holes are the dominant charge-carriers.

7.11. Finite temperatures

All we said so far applies to zero temperature. What happens at finite temperatures? And what is particularly important, what happens at about room temperature, at which most electronic devices are supposed to work?

For finite temperatures it is no longer valid to assume that all states up to the Fermi energy are filled and all states above that are empty. The demarcation line between filled and unfilled states will become less sharp.

Let us see first what happens to a metal. Its highest energy band is about half-filled at absolute zero; at higher temperatures some of the electrons will acquire somewhat higher energies in the band, but that is all. There will be very little change in the effective number of electrons. A metal will stay a metal at higher temperatures.

What will happen to an insulator? If there are many electrons per atom, then there are a number of completely filled bands that are of no interest. Let us concentrate our attention to the two highest bands (called *valence* and *conduction* bands) and take the zero of energy at the top of the valence band. Since at absolute zero the valence band is completely filled, the Fermi level must be somewhere above the top of the valence band. Assuming that it is about half-way between the bands (I shall prove this later), the situation is depicted in Fig. 7.17a for zero temperature and in Fig. 7.17b for finite temperature. Remember, when the Fermi function is less than 1, it means that

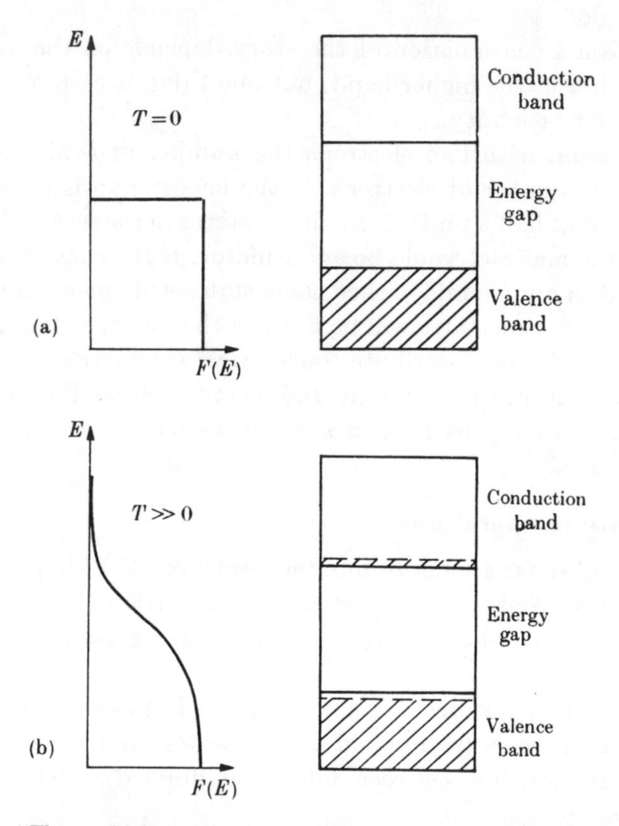

Fig. 7.17. (a) The two highest bands at $T = 0$ K. (b) The two highest bands at $T \gg 0$ K. There are electrons at the bottom of the conduction band, and holes at the top of the valence band.

the probability of occupation is less than 1; thus, some states in the valence band must remain empty. Similarly, when the Fermi function is larger than 0, it means that the probability of occupation is finite; that is, some electrons will occupy states in the conduction band.

We have come to the conclusion that at finite temperatures an insulator is no longer an insulator. There is conduction by electrons in the conduction band, and conduction by holes in the valence band. The actual amount of conduction depends on the energy gap. This can be appreciated if you remember that well away from the Fermi level the Fermi function varies exponentially; its value at the bottom of the conduction band and at the top of the valence band therefore depends critically on the width of the energy gap.

For all practical purposes diamond with an energy gap of 7 eV is an insulator, but silicon and germanium with energy gaps of 1·1 and 0·65 eV show noticeable conduction at room temperature. They are called semiconductors.

As you can see there is no profound difference in principle; insulators and semiconductors are distinguished only by the magnitudes of their respective energy gaps.

7.12. Concluding remarks

The band theory of solids is not an easy subject. The concepts are a little bewildering at first and their practical utility is not immediately obvious.

You could quite well pass examinations without knowing much about band theory, and you could easily become the head of a big electrical company without having any notion of bands at all. But if you ever want to create something new in solid-state electronic devices (which will be more and more numerous in your professional life) a thorough understanding of band theory is imperative. So my advice would be to go over it again and again until familiarity breeds comprehension.

I would like to add a few more words about the one-dimensional models we use so often. The sole reason for using one-dimensional models is mathematical simplicity, and you must appreciate that the results obtained are only qualitatively true. The real world is three-dimensional, and if we wish to get theoretical results that can be compared with those of experiments, no number of dimensions less than three will do.

Examples

1. Classify metals, semiconductors, an insulators on the basis of their band structure. Find out what you can about materials on the border lines of these classifications.
2. X-ray measurements show that electrons in the conduction band of lithium have energies up to 4·2 eV. What average effective mass should one assume in order to get the same result from free electron theory?

Assume one free electron per atom. The atomic weight of lithium is 6·94 and its density is 530 kg/m³.
3. Show, using the Feynman model, that the effective mass at the bottom of the band is inversely proportional to the width of the band.
4. Show from eqns. (7.3) and (7.5) that the group velocity of the electron is zero at $k = n\pi/a$.
5. In general, the reciprocal of the effective mass is a tensor whose components are given by eqn. (7.45). How would the classical equation of motion

$$m\frac{d\mathbf{v}}{dt} = \mathbf{F}$$

be modified?
6. The lowest energy bands in a solid are very narrow since there is hardly any overlap of the wave functions. In general, the higher up the band is, the wider it becomes. Assuming that the collision time is about the same for the valence and conduction bands, which would you expect to have higher mobility, an electron or a hole?

8. Semiconductors

Whatsoever things are true . . .
whatsoever things are pure . . .
think on these things.

PHILIPPIANS iv 8

8.1. Introduction

WITH the aid of band theory we have succeeded in classifying solids into metals, insulators, and semiconductors. We are now going to consider semiconductors (technologically the newest class) in more detail. Metals and insulators have been used for at least as long as we have been civilized, but semiconductors have found application only in the last century, and their more widespread application dates from the 1950s. During this brief period the electronics industry has been (to use a hackneyed word justifiably) revolutionized, first by the transistor, then by microelectronic circuitry. Each of these in succession, by making circuitry much cheaper and more compact, has led to the wider use of electronic aids such as computers in a way that is revolutionary in the social sense too.

Perhaps the key reason for this sudden change has been the preparation of extremely pure semiconductors, and hence the possibility of controlling impurity; this was a development of the 1940s and 50s. By crystal pulling, zone-refining, and epitaxial methods it is possible to prepare silicon and germanium with an impurity of only 1 part in 10^{10}. Compare this with long-established engineering materials such as steel, brass, or copper where impurities of a few parts per million are still virtually unattainable (and for most purposes, it must be admitted, not required). Probably the only other material that has ever been prepared with purity comparable to that of silicon and germanium is uranium, but people seem a little shy of quoting figures.

I shall now try to show why the important electrical properties of semiconductors occur and how they are influenced and controlled by small impurity concentrations. Next we shall consider what really came first, the preparation of pure material. We shall be ready then to discuss junction devices and integrated circuit technology.

8.2. Intrinsic semiconductors

The aim in semiconductor technology is to purify the material as much as possible and then to introduce impurities in a controlled manner. We shall call the pure semiconductor 'intrinsic' since its behaviour is determined by its intrinsic properties alone, and we shall call the semiconductor 'extrinsic' after external interference has changed its inherent properties. In devices it is mostly extrinsic semiconductors that are used, but it is better to approach our subject gradually and discuss intrinsic semiconductors first.

To be specific, let us think about silicon (though most of our remarks will be qualitatively true of germanium and other semiconductors). Silicon has the diamond crystalline structure; the four covalent bonds are symmetrically arranged. All the four valence electrons of each atom participate in the covalent bonds as we discussed before. But now, having learned band theory, we may express the same fact in a different way. We may say that all the electrons are in the valence band at 0 K. There is an energy gap of 1·1 eV above this before the conduction band starts. Thus to get an electron in a state in which it can take up kinetic energy from an electric field and can contribute to an electric current, we first have to give it a package of at least 1·1 eV of energy. This can come from thermal excitation, or by photon excitation quite independently of temperature.

Let us try to work out now the number of electrons likely to be free to take part in conduction at a temperature T. How can we do this? We have already solved this problem for the one-dimensional case: eqn. (7.51) gives us the effective number of electrons in a partly filled band; so all we need to do is to include the Fermi function to take account of finite temperature and to generalize the whole thing to three dimensions. It can be done but it is a bit too complicated. We shall do something else, which is less justifiable on strictly theoretical grounds but is physically much more attractive. It is really cheating because we use only those concepts of band theory that suit us, and instead of solving the problem honestly we shall appeal to approximations and analogies. It is a compromise solution that will lead us to easily manageable formulae.

First of all we shall say that the only electrons and holes that matter are those near the bottom of the conduction band and the top of the valence band respectively. Thus we may assume that

$$k_x a,\ k_y b,\ k_z c \ll 1 \tag{8.1}$$

and we may expand eqn. (7.34) to get the energy in the form

$$E = E_1 - 2A_x(1-\tfrac{1}{2}k_x^2a^2) - 2A_y(1-\tfrac{1}{2}k_y^2b^2) - 2A_z(1-\tfrac{1}{2}k_z^2c^2). \quad (8.2)$$

Using our definition of effective mass we can easily show from the above equation that

$$m_x^* = \frac{\hbar^2}{2A_xa^2}, \qquad m_y^* = \frac{\hbar^2}{2A_yb^2}, \qquad m_z^* = \frac{\hbar^2}{2A_zc^2}. \quad (8.3)$$

Substituting the values of A_xa^2, A_yb^2, and A_zc^2 from eqn. (8.3) back into eqn. (8.2) and condensing the constant terms into a single symbol, E_0, we may now express the energy as

$$E = E_0 + \frac{\hbar^2}{2}\left(\frac{k_x^2}{m_x^*} + \frac{k_y^2}{m_y^*} + \frac{k_z^2}{m_z^*}\right). \quad (8.4)$$

Taking further $E_0 = 0$, and assuming that everything is symmetric, that is

$$m_x^* = m_y^* = m_z^* = m^*, \quad (8.5)$$

we get

$$E = \frac{\hbar^2}{2m^*}(k_x^2 + k_y^2 + k_z^2). \quad (8.6)$$

This formula is identical to eqn. (6.2) obtained from the free electron model. Well, nearly identical. The mass in the denominator is not the real mass of an electron but the effective mass. But that is the only difference. Thus we are going to claim that electrons in the conduction band have a different mass but apart from that behave in the same way as free electrons. Hence the formula derived for the density of states (eqn. (6.10)) is also valid and we can use the same method to determine the Fermi level. So we shall have the total number of electrons by integrating.... Wait, we forgot about holes. How do we include them? Well, if holes are the same sort of things as electrons apart from having a positive charge, then everything we said about electrons in the conduction band should be true for holes in the valence band. The only difference is that the density of states must increase downwards for holes.

Choosing now the zero of energy at the top of the valence band we may write the density of states in the form

$$Z(E) = C_e(E - E_g)^{\frac{1}{2}}, \qquad C_e = 4\pi(2m_e^*)^{\frac{3}{2}}/h^3 \quad (8.7)$$

for electrons, and

$$Z(E) = C_{\mathrm{h}}(-E)^{\frac{1}{2}}, \qquad C_{\mathrm{h}} = 4\pi(2m_{\mathrm{h}}^{*})^{\frac{3}{2}}/h^{3} \tag{8.8}$$

for holes, both of them per unit volume. This is shown in Fig. 8.1, where E is plotted against $Z(E)$. You realize of course that the density of states has meaning only in the allowed energy band and must be identically zero in the gap between the two bands.

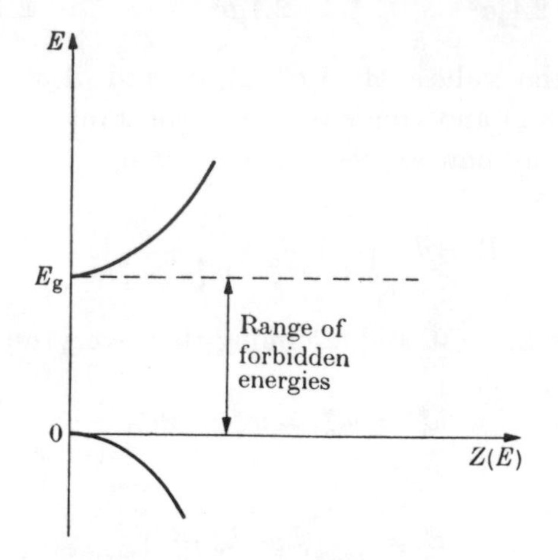

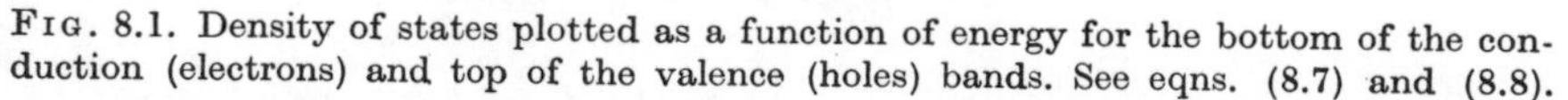
FIG. 8.1. Density of states plotted as a function of energy for the bottom of the conduction (electrons) and top of the valence (holes) bands. See eqns. (8.7) and (8.8).

Let us return now to the total number of electrons. To obtain that we must take the density of states, multiply by the probability of occupation (getting thereby the total number of occupied states), and integrate from the bottom to the top of the conduction band. So, formally, we have to solve the following integral

$$N_{\mathrm{e}} = \int_{\text{bottom of conduction band}}^{\text{top of conduction band}} (\text{density of states})(\text{Fermi function})\, \mathrm{d}E \tag{8.9}$$

There are several difficulties with this integral:

(i) Our solution for the density of states is valid only at the bottom of the band,

(ii) The Fermi function

$$F(E) = \left\{1 + \exp\left(\frac{E - E_{\mathrm{F}}}{kT}\right)\right\}^{-1} \tag{8.10}$$

is not particularly suitable for analytical integration.

(iii) We would need one more parameter in order to include the width of the conduction band.

We are saved from all these difficulties by the fact that the Fermi level lies in the forbidden band and in practically all cases of interest its distance from the band edge is large in comparison with kT (0·025 eV at room temperature). Hence,

$$E - E_F \gg kT \tag{8.11}$$

and the Fermi function may be approximated by

$$F(E) = \exp\{-(E - E_F)/kT\}, \tag{8.12}$$

as shown already in eqn. (6.18).

If the Fermi function declines exponentially, then the $F(E)Z(E)$ product will be appreciable only near the bottom of the conduction band as shown in Fig. 8.2. Thus we do not need to know the density

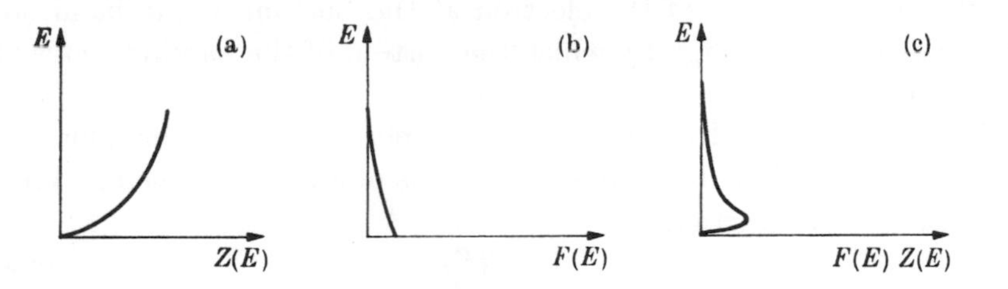

FIG. 8.2. (a) The density of states as a function of energy for the bottom of the conduction band. (b) The Fermi function for the same range of energies. (c) A plot of $F(E)Z(E)$ showing that the filled electron states are clustered together close to the bottom of the conduction band.

of states for higher energies (nor the width of the band) because the fast decline of $F(E)$ will make the integrand practically zero above a certain energy. But if the integrand is zero anyway, why not extend the upper limit up to infinity? We may then come to an integral that is known to mathematicians.

Substituting now eqns. (8.7) and (8.12) into eqn. (8.9) we get

$$N_e = C_e \int_{E_g}^{\infty} (E - E_g)^{\frac{1}{2}} \exp\{-(E - E_F)/kT\} \, dE. \tag{8.13}$$

Introducing now the new variable

$$x = (E - E_g)/kT \tag{8.14}$$

the integral takes the form

$$N_e = C_e(kT)^{\frac{3}{2}} \exp\{-(E_g - E_F)/kT\} \int_0^\infty x^{\frac{1}{2}} e^{-x}\, dx. \tag{8.15}$$

According to mathematical tables of high reputation†

$$\int_0^\infty x^{\frac{1}{2}} e^{-x}\, dx = \tfrac{1}{2}\sqrt{\pi} \tag{8.16}$$

leading to the final result

$$N_e = N_c \exp\{-(E_g - E_F)/kT\}, \tag{8.17}$$

where

$$N_c = 2(2\pi m_e^* kT/h^2)^{\frac{3}{2}}. \tag{8.18}$$

Thus we have obtained the number of electrons in the conduction band as a function of some fundamental constants, of temperature, of the effective mass of the electron at the bottom of the band, and of the amount of energy by which the bottom of the band is above the Fermi level.

We can deal with holes in an entirely analogous manner. The probability of a hole being present (that is of an electron being absent) is given by the function

$$1 - F(E), \tag{8.19}$$

which also declines exponentially along the negative E-axis. So we can choose the lower limit of integration as $-\infty$, leading to the result for the number of holes in the valence band:

$$N_h = N_v \exp(-E_F/kT), \tag{8.20}$$

where

$$N_v = 2(2\pi m_h^* kT/h^2)^{\frac{3}{2}}. \tag{8.21}$$

For an intrinsic semiconductor each electron excited into the conduction band leaves a hole behind in the valence band. Therefore the number of electrons should be equal to the number of holes (this would actually follow from the condition of charge neutrality too), that is

$$N_e = N_h. \tag{8.22}$$

Substituting now eqns. (8.17) and (8.20) into eqn. (8.22) we get

$$N_c \exp\{-(E_g - E_F)/kT\} = N_v \exp(-E_F/kT), \tag{8.23}$$

† Even better, you could work it out for yourself; it's not too difficult.

from which the Fermi level can be determined. With a little algebra we get

$$E_F = \frac{E_g}{2} + \tfrac{3}{4}kT \log_e \frac{m_h^*}{m_e^*}. \tag{8.24}$$

Since kT is small and the effective masses of electrons and holes are not very much different, we can say that the Fermi level is roughly halfway between the valence and conduction bands.

We know now everything we need about intrinsic semiconductors. Let us now look at the effect of impurities.

8.3. Extrinsic semiconductors

We shall continue to consider silicon as our specific example, but now with a controlled addition of a group V impurity (this refers to column five in the periodic table of elements), as, for example, antimony (Sb), arsenic (As), or phosphorus (P). If the impurity is less than, say, 1 in 10^6 silicon atoms, the lattice will be hardly different from that of a pure silicon crystal. Each group V atom will replace a silicon atom and use up four of its valence electrons for covalent bonding (Fig. 8.3a). There

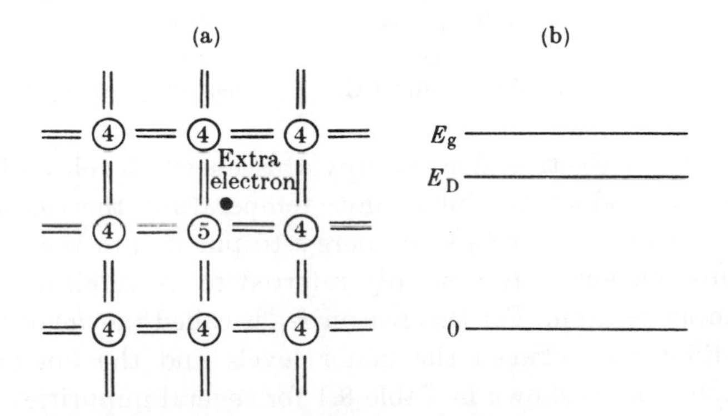

FIG. 8.3. (a) The extra electron 'belonging' to the group V impurity is much more weakly bound to its parent atom than the electrons taking part in the covalent bond. (b) This is equivalent to a *donor level* close to the conduction band in the band representation.

will, however, be a spare electron. It will no longer be so tightly bound to its nucleus as in a free group V atom, since the outer shell is now occupied (we might look at it this way) by eight electrons, the inert gas number; so the dangling spare electron cannot be very tightly bound. However, the impurity nucleus still has a net positive charge to distinguish it from its neighbouring silicon atoms. Hence we must suppose that the electron still has some affinity for its parent atom. Let us rephrase this somewhat anthropomorphic picture in terms of the band theory. We have said the energy gap represents the minimum energy required to ionize a silicon atom by taking one of its valency electrons. The electron belonging to the impurity atom clearly needs far less energy than this to become available for conduction. We would expect its energy level to be something like E_D in Fig. 8.3b. $(E_g - E_D)$ is typically of the order of 10^{-2} eV (see Table 8.1). At absolute zero

TABLE 8.1. *Energy levels of donor (group V) and acceptor (group III) impurities in Ge and Si. The energies given are the ionization energies, i.e. the distance of the impurity level from the band edge (in electron volts).*

	Impurity	Ge	Si
Donors	Antimony (Sb)	0·0096	0·039
	Phosphorous (P)	0·0120	0·045
	Arsenic (As)	0·0127	0·049
Acceptors	Indium (In)	0·0112	0·160
	Gallium (Ga)	0·0108	0·065
	Boron (B)	0·0104	0·045
	Aluminium (Al)	0·0102	0·057

temperature an electron does occupy this energy level, and it is *not* available for conduction. But at finite temperatures this electron needs no more than about 10^{-2} eV of energy to put it into the conduction band. This phenomenon is usually referred to as an electron *donated* by the impurity atom. For this reason E_D is called the *donor* level. The energy difference between the donor levels and the bottom of the conduction band is shown in Table 8.1 for several impurities.

Interestingly, a very rough model serves to give a quantitative estimate of the donor levels. Remember, the energy of an electron in a hydrogen atom (given by eqn. (4.18)) is

$$E = -me^4/8\epsilon_0^2h^2. \tag{8.25}$$

We may now argue that the excess electron of the impurity atom is held by the excess charge of the impurity nucleus; that is, the situation is like that in the hydrogen atom, with two minor differences.

(i) The dielectric constant of free space should be replaced by the dielectric constant of the material.

(ii) The free electron mass should be replaced by the effective mass of the electron at the bottom of the conduction band.

Thus this model leads to the following estimate

$$E_g - E_D = m^*e^4/8\epsilon^2h^2, \tag{8.26}$$

which for silicon (with a relative dielectric constant of 12 and effective mass about half the free electron mass) gives a value of about 0·05 eV for $(E_g - E_D)$, which is not far from what is actually measured.

If instead of a group V impurity we had some group III atoms, as, for example, indium (In), aluminium (Al), or boron (B), there would be an electron missing from one of the covalent bonds (see Fig. 8.4). If one electron is missing, there must be a hole present.

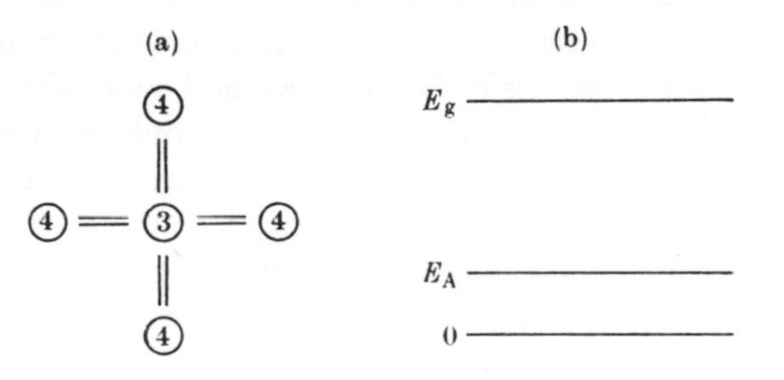

FIG. 8.4. (a) In the case of a group III impurity one bonding electron is missing—there is a 'hole' which any valence electron with a little surplus energy can fall into. (b) This shows in the band representation as an *acceptor level* just above the valence band edge.

Before going further let me say a few words about holes. You might have been slightly confused by our rather inconsistent references to holes. To clear this point—there are three equivalent representations of holes, and you can always (or nearly always) look at them in the manner most convenient under the circumstances.

You may think of a hole as a full-blooded positive particle moving around in the crystal, or as an electron missing from the top of the valence band, or as the actual physical absence of an electron from a place where it would be desirable to have one.

In the present case the third interpretation is the most convenient one—to start with. A group III atom has three valence electrons. So when it replaces a silicon atom at a certain atomic site it will try to contribute to bonding as much as it can. But it possesses only three electrons for four bonds. Any electron wandering around would thus be welcome to help out, or the impurity nucleus might actually consider stealing an electron from the next site.

The essential point is that a low energy state is available for electrons. Not as low an energy as at the host atom but low enough to come into consideration when an electron has acquired some extra energy and feels an urge to jump somewhere. Therefore (changing now to band theory parlance) the energy levels due to group III impurities must be just above the valence band. Since these atoms accept electrons so willingly they are called *acceptors* and the corresponding energy levels are referred to as *acceptor levels*.

A real material will usually have both electron donors and acceptors present (not necessarily group V and group III elements; these were chosen for simplicity of discussion and because they are most often used in practice). However, usually one type of impurity exceeds the other, and we can talk of impurity semiconductors as n- (negative carrier) or p- (positive carrier) types according to whether the dominant charge carriers are electrons or holes. If we had some silicon with 10^{20} atoms per cubic metre of trivalent indium, it would be a p-type semiconductor. If we were somehow to mix in 10^{21} atoms per cubic metre of pentavalent phosphorous, the spare phosphorous electrons would not only get to the conduction band but would also populate the acceptor levels, thus obliterating the p characteristics of the silicon and turning it into an n-type semiconductor.

Let us calculate now the number of electrons and holes for an extrinsic semiconductor. We have in fact already derived formulae for them, as witnessed by eqns. (8.17) and (8.20); but they were for intrinsic semiconductors. Did we make any specific use of the fact that we were considering intrinsic semiconductors? Perhaps not. We said that only electrons at the bottom of the band matter, and we also said that the bottom of the band is many times kT away from the Fermi level in energy but this could all be equally valid for extrinsic semiconductors. It turns out that these approximations are valid apart from certain exceptional cases (we shall meet one exception when we consider devices).

But how can we determine the Fermi level? It is certainly more difficult for an extrinsic semiconductor. We have to consider now all the donors and acceptors. The condition is that the crystal must be electrically neutral, that is, the net charge density must be zero. Let us take a count of what sort of charges we may meet in an extrinsic semiconductor. There are our old friends, electrons and holes; then there are the impurity atoms that donated an electron to the conduction band, and are left with a positive charge; and finally there are the acceptor atoms that accepted an electron from the valence band and thus have a negative charge. Hence the formula for overall charge neutrality is

$$N_e + N_A^- \rightleftharpoons N_h + N_D^+, \tag{8.27}$$

where N_A^- is the number of *ionized* acceptor atoms (which accepted an electron from the valence band) and N_D^+ is the number of *ionized* donor atoms (which donated an electron to the conduction band). We have written eqn. (8.27) with the $\rightleftharpoons$ sign used by chemists to show that it is a dynamic equilibrium, rather than a once and for all equation.

How can we find the number of ionized impurities, N_A^- and N_D^+, from the actual number of impurity atoms N_A and N_D? Looking at Fig. 8.3b you may recollect that the N_D donor electrons live at the level E_D (at 0 K) all ready and willing to become conduction electrons if somehow they can acquire $(E_g - E_D)$ joules of energy. N_D^+ is hence a measure of how many of them have gone. In other words, if we multiply N_D by the probability of an electron *not* being at E_D we should get N_D^+, that is

$$N_D^+ = N_D\{1 - F(E_D)\}. \tag{8.28}$$

For acceptor atoms the argument is very similar. The probability of an electron occupying a state at the energy level E_A is $F(E_A)$. Therefore the number of ionized acceptor levels is

$$N_A^- = N_A F(E_A). \tag{8.29}$$

We are now ready to calculate the position of the Fermi level in any semiconductor whose basic properties are known, that is, if we know N_A and N_D, the energy gap, E_g, and the effective masses of electrons and holes. Substituting for N_e, N_h, N_A^-, N_D^+ from eqns. (8.17), (8.20), (8.28), and (8.29) into eqn. (8.27) we get an equation that can be solved for E_F. It is a rather cumbersome equation but can always be solved with the aid of a computer. Fortunately, we seldom need to use all the terms, since, as mentioned above, the dominant impurity usually swamps the others. For example, in an n-type semiconductor, usually $N_e \gg N_h$ and $N_D^+ \gg N_A^-$ and eqn. (8.27) reduces to

$$N_e \cong N_D^+. \tag{8.30}$$

This, of course, implies that all conduction electrons come from the donor levels rather than from host lattice bonds. Substituting eqn.

(8.17) and (8.28) into eqn. (8.30) we get

$$4{\cdot}8\times10^{21}\left(\frac{m_e^*}{m_0}T\right)^{\frac{3}{2}}\exp\left(-\frac{E_g-E_F}{kT}\right)\cong N_D\left(1+\exp\frac{E_F-E_D}{kT}\right)^{-1}, \tag{8.31}$$

where we have put in the numerical values of the constants in MKS units.†

For a particular semiconductor eqn. (8.31) is easily soluble and we may plot E_F as a function of N_D or of temperature. Let us first derive a formula for the simple case when $(E_F-E_D)/kT$ is a large negative number. Eqn. (8.31) then reduces to

$$(\text{constant})\exp\frac{E_F}{kT}\cong N_D, \tag{8.32}$$

that is, E_F increases with the logarithm of N_D. So we have already learned that the position of the Fermi level moves upwards, and varies rather slowly with impurity concentration.

Let us consider now a slightly more complicated situation where the above approximation does not apply. Take silicon at room temperature with the data

$$E_g = 1{\cdot}15\text{ eV},\qquad E_g-E_D = 0{\cdot}049\text{ eV},\qquad N_D = 10^{22}\text{ m}^{-3} \tag{8.33}$$

We may get easily the solution by introducing the notation

$$x = \exp\frac{E_F}{kT} \tag{8.34}$$

reducing thereby eqn. (8.31) to the form

$$Ax = N_D/(1+Bx), \tag{8.35}$$

which leads to a quadratic equation in x, giving finally for the Fermi level, $E_F = 0{\cdot}97$ eV. Thus, in the practical case of an extrinsic semiconductor the Fermi level is considerably above the middle of the energy gap.

Let us consider now the variation of Fermi level with temperature. This is somewhat complicated by the fact that E_g and E_D are dependent on lattice dimensions, and hence both change with temperature, but we shall ignore this effect for the moment.

† Eqn. (8.31) is not, however, valid at the limit of *no* impurity, because holes cannot then be neglected; nor is it valid when N_D is very large, because some of the approximations (e.g. eqn. (8.11)) are then incorrect, and in any case many impurity atoms getting close enough to each other will create their own impurity band.

At very low temperatures (a few degrees absolute) the chance of excitation across the gap is fantastically remote compared with the probability of ionization from a donor level (which is only remote). Try calculating this for a 1-eV value of E_g and a 0·05 eV value of E_g-E_D. You should find that with 10^{28} lattice atoms per cubic metre, of which only 10^{21} are donors, practically all the conduction electrons are from the latter. Thus, at low temperature the material will act like an intrinsic semiconductor whose energy gap is only E_g-E_D. So we can argue that the Fermi level must be about halfway within this 'gap', i.e.

$$E_F \simeq \tfrac{1}{2}(E_g+E_D). \tag{8.36}$$

At very high temperatures practically all the electrons from the impurity atoms will be ionized but, because of the larger reservoir of valency electrons, the number of carriers in the conduction band will be much greater than N_D. In other words, the material (now a fairly good conductor) will behave like an intrinsic semiconductor with the Fermi level at about $E_g/2$. For larger impurity concentration the intrinsic behaviour naturally comes at a higher temperature. Thus a sketch of E_F against temperature will resemble Fig. 8.5 for an n-type semiconductor. The relationship for a p-type semiconductor is entirely analogous and is shown in Fig. 8.6.

There is just one further point to note about the variation of energy gap with temperature. We have seen in the Ziman model of the band

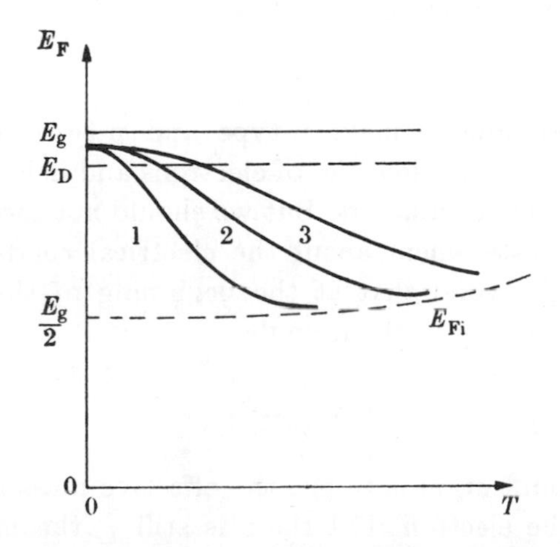

FIG. 8.5. The variation of the Fermi level as a function of temperature for an n-type semiconductor. The curves 1, 2, and 3 correspond to increasing impurity concentrations. E_{Fi} is the intrinsic Fermi level (plotted from eqn. (8.24) for $m_h > m_e$) to which all curves tend at higher temperatures.

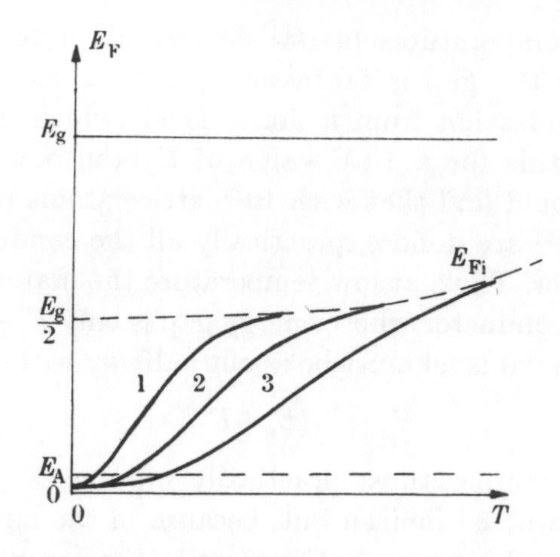

FIG. 8.6. The variation of the Fermi level as a function of temperature for a p-type semiconductor. The curves 1, 2, and 3 correspond to increasing impurity concentrations. E_{Fi} is the intrinsic Fermi level to which all curves tend at higher temperatures.

structure that the interband gap is caused by the interaction energy when the electrons' de Broglie half-wavelength is equal to the lattice spacing. It is reasonable to suppose that this energy would be greater at low temperatures for the following reason. At higher temperatures the thermal motion of the lattice atoms is more vigorous; the lattice spacing is thus less well defined and the interaction is weaker. This, qualitatively, is the case; for example, in germanium the energy gap decreases from about 0·75 eV at 4 K to 0·67 eV at 300 K.

8.4. Scattering

Having learned how to make n-type and p-type semiconductors and how to determine the densities of electrons and holes we now know quite a lot about semiconductors. But we should not forget that so far we have made no statement about the electrical conductivity, more correctly nothing beyond that at the beginning of the course (eqn. (1.10)), where we produced the formula

$$\sigma = \frac{e^2}{m}\tau N_e. \tag{8.37}$$

An obvious modification is to put the effective mass in place of the actual mass of the electron. But there is still τ, the mean free time between collisions. What will τ depend on and how?

We have now asked one of the most difficult questions in the theory of solids. As far as I know no one has managed to derive an expression for τ starting from first principles (that is, without the help of experimental results).

Let us first see what happens at absolute zero temperature. Then all the atoms are at rest;† so the problem seems to be how long can an electron travel in a straight line without colliding with a stationary atom. Well, why would it collide with an atom at all? The electron hasn't really any desire to bump into an atom. As we know from the Feynman model the electron does something quite different. It sits in an energy level of a certain atom then tunnels through the adverse potential barrier and takes a seat at the next atom, and again at the next atom. So it just walks across the crystal without any collision whatsoever. The mean free path is the length of the crystal. So it is *not* the presence of the atoms that causes the collisions. What then? The imperfections. If the crystal were perfect we should have nice periodic solutions (as in eqn. (7.33)) for the wave function and there would be equal probability for an electron being at any atom. It could thus start at any atom and could wriggle through the crystal to appear at any other atom. But the crystal is not perfect. The ideal periodic structure of the atoms is upset, partly by the thermal motion of the atoms, and partly by the presence of impurities, to mention only the two most important effects. So, strictly speaking, the concept of collisions as visualized for gas molecules makes little sense for electrons. The reason why nearly everybody clings to this picture is that no better one is available and we hate to be left without *some* physical picture.

It is a pretty difficult business to derive even approximate formulae for τ. We shall not make the attempt. There is not much point in quoting the formulae either, for it is most unlikely that you will ever need them. It is, however, worth noting the temperature-dependence of these 'collision' processes. According to simple theories

$$\tau_{\text{thermal}} \sim T^{-\frac{3}{2}} \quad \text{and} \quad \tau_{\text{ionized impurity}} \sim T^{\frac{3}{2}}, \tag{8.38}$$

which is not quite borne out in practice but is not far from the truth. It is certainly correct to say that at high temperatures the thermal effects dominate and at low temperatures the impurity effects dominate. When both of them are present the resultant collision time may be

† For the purpose of the above discussion we can assume that the atoms are at rest, but that can never happen in an actual crystal. If the atoms were at rest then we would know both their positions and velocities at the same time, which contradicts the uncertainty principle. Therefore even at absolute zero temperature the atoms must be in some motion.

obtained from the equation

$$\frac{1}{\tau} = \frac{1}{\tau_{\text{thermal}}} + \frac{1}{\tau_{\text{ionized impurity}}}. \tag{8.39}$$

If a high value of τ is required, then one should use a very pure material and work at low temperatures. A high value of τ means high mobility and hence high average velocity for the electrons.

We do not know yet whether high electron velocities in crystals will have many useful applications,† but since fast electrons in vacuum give rise to interesting phenomena, it might be worth while making an effort to obtain high carrier velocities in semiconductors.

Everything we have said about electrons is also valid for holes. Since the currents are additive, the total conductivity is given as

$$\sigma = \frac{e^2\tau_e N_e}{m_e^*} + \frac{e^2\tau_h N_h}{m_h^*}. \tag{8.40}$$

8.5. A relationship between electron and hole densities

Let us now return to eqns. (8.17) and (8.20). These were derived originally for intrinsic semiconductors, but they are valid for extrinsic semiconductors as well. Multiplying them together we get

$$N_e N_h = 4\left(\frac{2\pi kT}{h^2}\right)^3 (m_e^* m_h^*)^{\frac{3}{2}} \exp\left(-\frac{E_g}{kT}\right). \tag{8.41}$$

It is interesting to note that the Fermi level has dropped out and only the 'constants' of the semiconductor are contained in this equation. Thus, for a given semiconductor (i.e. for known values of m_e^*, m_h^*, and E_g) and temperature we can define the product $N_e N_h$ exactly, *whatever the Fermi energy and hence whatever the impurity density*. In particular, for an intrinsic material, where $N_e = N_h = N_i$, we get

$$N_e N_h = N_i^2. \tag{8.42}$$

Let us think over the implications. We start with an intrinsic semiconductor; so we have equal numbers of electrons and holes. Now add some donor atoms. The number of electrons must then increase but according to eqn. (8.42) the product must remain constant, that is, if the number of electrons increases, the number of holes must decrease. At first this seems rather odd. One would think that the number of electrons excited thermally from the valence band into the conduction

† Two present applications are the Gunn effect and very high frequency transistors, both to be discussed in Chapter 9.

band (and thus the number of holes left behind) would depend on temperature only, and be unaffected by the presence of donor atoms. This is not so. By increasing the concentration of donors, the total number of electrons in the conduction band is increased but the number of electrons excited across the gap is decreased (not only in their relative proportion but in their absolute number too). Why?

We can obtain a qualitative answer to this question by considering the 'dynamic equilibrium' mentioned briefly before. It means that electron-hole pairs are constantly created and annihilated and there is equilibrium when the rate of creation equals the rate of annihilation (the latter event is more usually referred to as 'recombination').

Now it is not unreasonable to assume that electrons and holes can find each other more easily if there are more of them present; so the rate of recombination must be proportional to the densities of holes and electrons. For an intrinsic material we may write

$$r_{\text{intrinsic}} = aN_{\text{i}}^2, \qquad g_{\text{intrinsic}} = aN_{\text{i}}^2, \tag{8.43}$$

where a is a proportionality constant, and r and g are the rates of recombination and creation respectively.

Now we may argue that by adding a small amount of impurity neither the rate of creation nor the proportionality constant should change. So for an extrinsic semiconductor

$$g_{\text{extrinsic}} = aN_{\text{i}}^2 \tag{8.44}$$

is still valid.

The rate of recombination should, however, depend on the actual densities of electrons and holes, that is

$$r_{\text{extrinsic}} = aN_{\text{e}}N_{\text{h}}. \tag{8.45}$$

From the equality of eqns. (8.44) and (8.45) we get the required relationship

$$N_{\text{i}}^2 = N_{\text{e}}N_{\text{h}}. \tag{8.46}$$

So we may say that as the density of electrons is increased above the intrinsic value, the density of holes must decrease below the intrinsic value in order that the rate of recombination of electron–hole pairs may remain at a constant value equal to the rate of thermal creation of pairs.

Those of you who have studied chemistry may recognize this relationship as a particular case of the *law of mass action*. This can be illustrated by a chemical reaction between A and B giving rise to a compound AB, viz.

$$\text{A}+\text{B} \rightleftharpoons \text{AB}. \tag{8.47}$$

If we represent the molecular concentration of each component by writing its symbol in square brackets, the quantity

$$[A][B][AB]^{-1} \tag{8.48}$$

is a constant at a given temperature. Now our 'reaction' is

$$\text{electron} + \text{hole} \rightleftharpoons \text{bound electron.} \tag{8.49}$$

As the number of bound electrons (cf. [AB]) is constant, this means that

$$[\text{electron}][\text{hole}] = N_e N_h \tag{8.50}$$

will also be constant.

8.6. Non-equilibrium processes

In our investigations so far the semiconductor was always considered to be in thermal equilibrium. Let us look briefly at a few cases where the equilibrium is disturbed.

The simplest way of disturbing the equilibrium is to shine electromagnetic waves (in practice these are mostly in the visible range) upon the semiconductor. As a result *photoemission* may occur, as in metals, but more interestingly, the number of carriers available for conduction may significantly increase (by as much as a factor of 100). This case is called *photoconduction*.

The three possible processes of producing carriers for conduction are shown in Fig. 8.7; (i) creating an electron–hole pair, that is exciting an electron from the valence band into the conduction band; (ii) exciting an electron from an impurity level into the conduction band; (iii)

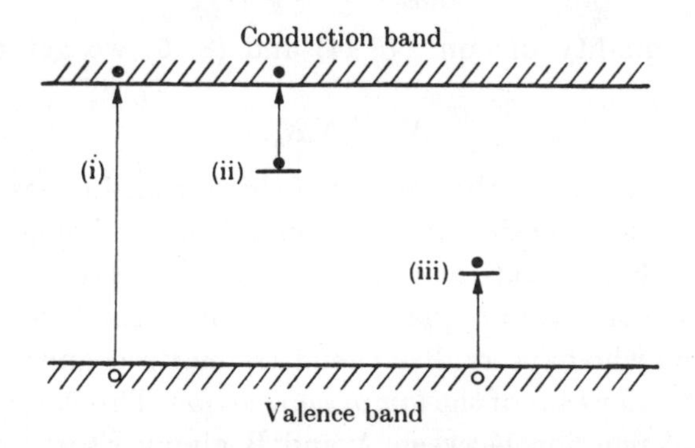

FIG. 8.7. Three methods of obtaining free carriers by illumination: (i) band-to-band transitions yielding an electron-hole pair, (ii) ionization of donor atoms, (iii) ionization of acceptor atoms.

exciting an electron from the top of the valence band into an impurity level, and thus leaving a hole behind.

The extra carriers are available for conduction as long as the semiconductor is illuminated. What happens when the light is switched off? The number of carriers must fall gradually to the equilibrium value. The time in which the extra density is reduced by a factor e is called the *lifetime* of the carrier and is generally denoted by τ (and is thus quite often confused with the collision time). It is an important parameter in the design of many semiconductor devices.

Assume now that only part of the semiconductor is illuminated; we shall then have a region of high concentration in connection with regions of lower concentration. This is clearly an unstable situation, and by analogy with gases we may expect the carriers to move away from places of high concentration towards places of lower concentration. The analogy is incidentally correct; this motion of the carriers has been observed, and can be described mathematically by the usual *diffusion equation*

$$J = eD\nabla N, \tag{8.51}$$

where D is the diffusion coefficient. This is quite plausible physically; it means that if there is a density gradient, a current must flow.

Equation (8.51) is equally valid for holes and electrons, though in a practical case the signs should be chosen with great care.

8.7. Real semiconductors

All our relationships obtained so far have been based on some idealized model. Perhaps the greatest distortion of reality came from our insistence that all semiconductors crystallize in a simple cubical structure. They don't. The most commonly used semiconductors, silicon and germanium, crystallize in the diamond structure, and this makes a significant difference.

Plotting the E vs k_x curve (Fig. 8.8) for the conduction band of silicon, for example, it becomes fairly obvious that it bears no close resemblance to our simple $E = E_0 - 2A_x \cos k_x a$ curve, which had its minimum at $k_x = 0$. Even worse, the $E(k_y)$ curve would be very different from the one plotted. The surfaces of constant energy in k-space are *not* spheres.

The situation is not much better in the valence band where the constant energy surfaces are nearly spheres but—I regret to say—there are three different types of holes present. This is shown in Fig. 8.8, where the letters h, l, and s stand for *heavy*, *light*, and *split-off* bands

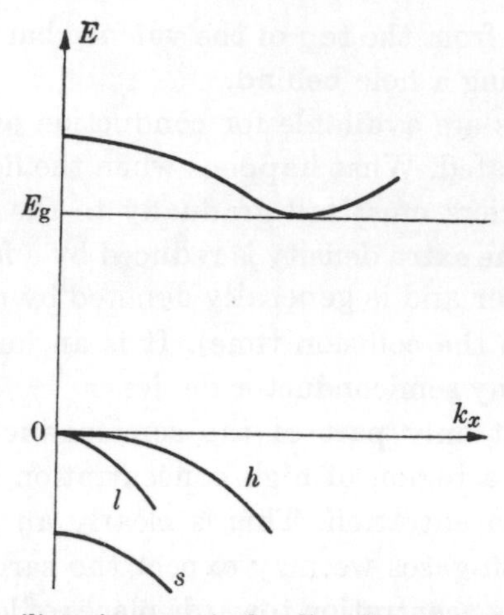

FIG. 8.8. E–k curve for silicon in a particular direction. Note that the minimum of the conduction band is not at $k = 0$ and there are three different types of holes. The situation is similar in germanium which is also an indirect-gap (minimum of conduction band not opposite to maximum of valence band) semiconductor.

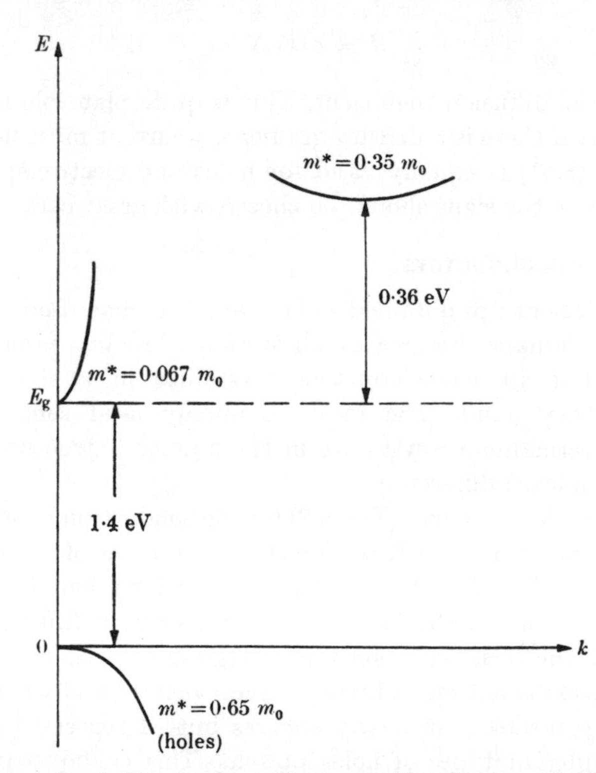

FIG. 8.9. Energy band diagram for gallium arsenide showing subsidiary valley in conduction band.

respectively. What does it mean to have three different types of holes? Well, just think of them as holes painted red, blue, and green. They coexist peacefully, though occasionally, owing to collisions, a hole may change its complexion.

Does this mean that all we have said so far is wrong? Definitely not. Does it mean that considerable modifications are needed? No, for most purposes not even that. We can get away with our simple model because, in general, only average values are needed. It is nice to know that there are three different types of holes in the valence band of silicon but for device operation only the average effective mass and some sort of average collision time are needed.

The picture is not as black as it seems. In spite of the anisotropy in the $E(k)$ curves the grand total is isotropic. I mean that by performing all the relevant averaging processes in silicon (still for a single-crystal material) the final result is an isotropic effective mass and isotropic collision time. Measuring the conductivity in different directions will thus always give the same result†.

Another important deviation from the idealized band structure occurs in a number of III-V compounds, where a subsidiary valley appears in the conduction band (shown for GaAs in Fig. 8.9). For most purposes (making a gallium arsenide transistor, for example) the existence of this additional valley can be ignored but it acquires special significance at high electric fields. It constitutes the basis for the operation of a new type of device, which we shall discuss later among semiconductor devices. You must realize, however, that this is the exception rather than the rule. The details of the band structure generally do not matter. For the description and design of the large majority of semiconductor devices our model is quite adequate.

8.8. Measurement of semiconductor properties

The main properties to be measured are (i) mobility, (ii) Hall coefficient, (iii) effective mass, (iv) energy gaps (including the distance of any impurity layer from the band edge), and (v) carrier lifetime.

Mobility. This quantity was defined as the carrier drift velocity for unit field

$$\mu = v_{\mathrm{D}}/\mathscr{E}. \tag{8.52}$$

† This is not true for graphite, for example, which has very different conductivities in different directions, but fortunately single crystal graphite is not widely used in semiconductor devices. Polycrystalline graphite has, of course, been used ever since the birth of the electrical and electronics industries (for brushes and microphones) but the operational principles of these devices are so embarrassingly simple that we can't possibly discuss them among the more sophisticated semiconductor devices.

Thus the most direct way of measuring mobility is to measure the drift velocity caused by a known d.c. electric field. Since the electric field is constant in a conductor (and in a semiconductor too), it can be deduced from measurements of voltage and distance. How can we measure the drift velocity? Well, in the same way we always measure velocity; by measuring the time needed to get from point A to C. But can we follow the passage of carriers? Not normally. When in the circuit of Fig. 8.10 we close the switch the electrons acquire an ordered motion *everywhere*.

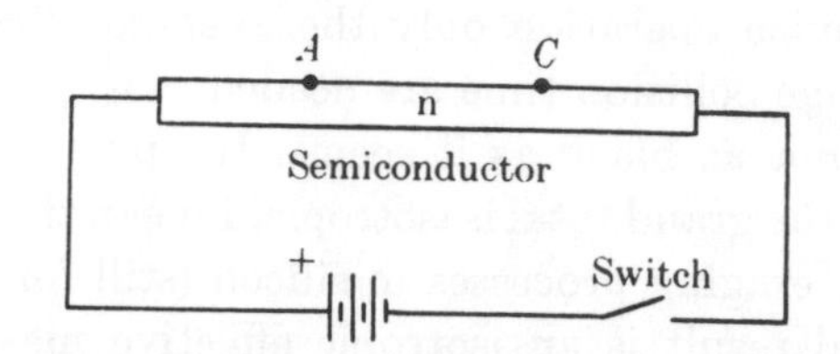

FIG. 8.10. Current flow in an n-type semiconductor.

Those that happen to be at point C when the switch is closed will arrive at point A some time later, but we have no means of learning when. From the moment the switch is closed the flow of electrons is uniform at both points, C and A. What we need is a circuit in which carriers can be launched at one point and detected at another point. A circuit that can do this was first described by Haynes and Shockley and is shown in Fig. 8.11a. When S is open there is a certain current flowing across the resistor R. At t_1 the switch is closed and according to the well-known laws of Kirchhoff there is a sudden increase of current (and voltage, as shown in Fig. 8.11b) through R. But that is not all. The contact between the metal wire and the n-type semiconductor is a rather special one. It has the curious property of being able to *inject holes*. We shall say more about injection later, but for the time being please accept that holes appear at point A, when S is closed. Under the influence of the battery B_1 the holes injected at A will move towards C. When they arrive at C (say at time t_2) there is a new component of current that must flow across R. The rise in current (and in voltage) will be gradual because some holes have a velocity higher than the average, but after a while a steady state develops. Now we know the distance between points A and C and we know fairly accurately the time needed by the holes to get from A to C; the drift velocity can thus be determined. The electric field can easily be obtained so we have managed to measure the mobility.

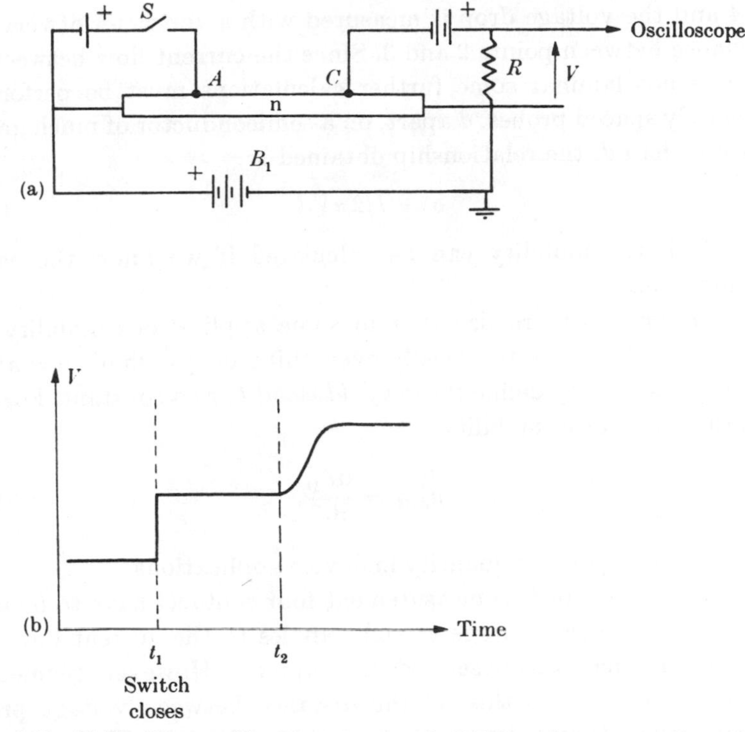

FIG. 8.11. (a) The Haynes–Shockley experiment. (b) The voltage across R as a function of time. The switch S is closed at $t = t_1$. The holes drift from A to C in a time $t_2 - t_1$.

A less direct way of determining the mobility is to measure the conductivity and use the relationship

$$\mu = \sigma/Ne. \tag{8.53}$$

A method often used in practice is the so-called 'four-point probe' arrangement shown in Fig. 8.12. The current is passed from contact

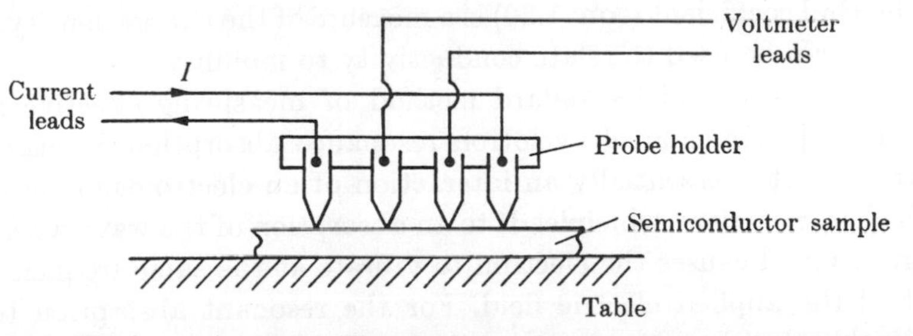

FIG. 8.12. The four-point probe. The probes are sharply pointed and held rigidly in a holder which can be pressed with a known force on to the semiconductor. A typical spacing is 1 mm between probes.

1 to 4 and the voltage drop is measured with a voltmeter of very high impedance between points 2 and 3. Since the current flow between the probes is not laminar some further calculations must be performed. For equally spaced probes, d apart, on a semiconductor of much greater thickness than d, the relationship obtained is

$$\sigma = I/2\pi Vd \tag{8.54}$$

from which the mobility can be calculated if we know the carrier concentration.

It is important to realize that in some applications mobility is a function of field. Since practically everything obeys Ohm's law at low enough fields we may define the *low field mobility* as a constant. For high fields the differential mobility

$$\mu_{\mathrm{diff}} = \frac{\mathrm{d}v_{\mathrm{D}}}{\mathrm{d}\mathscr{E}} \tag{8.55}$$

is usually the important quantity in device applications.

Hall coefficient. For this measurement four contacts have to be made so as to measure the voltage at right angles to the current flow. The basic measurement was described in Chapter 1. However, geometrical factors also come into this. If the distance between voltage probes is greater than that between the current probes the Hall voltage is reduced. Again, this reduction factor is calculable by detailed consideration of the patterns of current flow.

Very often conductivity and the Hall coefficient are measured by an ingenious method due to van der Pauw†. Four 'dot' contacts are made on the surface of the semiconductor. By measuring the voltage across two of them when a current is passed through the other pair, the conductivity can be found. The changes in the voltage-to-current ratio when a magnetic field is applied enable the Hall coefficient to be found.

The Hall coefficient (eqn. 1.20) is a measure of the charge density, and hence it can be used to relate conductivity to mobility.

Effective mass. The standard method of measuring effective mass uses the phenomenon of cyclotron resonance absorption discussed in Chapter 1. It is essentially an interaction of an electromagnetic wave with charge carriers, which leads to an absorption of the wave when the magnetic field causes the electron to vibrate at the same frequency as that of the applied electric field. For the resonant absorption to be noticeable the electron must travel an appreciable part of the period

† Van der Pauw, *Philips Res. Rep.*, *13*, 1, (1958).

without collisions; thus a high-frequency electric field, a high-intensity magnetic field, and low temperatures are used. These measurements are commonly made in the microwave region (10^{10} Hz) at liquid helium temperature (about 4 K) or in the infrared (about 10^{13} Hz) at liquid nitrogen temperature (77 K).

In the apparatus for a microwave measurement shown diagrammatically in Fig. 8.13 the sample is enclosed in a waveguide in a Dewar

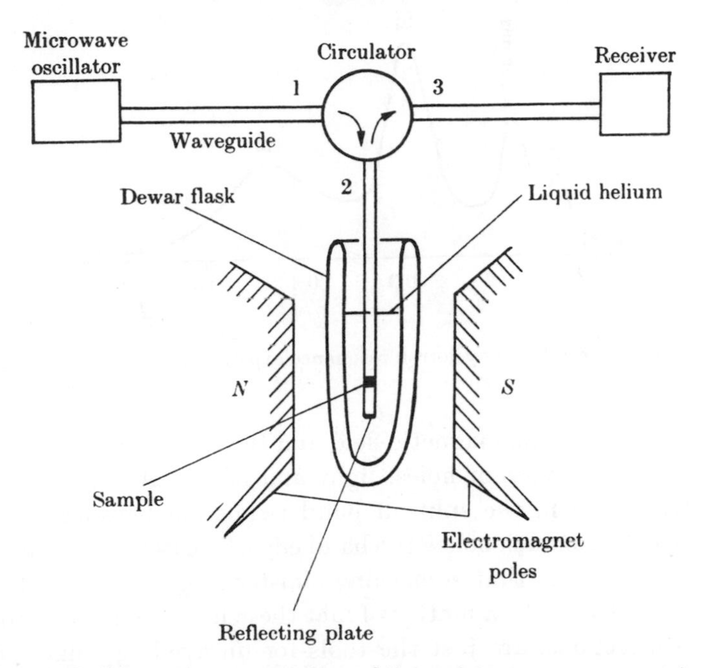

FIG. 8.13. Schematic representation of the cyclotron resonance experiment.

flask filled with liquid helium which is placed between the poles of a large electromagnet. The microwave signal is fed in through a circulator.† Thus the signal entering arm (2) is reflected by the reflecting plate at the end of the waveguide having passed through the semiconductor in each direction and ends up in the receiver connected to arm (3). Employing a wave of fixed frequency and a variable magnetic field, the effective mass is given (eqn. (1.69)) by

$$m^* = eB/\omega_c, \tag{8.56}$$

† The circulator has the following magical properties: a signal fed into arm (1) goes out entirely by arm (2) and a signal fed into arm (2) leaves the circulator by arm (3).

where B is the magnetic field corresponding to an absorption of signal. There will generally be several absorption peaks, corresponding to the various holes and electrons present. The experimentally obtained absorption curve for germanium is shown in Fig. 8.14 for a certain

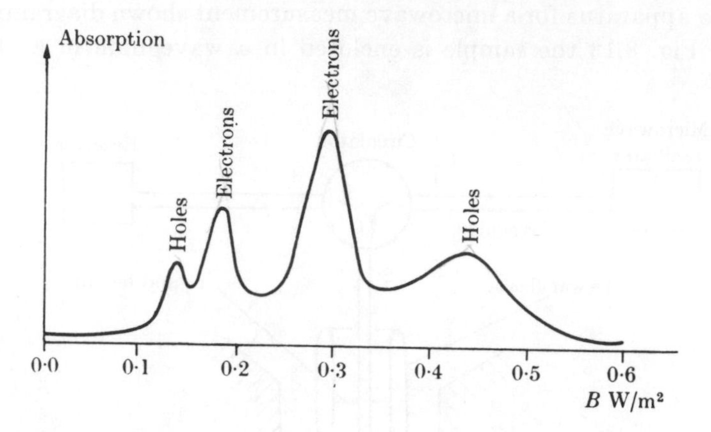

FIG. 8.14. Result of a cyclotron resonance experiment for germanium.

orientation between the magnetic field and the crystal axes. As may be seen there are two types of holes, light and heavy. A third resonance peak for the holes in the split-off band is missing because there are hardly any holes so much below the band edge. The two resonance peaks for electrons indicate that something sinister is going on in the conduction band as well. As a matter of fact these measurements, repeated in various directions, are just the tools for unravelling what the real E–k curve looks like. In addition, from the amplitude and width of the peaks, information about the density and collision times of the various carriers can be obtained.

Energy gap. A simple way to measure the energy gap between the valence and conduction bands is to see how the conductivity varies with temperature. For any semiconductor the conductivity is given by

$$\sigma = (N_e\mu_e + N_h\mu_h)e, \tag{8.57}$$

which is the same as eqn. (8.40). For an intrinsic material, we have from eqn. (8.41)

$$N_e = N_h = N_i = \text{constant} \times T^{\frac{3}{2}} \exp\left(\frac{-E_g}{2kT}\right). \tag{8.58}$$

Combining eqn. (8.57) with eqn. (8.58) we get

$$\sigma = \text{constant} \times e(\mu_c + \mu_h) T^{\frac{3}{2}} \exp\left(-\frac{E_g}{2kT}\right)$$

$$= \sigma_0 \exp\left(-\frac{E_g}{2kT}\right) \qquad (8.59)$$

We shall ignore the $T^{\frac{3}{2}}$ variation, which will almost always be negligible compared with the exponential temperature variation. Hence a plot of $\log_e \sigma$ versus $1/T$ will have a slope of $-E_g/2k$, which gives us E_g. Also in eqn. (8.59) we have ignored the variation of E_g with temperature; this simplification is not always justified. Leaving this aside for a moment, let us now consider what happens with an impurity semiconductor. We have discussed the variation of the Fermi level with temperature and concluded that at high temperatures semiconductors are intrinsic in behaviour, and at low temperatures they are pseudo-intrinsic with an energy gap equal to the gap between the impurity level and the band edge. Thus we would expect two definite straight-line regions with greatly different slopes on the plot of $\log_e \sigma$ against $1/T$, as illustrated in Fig. 8.15. In the region between these slopes the temperature is high enough to ionize the donors fully but not high enough to ionize an appreciable number of electrons from the host lattice. Hence, in this middle temperature range the carrier density will not be greatly influenced by temperature and the variations in mobility and the $T^{\frac{3}{2}}$ factor that we neglected will determine the shape of the curve.

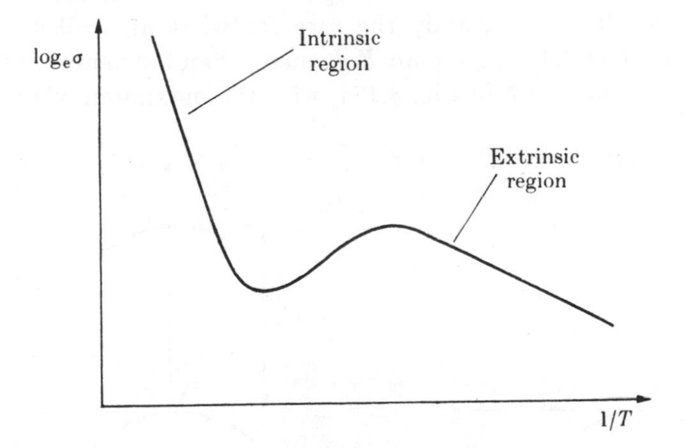

Fig. 8.15. Typical log conductivity–reciprocal temperature curve for an extrinsic semiconductor.

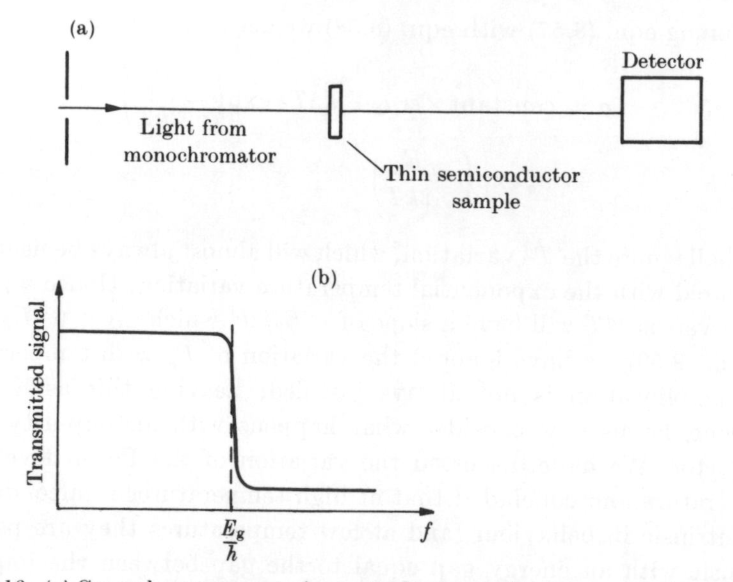

FIG. 8.16. (a) General arrangement of an optical transmission measurement and (b) the result for a direct-gap semiconductor.

An apparently simpler method of measuring the energy gap is to study optical transmission. The light is shone through a thin slice of semiconductor (Fig. 8.16a) and the transmission coefficient is plotted as a function of optical frequency (f). If $hf < E_g$ there will be some small transmission loss owing to various causes. But when $hf > E_g$ it becomes possible for electrons to be optically excited into the conduction band. Thus there should be a sharp change in transmission, as illustrated in Fig. 8.16b. This is certainly the case for what are called *direct-gap semiconductors*. These have an E–k energy band structure, illustrated in Fig. 8.9 and again in Fig. 8.17a, with the maximum of the valence

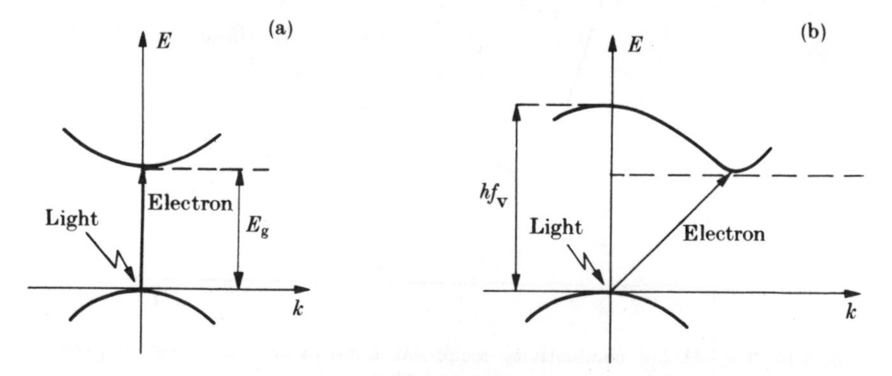

FIG. 8.17. Photon absorption by (a) a direct- and (b) an indirect-gap semiconductor.

band at the same k value as the minimum of the conduction band. GaAs and CdS are examples of direct-gap semiconductors.

How does an electron make a transition when excited by a photon? We have not yet studied this problem in any detail. All we have said so far is that if an electron receives the appropriate amount of energy it can be excited to a state of higher energy. This is not true in general because, as in macroscopic collision processes, not only energy but momentum as well should be conserved. So, strictly speaking, we should introduce the momenta of photons, bound electrons and lattice vibrations. This is unfortunately beyond the scope of this course. I have to ask you to believe that photon-excited electrons like to move vertically (well, nearly vertically) up in the E–k diagram, as shown in Fig. 8.17a.

In an indirect-gap semiconductor (like Si or Ge) the condition $hf = E_g$ is not sufficient to cause vertical transitions (Fig. 8.17(a)). It turns out that transitions are still possible (that is, the electrons are able to move sideways as well) with the help of lattice vibrations but the process is much less likely to occur. Therefore the transmission of light through a slice of indirect-gap semiconductor varies rather slowly as a function of frequency until f_v is reached, when vertical transitions become possible.

Whether or not a material is a direct-gap material is of increasing importance in the development of optical semiconductor devices—semiconductor lasers are all direct-gap materials.

Carrier lifetime. We are usually interested in *minority* carrier lifetime. The reason is simply that owing to injection or optical generation the minority carrier density may be considerably above the thermal equilibrium value, whereas the change in the density of majority carriers is generally insignificant. Consider, for example, silicon with 10^{22} fully ionized impurities per cubic metre. Then, as N_i for silicon is about $10^{16}/\text{m}^3$ at room temperature (eqn. (8.41)), N_h will be about $10^{10}/\text{m}^3$. Now suppose that in addition 10^{15} electron–hole pairs per cubic metre

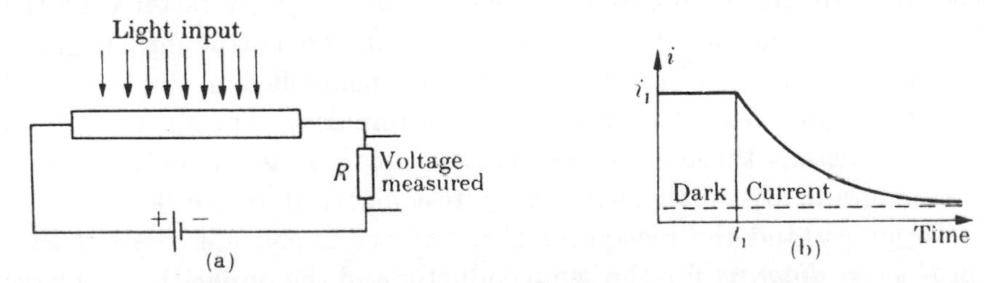

FIG. 8.18. (a) Photoconduction experiment in a semiconductor (b) when the light is switched off the current decays to its dark current value.

are created by input light. The hole density in the silicon will then increase by a huge factor, 10^5, but the change in electron density will be an imperceptible 10^{-5}%. Thus to 'see' the change of hole current is relatively simple; the only trick is to make a junction that lets through the holes but restricts the electron flow to a low value. (This is again something we shall discuss later.) Thus, the current flowing in the circuit of Fig. 8.18a consists mainly of holes created by the input light. If the light is switched off at $t = t_1$, the current (and so the voltage) across the resistance R declines (Fig. 8.18b) and from the curve the lifetime of the holes can be determined.

8.9. Preparation of pure controlled-impurity single crystal semiconductors

Crystal growth from the melt. This is the simplest way of preparing a single crystal. The material is purified by chemical means, perhaps to an impurity concentration of a few parts per million, then melted in a crucible of the shape shown in Fig. 8.19. The crucible is slowly

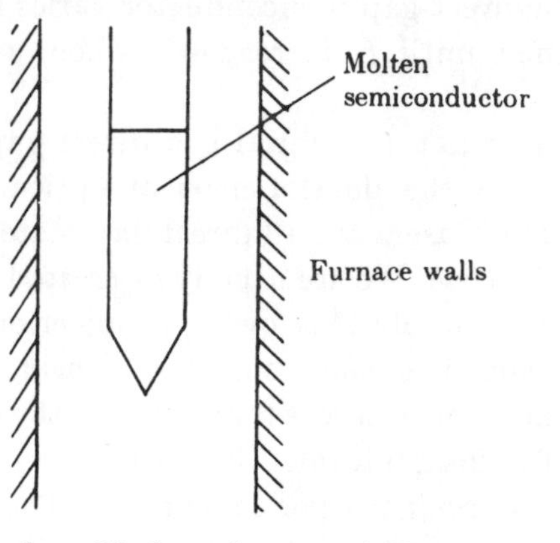

FIG. 8.19. A form of crucible for melt-grown single crystals.

cooled down. As the pointed end tends to cool slightly faster than the bulk of the material, the crystal 'seeds' at the bottom, then grows through the melt. If conditions are well controlled a single crystal growth is obtained. It is found that the impurity concentration is no longer constant throughout the crystal, but there is a definite concentration gradient, usually with the purest material at the bottom.

To understand the reason for this we have to consider the *metallurgical phase diagram* for the semiconductor and the impurity. You have probably come across the phase diagram for copper and zinc, stretching from 100% copper, 0% zinc to 0% copper, 100% zinc, with brasses in the middle, and curves representing liquidus and solidus lines, with temperature as the ordinate. We do not need to consider such a range of composition, since we are considering only a minute amount of impurity in silicon. We need only look at the region close to pure Si, where there will be no complications of eutectics but only the liquidus and solidus lines, shown diagrammatically in Fig. 8.20. The temperature separations of these lines will be only a few degrees.

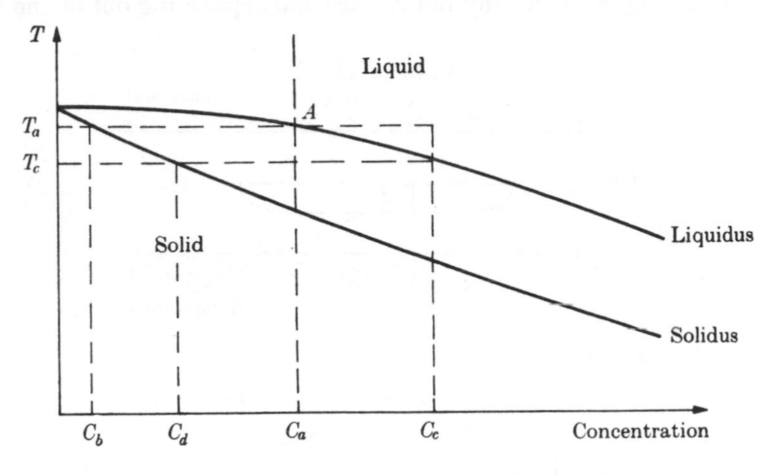

FIG. 8.20. Illustration of change of composition on freezing.

Suppose there is initially an impurity content C_a (Fig. 8.20). As the melt cools down it stays liquid until it reaches the temperature T_a. At this temperature there can exist liquid of composition C_a and solid of composition C_b. Solid of the latter composition is the first to crystallize out. As this is purer material and becomes lost to the rest of the melt once it solidifies, the remainder has a higher impurity concentration, say C_c. Thus no more solidification occurs until the temperature T_c is reached, when more solid impurity concentration C_d comes out. And so it goes on. Of course, if the cooling is slow this is a continuous process and the impurity concentration of the solid, still starting at C_b smoothly increases up the crystal.

It is usual to describe this process in terms of the *distribution coefficient k*, defined as the concentration of impurity in the solid phase divided by the concentration in the liquid phase, both measured close to the phase boundary. For the case of Fig. 8.20

$$k = C_b/C_a. \tag{8.60}$$

If it is assumed that k does not change during solidification, it is a simple matter to find the impurity concentration gradient of the crystal.

Zone refining. The different concentrations of impurity in the solid and liquid phase can be exploited in a slightly different way. We start with a fairly uniform crystal, melt a slice of it, and arrange for the *molten zone* to travel along the crystal length. This can be done by putting it in a refractory boat and dragging it slowly through a furnace, as shown in Fig. 8.21. At any point, the solid separating out at the back

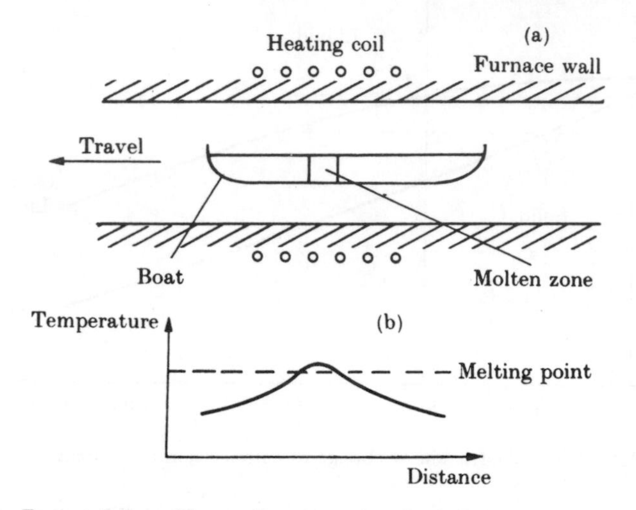

FIG. 8.21. Zone refining. The molten zone moving through the crystal sweeps the impurities to the far end.

of the zone will be k times as impure as the melted material which, as $k < 1$, is an improvement. By a fairly simple piece of algebra it can be shown that the impurity concentration in the solid, $C_s(x)$, after the zone has passed down the crystal (of length l) once is

$$C_s(x) = C_0\{1-(1-k)\exp(-kx/z)\}, \tag{8.61}$$

where C_0 is the initial concentration and z is the length of the molten zone. Clearly, at the end of the crystal that is melted first the value of impurity concentration will be

$$C_s(0) = k^n C_0 \tag{8.62}$$

if this process is repeated n times. Since k is typically 0·1, it is possible to drive most of the impurity to a small volume at the far end with relatively few passes.

This very simple idea is the basis of the great success of semiconductor engineering. As we have said before impurities can be reduced to few parts in 10^{10}, and then they are usually limited by impurities picked up from reactions with the boat.

Floating zone purification. This latter problem showed up rather strongly when the semiconductor industry went over from germanium (melting point 937°C) to silicon (melting point 1415°C). The solution was the *floating zone method*, which dispensed with the boat altogether. In this method the crystal is held vertically in a rotating chuck (Fig. 8.22). It is surrounded at a reasonable distance by a cool silica envelope,

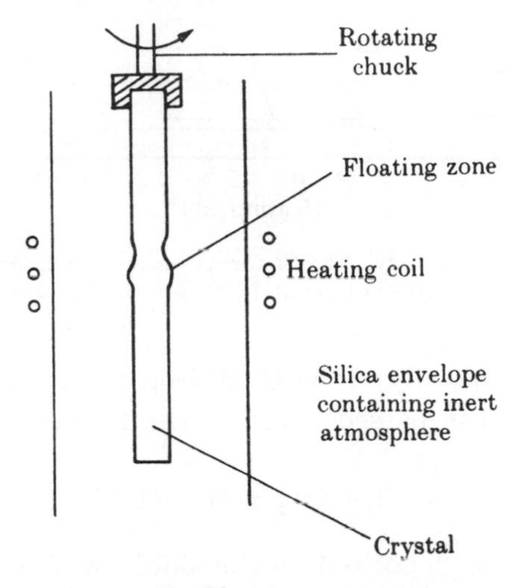

FIG. 8.22. Floating zone refining.

so that it can be kept in an inert atmosphere, then outside this is a single-turn coil of water-cooled copper tubing. A large high-frequency current (several MHz) is passed through the coil, and the silicon crystal is heated to melting point by the *eddy currents* induced in it. The coil is slowly moved up the crystal so that the molten zone passes along its length. This technique can be used only for fairly small crystals because the weight has to be supported by the surface tension of the molten zone.

Epitaxial growth. The processes of growing and refining single crystals made possible the advent of the transistor in the 1950's. The next stage has been the *planar* techniques, starting in about 1960, that have led to the development of integrated circuits, which promise to have an even greater impact on electronics than the transistor. We shall discuss the essentials of this process when we consider actual devices later, but now I shall just describe the *epitaxial growth* method of material preparation, which is eminently compatible with the manufacture of integrated microelectronic circuits.

'Epitaxial' is derived from a Greek word meaning 'arranged upon'.† There are several ways in which such growth can be carried out. To deposit silicon epitaxially from the *vapour phase* the arrangement of Fig. 8.23 can be used. Wafers of single-crystal silicon are contained in a

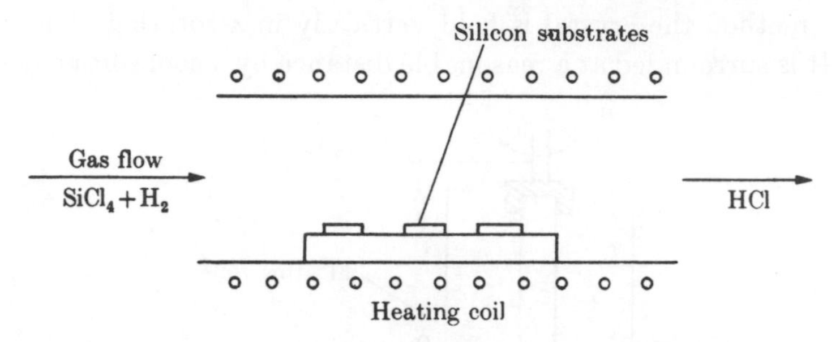

FIG. 8.23. Vapour phase epitaxy. The Si forms on the single crystal substrates at a temperature of about 1200°C in the furnace.

tube furnace at (typically) 1250°C. Silicon tetrachloride vapour in a stream of hydrogen is passed through the furnace and the chemical reaction

$$SiCl_4+2H_2 \rightleftharpoons Si+4HCl \tag{8.63}$$

takes place. The Si is deposited on the silicon wafers as a single crystal layer following the crystal arrangement of the substrate. Sometimes the silane reaction

$$SiH_4 \rightleftharpoons Si+2H_2 \tag{8.64}$$

is preferred, since it gives no corrosive products.

The epitaxial layer can be made very pure by controlling the purity of the chemicals; or more usefully it can be deliberately doped to make it n- or p-type by bubbling the hydrogen through a weak solution of (for example) phosphorous trichloride or boron trichloride respectively before it enters the epitaxy furnace. In this way epitaxial layers of about 2 μm to 20 μm thick can be grown to a known dimension and a resistivity that is controllable to within 5% from batch to batch.

† My friends who speak ancient Greek tell me that *epitactic* should be the correct adjective. Unfortunately, *epitaxial* has gained such a wide acceptance among technologists having no Greek-speaking friends that we have no alternative but to follow suit.

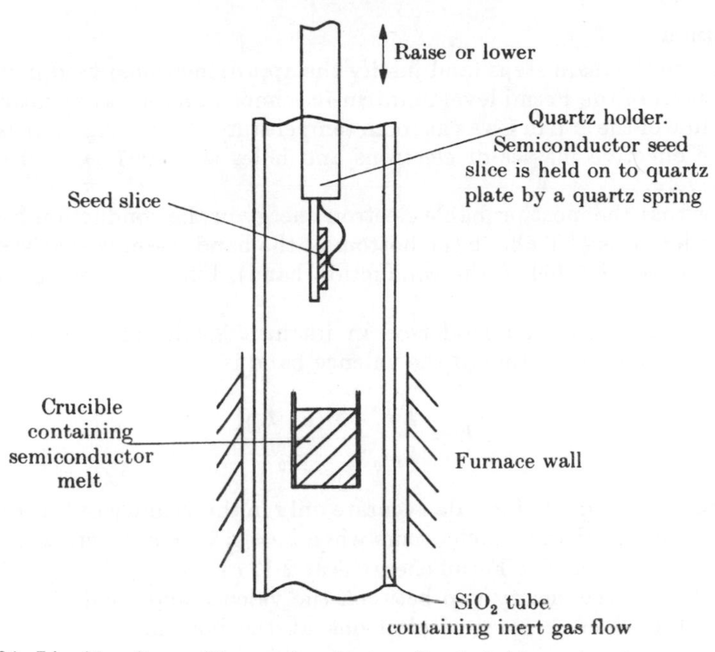

FIG. 8.24. Liquid epitaxy. The semiconductor slice is held on the plate by a quartz spring clip and lowered into the molten semiconductor alloy. By correct cooling procedures the pure semiconductor is encouraged to precipitate onto the surface of the slice.

Liquid epitaxy has also been used, mainly with compound semiconductors. The substrate crystal is held above the melt on a quartz plate and dipped into the molten semiconductor (Fig. 8.24). By accurately controlling the cooling rate a single-crystal layer can be grown epitaxially on the crystal.

TABLE 8.2. *The more important properties of a few semiconductors at room temperature*

Semi-conductor	Energy gap E_g (eV)	Effective masses m_e^*/m_0	m_h^*/m_0	Mobilities, $m^2V^{-1}sec^{-1} \times 10^{-4}$ μ_e	μ_h
Ge	0·67	0·12	0·04 0·28 0·08	3800	1800
Si	1·11	0·26	0·16 0·50 0·24	1500	500
GaAs	1·40	0·067	0·65	8500	400
GaP	2·25	0·35	0·5	150	140
InP	1·30	0·08	0·2	4600	150
InSb	0·17	0·013	0·18	70000	1000
CdS	2·5	0·27	0·07	340	18

Examples

1. Indicate the main steps (and justify the approximations) used in deriving the position of the Fermi level in intrinsic semiconductors. How near is it to the middle of the gap in GaAs at room temperature? The energy gap is 1·4 eV and the effective masses of electrons and holes are $0{\cdot}067\ m_0$ and $0{\cdot}65\ m_0$ respectively.
2. Show that the most probable electron energy in the conduction band of a semiconductor is $\frac{1}{2}kT$ above the bottom of the band (assume that the Fermi level is several kT below the conduction band). Find the average electron energy.
3. In a one-dimensional model of an intrinsic semiconductor the energy measured from the bottom of the valence band is

$$E = \frac{\hbar^2 k_1^2}{3m_0} + \frac{\hbar^2 (k-k_1)^2}{m_0}.$$

This is an approximate formula accurate only in the vicinity of the minimum of the conduction band which occurs when $k = k_1 = \pi/a$, where a the lattice spacing is 0·314 nm. The Fermi energy is at 2·17 eV.

Calculate (i) the energy gap between the valence and conduction bands, and (ii) the effective mass of electrons at the bottom of the conduction band.

Assume that the Fermi level is halfway between the valence and conduction bands.

4. The variation of the resistivity of intrinsic germanium with temperature is given by the following table:

T(K)	385	455	556	714
$\rho(\Omega\text{m})$	0·028	0·0061	0·0013	0·000274

It may be assumed, as a rough approximation, that the hole and electron mobilities both vary as $T^{-\frac{3}{2}}$ and that the forbidden energy gap, E_g, is independent of temperature.

(i) Determine the value of E_g.

(ii) At about what wavelength would you expect the onset of optical absorption?

5. What is the qualitative difference between the absorption spectra of a direct gap and that of an indirect gap semiconductor?
6. What values of carrier concentrations, N_e and N_h, give minimum conductivity at a given temperature? Determine N_e/N_h if the collision times for electrons and holes are equal and $m_e^*/m_h^* = \frac{1}{2}$.
7. A silicon crystal is doped with indium for which the electron acceptor level is 0·16 eV above the top of the valence band. For silicon the energy gap is 1·10 eV and the effective masses of holes and electrons are $0{\cdot}39\ m_0$ and $0{\cdot}26\ m_0$ respectively. What impurity density would cause the Fermi level to coincide with the impurity level at 300 K and what fraction of the acceptor levels will be filled? What are the majority and minority carrier concentrations in the crystal?

8. (i) In a certain n-type semiconductor a fraction α of the donor atoms is ionized. Derive an expression for $E_F - E_D$, where E_F and E_D are the Fermi level and donor level respectively.

(ii) In a certain p-type semiconductor a fraction β of the acceptor atoms is ionized. Derive an expression for $E_F - E_A$ where E_A is the acceptor level.

(iii) Assume that both of the above materials were prepared by doping the same semiconductor, and that $E_D = 1{\cdot}1$ eV and $E_A = 0{\cdot}1$ eV, where energies are measured from the top of the valence band. By various measurements at a certain temperature T we find that $\alpha = 0{\cdot}5$ and $\beta = 0{\cdot}05$. When a p–n junction is made, the built-in voltage measured at the same temperature is found to be 1·05 eV. Determine T.

9. The rate of recombination (equal to the rate of generation) of carriers in an extrinsic semiconductor is given by eqn. (8.45). If the minority carrier concentration in an n-type semiconductor is above the equilibrium value by an amount $(\delta N_h)_0$ at $t = 0$, show that this extra density will reduce to zero according to the relationship

$$\delta N_h = (\delta N_h)_0 \exp(-t/\tau_p)$$

where

$$\tau_p = \frac{1}{aN_e}.$$

(Hint: The rate of recombination is proportional to the *actual* density of carriers, while the rate of generation remains constant.)

10. Derive the continuity equation for minority carriers in an n-type semiconductor.

(Hint: Take account of recombination of excess holes by introducing the lifetime, τ_p.)

11. Fig. 8.14 shows the result of a cyclotron resonance experiment with Ge. Microwaves of frequency 24,000 MHz were transmitted through a slice of Ge and the absorption was measured as a function of a steady magnetic field applied along a particular crystalline axis of the single-crystal specimen. The ordinate is a linear scale of power absorbed in the specimen. The total power absorbed is always a very small fraction of the incident power.

(i) How many distinct types of charge carriers do there appear to be in Ge from this data?

(ii) How many types of charge carriers are really there? How would you define whether they are 'real' charge carriers or not?

(iii) What are the effective masses for this particular crystal direction? Can this effective mass be directly interpreted for electrons? For holes?

(iv) If the figure were not labelled would you be able to tell which peaks referred to electrons?

(v) Estimate the collision times of the holes (Hint: Use eqn. 1.66).

(vi) Estimate the relative number density of the holes.

(vii) In Section 8.7 we talked about *three* different types of holes. Why are there only two resonance peaks for holes?

9. Principles of semiconductor devices

Les mystères partout coulent comme les sèves,

BAUDELAIRE *Les Sept Vieillards*

This thing with knobs and a pretty light.

A. WESKER *Chips with everything*

9.1. Introduction

YOU have bravely endured lengthy discussions on rather abstract and occasionally nebulous concepts in the hope that something more relevant to the practice of engineering will emerge. Well, here we are; at last we are going to discuss various semiconductor devices. It is impossible to include all of them, for there are so many nowadays. But if you follow carefully (and if everything we have discussed so far is at your fingertips) you will stand a good chance of understanding the operation of all existing devices and—I would add—you should be in a very good position to understand the operation of semiconductor devices to come in the near future. This is because human ingenuity has rather narrow limits. Hardly anyone ever produces a new idea. It is always some combination of old ideas that leads to reward. Revolutions are few and far between. It is steady progress that counts.

9.2. The p–n junction in equilibrium

Not unexpectedly, when we want to produce a device we have to put things together. This is how we get the simplest semiconductor device, the p–n junction, which consists of a p- and an n-type material in contact (Fig. 9.1a). Let us imagine now that we literally put the two pieces together.† What happens when they come into contact? Remember, in the n-type material there are lots of electrons, and holes abound in the p-type material. At the moment of contact the electrons will rush over into the p-type material and the holes into the n-type material. The reason is, of course, diffusion; both carriers make an attempt to occupy uniformly the space available. Some electrons moving towards the left collide head-on with the onrushing holes and recombine, but others will be able to penetrate farther into the p-type material. How far? Not very far; or, to put it another way, not *many* get very far because their efforts are frustrated by the appearance of an electric

† This is *not* how junctions are made.

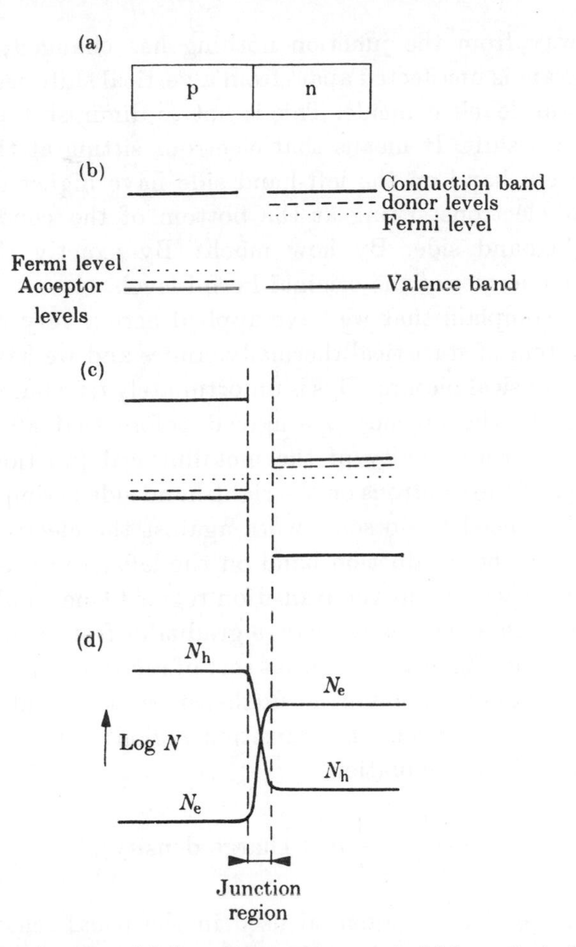

FIG. 9.1. The p–n junction. (a) A p- and an n-type material in contact, (b) the energy diagrams before contact, (c) the energy diagrams after contact, (d) electron and hole densities.

field. The electrons leave positively charged donor atoms behind, and similarly there are negatively charged acceptor atoms left in the p-type material when the holes move out. This charge imbalance will give rise to an electric field that will increase until equilibrium is reached.

Having reached equilibrium we can now apply a theorem mentioned before when discussing metal–metal junctions. We said that whenever two or more materials are in thermal equilibrium their respective Fermi levels must agree.

The Fermi levels before contact are shown in Fig. 9.1b and after contact in Fig. 9.1c. Here we assumed that some (as yet unspecified) distance away from the junction nothing has changed; that is, the energy diagram is unaffected apart from a vertical shift needed to make the two Fermi levels coincide. This is not to diminish the significance of the vertical shift. It means that electrons sitting at the bottom of the conduction band on the left-hand side have higher energies than their fellow electrons sitting at the bottom of the conduction band at the right-hand side. By how much? By exactly the difference between the energies of the original Fermi levels.

You may complain that we have applied here a very profound and general theorem of statistical thermodynamics and we have lost in the process the physical picture. This is unfortunately true but nothing stops us returning to the physics. We agreed before that an electric field would arise in the vicinity of the metallurgical junction. Thus, the lower energy of the electrons on the right-hand side is simply due to the fact that they need to do some work against the electric field before they can reach the conduction band on the left-hand side.

What can we say about the transition region? One would expect the electron and hole densities to change gradually from high to low densities as shown in Fig. 9.1d. But what sort of relationship will determine the density at a given point? And furthermore, what will be the profile of the conduction band in the transition region? They can all be obtained from Poisson's equation

$$\frac{d^2U}{dx^2} = \frac{1}{\epsilon}\text{(net charge density)},$$

where U is the electric potential used in the usual sense.† Since the density of mobile carriers depends on the actual variation of potential in the transition region, this is not an easy differential equation to solve. Fortunately, a simple approximation may be employed that leads quickly to the desired result.

As may be seen in Fig. 9.1d, the density of mobile carriers rapidly decreases in the transition region. We are therefore nearly right if we maintain that the transition region is completely depleted of mobile carriers. (For this reason the transition region is often called the 'depletion' region.) Hence we may assume the net charge densities are approximately of the form shown in Fig. 9.2, where $-x_p$ and x_n are the

† We are in a slight difficulty here because up to now potential meant the potential energy of the electron, denoted by V. The relationship between the two quantities is $eU = V$, which means that if you confuse the two things you'll be wrong by a factor of 10^{19}.

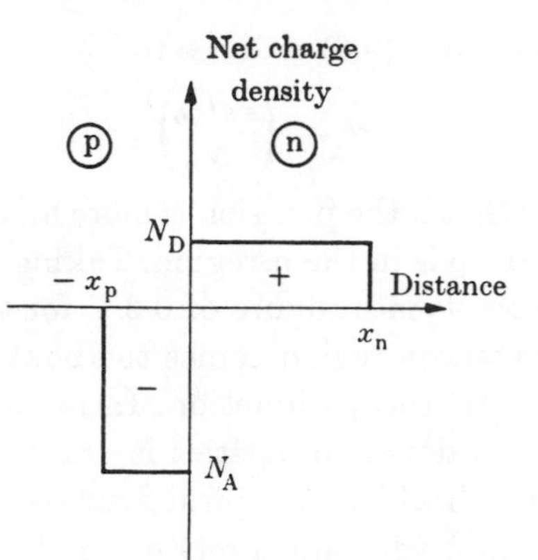

FIG. 9.2. Net charge densities in the transition region of a p–n junction.

widths of the depletion regions in the p- and n-type materials respectively. Charge conservation is expressed by the condition

$$N_A x_p = N_D x_n. \tag{9.1}$$

Poisson's equation for the region $-x_p$ to 0 reduces now to the form

$$\frac{d^2U}{dx^2} = \frac{eN_A}{\epsilon}, \tag{9.2}$$

which, integrated twice, gives the solution

$$U(-x_p) - U(0) = eN_A x_p^2/2\epsilon. \tag{9.3}$$

Performing the analogous integration for the 0 to x_n region, we get

$$U(0) - U(x_n) = eN_D x_n^2/2\epsilon. \tag{9.4}$$

Hence the total potential difference across the transition region is

$$\begin{aligned} U_0 &= U(-x_p) - U(x_n) \\ &= e(N_A x_p^2 + N_D x_n^2)/2\epsilon. \end{aligned} \tag{9.5}$$

This is the 'built-in' voltage between the p- and n-regions, often referred to as the contact potential. A typical figure for it is 0·3 V.

The total width of the transition region may be expressed as

$$w = x_p + x_n = \left(\frac{2\epsilon}{e}\right)^{\frac{1}{2}}\left[\left\{\frac{U(-x_p)-U(0)}{N_A}\right\}^{\frac{1}{2}} + \left\{\frac{U(0)-U(x_n)}{N_D}\right\}^{\frac{1}{2}}\right], \tag{9.6}$$

which, after a little algebra, may be put in the form

$$w = \left\{\frac{2\epsilon U_0}{e(N_A+N_D)}\right\}^{\frac{1}{2}}\left\{\left(\frac{N_A}{N_D}\right)^{\frac{1}{2}} + \left(\frac{N_D}{N_A}\right)^{\frac{1}{2}}\right\}. \tag{9.7}$$

If, say, $N_A \gg N_D$, eqn. (9.6) reduces to

$$w = \left(\frac{2\epsilon U_0}{eN_D}\right)^{\frac{1}{2}}, \tag{9.8}$$

which shows clearly that if the p-region is more highly doped, practically all of the potential drop is in the n-region. Taking for the donor density $N_D = 10^{21}/\text{m}^3$ and the typical figure of 0·3 V for the contact potential, the width of the transition region comes to about 0·18 μm. Remember, this is the value for an abrupt junction. In practice, the change from acceptor impurities to donor impurities is gradual, and the transition region is therefore much wider. A typical figure is about 1 μm. Thus in a practical case we cannot very much rely on the formulae derived above but if we have an idea how the acceptor and donor concentrations vary, similar equations can be derived.

From our simple model (assuming a depletion region) we obtained a quadratic dependence of the potential energy in the transition region. More complicated models give somewhat different dependence but they all agree that the variation is monotonic. Our energy diagram is thus as shown in Fig. 9.3. The energy difference between the bands on the p- and n-sides is eU_0.

We can describe now the equilibrium situation in yet another way. The electrons sitting at the bottom of the conduction band at the p-side will roll down the slope because they lower their energy this way. So there will be a flow of electrons from left to right proportional to the density of electrons in the p-type material

$$I_{e(\text{left to right})} \sim N_{ep}. \tag{9.9}$$

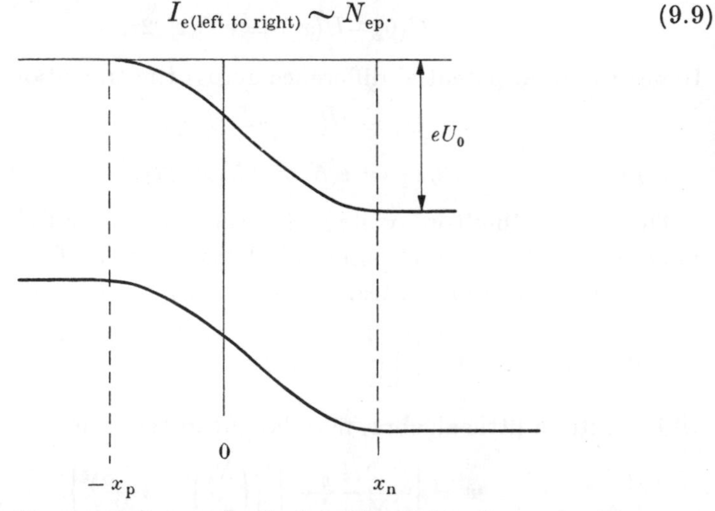

FIG. 9.3. The energy diagram for the transition region of a p–n junction.

The electrons in the n-type material, being the majority carriers, are very numerous. So although most of them will be sitting at the bottom of the conduction band there will still be a considerable number with sufficient energies to cross to the p-side. Assuming Boltzmann statistics this number is given by

$$N_{en} \exp(-eU_0/kT). \tag{9.10}$$

Substituting N_{en} from eqn. (8.17) we get

$$N_c \exp\{-(E_g+eU_0-E_F)/kT\} \tag{9.11}$$

Hence the electron current from right to left is given by

$$I_{e(\text{right to left})} \sim N_c \exp\{-(E_g+eU_0-E_F)/kT\}. \tag{9.12}$$

In equilibrium the current flowing from left to right should equal the current flowing from right to left; that is,

$$N_{ep} = N_c \exp\{-(E_g+eU_0-E_F)/kT\}. \tag{9.13}$$

Eqn. (9.13) gives nothing new. If we expressed the electron density in the p-type material with the aid of the Fermi level then we could show from eqn. (9.13) that eU_0 should be equal to the difference between the original Fermi levels, which we already knew. But although eqn. (9.13) doesn't give any new information we shall see in a moment that by describing the equilibrium in terms of currents flowing in opposite directions, the rectifying properties of the p–n junction can be easily understood.

We could go through the same argument for holes without much difficulty provided we can imagine particles rolling uphill, because for holes that is the way to lower their energy. The equations would look much the same and I shall not bother to derive them.

9.3. Rectification

Let us now apply a voltage as shown in Fig. 9.4. Since there are much fewer carriers in the transition region, we may assume that all the applied voltage will drop across the transition region. Then, depending only on the polarity, the potential barrier between the p and n regions

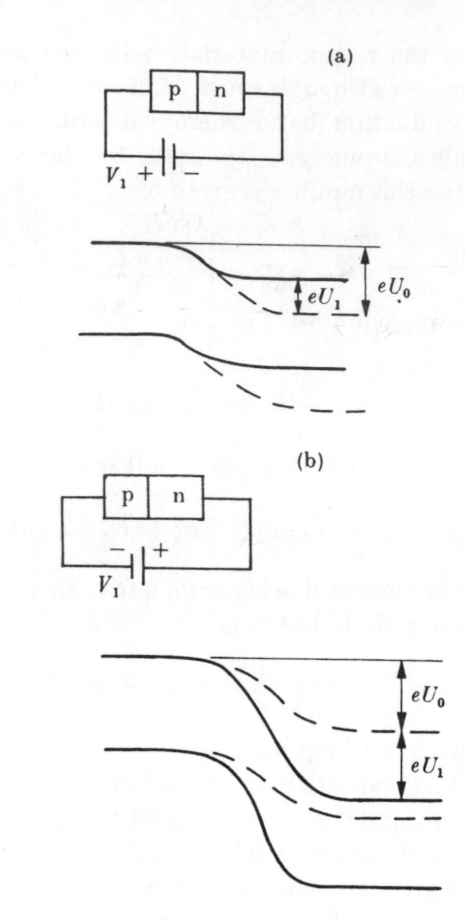

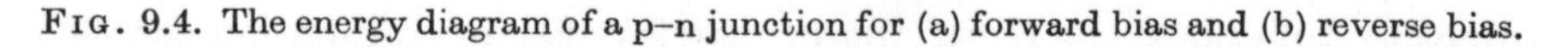

FIG. 9.4. The energy diagram of a p–n junction for (a) forward bias and (b) reverse bias.

will decrease or increase. If the p-side is made positive the potential barrier is *reduced* and we talk of *forward bias*. The opposite case is known as *reversed bias*; the p side is then negative and the potential barrier is increased.

It is fairly obvious qualitatively that the number of electrons flowing from left to right is not affected in either case. The same number of electrons will still roll down the hill as in equilibrium. But the flow of electrons from right to left is seriously affected. For reverse bias it will be reduced and for forward bias it will significantly increase. So we can see qualitatively that the total current flowing for a voltage U_1 will differ from the current flowing at a voltage $-U_1$. This is what is meant by rectification.

It is not difficult to derive the mathematical relationships; we have practically everything ready. The current from left to right is the same; let us denote it by I_0. The current from right to left may be obtained by putting $e(U_0 - U_1)$ in place of eU_0 in eqn. (9.12). (This is because we now want the number of electrons having energies in excess of $e(U_0 - U_1)$, etc.) At $U_1 = 0$, this current is equal to I_0 and increases exponentially with U_1; that is

$$I_{\mathrm{e(right\ to\ left)}} = I_0 \exp eU_1/kT. \tag{9.14}$$

Hence the total current

$$\begin{aligned} I_\mathrm{e} &= I_{\mathrm{e(right\ to\ left)}} - I_{\mathrm{e(left\ to\ right)}} \\ &= I_0(\exp eU_1/kT - 1), \end{aligned} \tag{9.15}$$

which is known as the *rectifier equation*;† it is plotted in Fig. 9.5. For

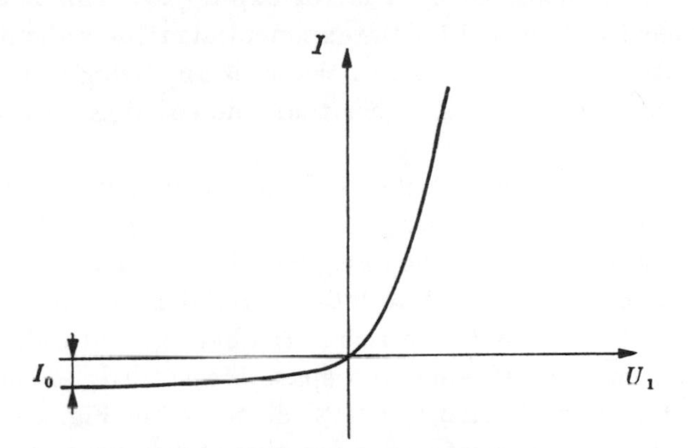

FIG. 9.5. The current as a function of applied voltage for a p–n junction.

negative values of U_1, I_e tends to I_0 and there is the exponential increase of current with forward voltage. It is worth noting that in spite of the simple reasoning this equation is qualitatively true for real diodes.

So if we plot a graph of log I_e versus applied forward bias voltage, we get a pretty good straight line for most rectifiers. (Above about 1 volt bias, the final '-1' in the rectifier equation can be neglected.) However there are two snags:

(1) the slope of the line is $\dfrac{e}{mkT}$ not $\dfrac{e}{kT}$, where m is a number usually lying between 1 and 2;

† Adding the hole current would increase I_0 but the form of the equation would not change because the same exponential factor applies to the hole density in the p-type material.

(2) the current intercept when the graph is extrapolated back to zero voltage gives log I_0. But this value of I_0 is several orders of magnitude less than the value of I_0 obtained by measuring the reverse current (Fig. 9.5). Explaining this away is beyond the scope of this course: it is necessary to take into account recombination and generation of carriers in the depletion region. A good account is given in the books by J. P. McKelvey and A. S. Grove cited in the further reading list.

9.4. Injection

In thermal equilibrium the number of electrons moving towards the left is equal to the number of electrons moving towards the right. However, when a forward bias is applied, the number of electrons poised to move left is increased by a factor exp U_1/kT. This is quite large; for an applied voltage of 0·1 V the exponential factor is about 55 at room temperature. Thus the number of electrons appearing at the boundary of the p-region is 55 times higher than the equilibrium concentration of electrons there.

What happens to these electrons? When they move into the p-region they become minority carriers, rather like immigrants travelling to a new country suddenly become foreigners. But, instead of mere political friction, the electrons' ultimate fate is annihilation. They are slain by heroic holes who themselves perish in the battle.† Naturally, to annihilate all the immigrants, time and space are needed; so some of them get quite far inside foreign territory, as shown in Fig. 9.6, where the density of electrons is plotted as a function of distance in a p–n junction under forward bias. The electron density declines, but not very rapidly. A typical distance is about 1 mm, which is about a thousand times larger than the width of the transition region.

Let us go back now to the plight of the holes. They are there in the p-type material to neutralize the negative charge of the acceptor atoms. But how will space charge neutrality be ensured when electrons are injected? It can be done in only one way; wherever the electron density is increased, the hole density must increase as well. And this means that new holes must move in from the contacts. Thus, as electrons move in from the right, holes must move in from the left to ensure charge

† I don't think that the pacifist interpretation of the recombination of electrons and holes (they get married, and live happily ever after) can bear closer scrutiny. When an electron and a hole recombine, they disappear from the stage and that's that. I would, however, be willing to accept the above interpretation for excitons which may be looked upon as electron–hole pairs bound together, but that's another story.

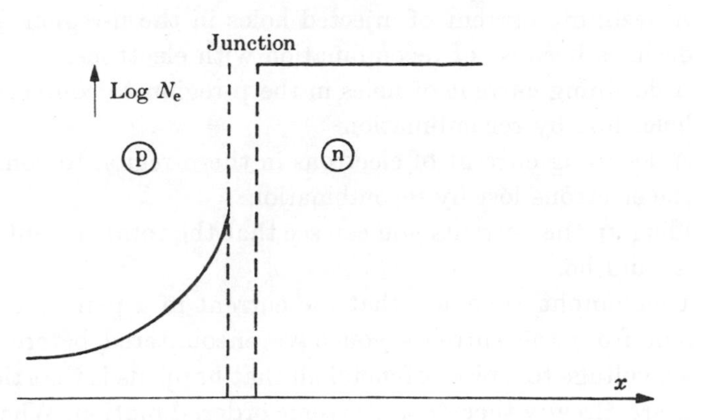

FIG. 9.6. The electron distribution in a forward biased p–n junction.

neutrality. Hence the current of electrons and holes will be made up of six constituents as shown in Fig. 9.7:

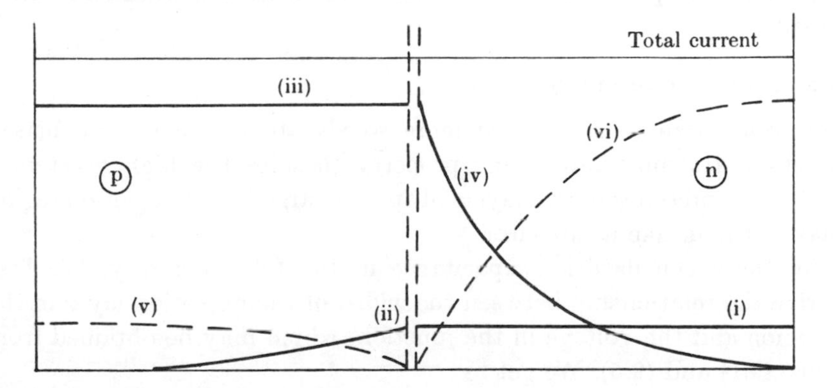

FIG. 9.7. The current distribution in a forward biased p–n junction.

(i) The electron current flowing in the n-type material and providing the electrons to be injected into the p-type material. It is constant in the n-region.

(ii) A declining current of injected electrons in the p-region. The current declines because the number of electrons becomes less and less as they recombine with the holes.

(iii) The current of holes in the p-region to provide the holes to be injected into the n-region. We have not discussed this because the injection of holes is entirely analogous to the injection of electrons.

(iv) A declining current of injected holes in the n-region. The current declines because of recombination with electrons.

(v) A declining current of holes in the p-region to compensate for the holes lost by recombination.

(vi) A declining current of electrons in the n-region to compensate for the electrons lost by recombination.

Adding up the currents you can see that the total current is constant, as it should be.

Let me emphasize again that the current in a p–n junction is quite different from the currents you have encountered before. When you apply a voltage to a piece of metal all that happens is that the electrons, which are already there, acquire some ordered motion. When a forward bias is applied to a p–n junction minority carriers get injected into both regions. These minority carriers were *not* there originally in such a high density; they came as a consequence of the applied voltage.

The distinction between ordinary conduction and minority carrier injection is important. It is the latter which makes transistor action possible.

9.5. Junction capacity

I would like to say a few more words about the reverse biased junction. Its most interesting property (besides the high resistance) is that the presence of two layers of space charge in the depletion region makes it look like a capacitor.

We may calculate its capacitance in the following way. We first derive the relationship between the width of the depletion layer in the n-region and the voltage in the junction, which may be obtained from eqns. (9.1) and (9.5). We get

$$x_{\mathrm{n}} = \left\{\frac{2\epsilon U_0 N_{\mathrm{A}}}{eN_{\mathrm{D}}(N_{\mathrm{D}}+N_{\mathrm{A}})}\right\}^{\frac{1}{2}}, \tag{9.16}$$

where U_0 is the 'built-in' voltage. For reverse bias the only difference is that the barrier becomes larger, that is U_0 should be replaced by U_0+U_1 where U_1 is the applied voltage in the reverse direction, yielding

$$x_{\mathrm{n}} = \left\{\frac{2\epsilon(U_0+U_1)N_{\mathrm{A}}}{eN_{\mathrm{D}}(N_{\mathrm{D}}+N_{\mathrm{A}})}\right\}^{\frac{1}{2}}. \tag{9.17}$$

The total charge of the donor atoms is

$$Q = eN_{\mathrm{D}}x_{\mathrm{n}} = \left\{2\epsilon e(U_0+U_1)\frac{N_{\mathrm{A}}N_{\mathrm{D}}}{N_{\mathrm{A}}+N_{\mathrm{D}}}\right\}^{\frac{1}{2}}. \tag{9.18}$$

Now a small increase in voltage will add charges at the boundary– as it happens in a real capacitance. We may, therefore, define the capacitance of the junction (per unit area) as

$$C = \frac{\partial Q}{\partial U_1} = \left\{\frac{\epsilon e}{2(U_0+U_1)}\frac{N_{\mathrm{A}}N_{\mathrm{D}}}{N_{\mathrm{A}}+N_{\mathrm{D}}}\right\}^{\frac{1}{2}}. \tag{9.19}$$

We can now put down an equivalent circuit for the junction and assign a physical function to the three elements (Fig. 9.8). R_1 is simply the ohmic resistance of the 'normal', i.e. not depleted, semiconductor. $R(U_1)$ is the junction resistance. It is very small in the forward direction

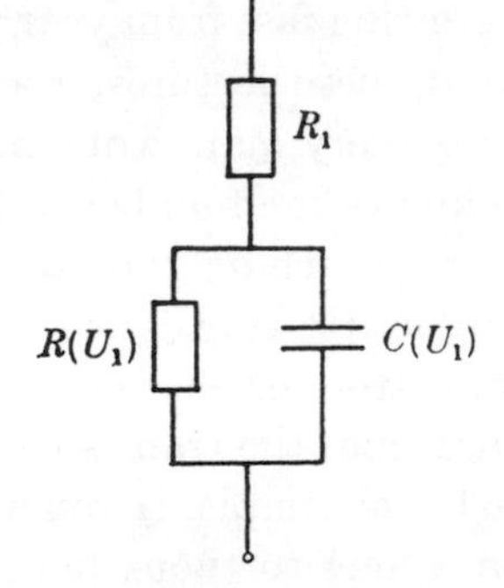

FIG. 9.8. The equivalent circuit of a p–n junction.

(0·1–10 Ω typically) and large in the reverse direction (10^6–10^8 Ω typically). $C(U_1)$ is the capacitance that varies with applied voltage, given by eqn. (9.19). Clearly, this equation loses validity for strong forward bias because the depletion layer is then flooded by carriers.

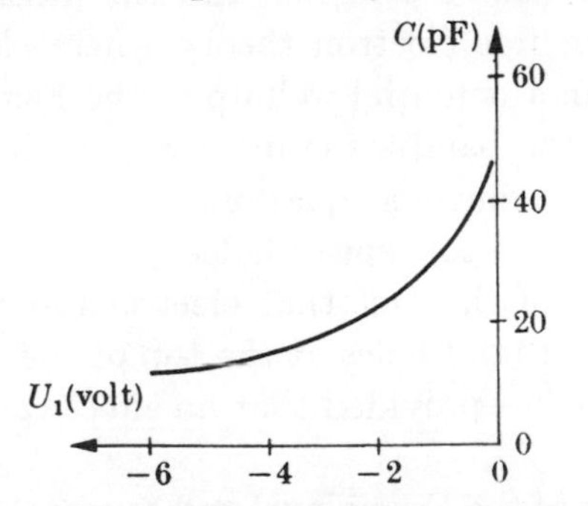

FIG. 9.9. Capacitance of a commercially available junction diode as a function of applied voltage.

This is actually borne out by our equivalent circuit, which shows that the capacity is shorted-out when $R(U_1)$ becomes small. With a reversed biased diode, $C(U_1)$ is easily measured, and can be fitted to an equation like (9.19) very well. The characteristic of a commercial diode is plotted in Fig. 9.9. Here C is proportional to $(U_1+0{\cdot}8)^{-\frac{1}{2}}$ whence the constant 0·8 V may be identified with the 'built-in' voltage.

9.6. The transistor

Here we are. We have arrived at last at the transistor, the most famous electronic device of the century. It has a history of little more than thirty years (it was discovered by Bardeen, Brattain, and Shockley at Bell Telephone Laboratories in 1948) but in that time it conquered the world. It made possible both the pocket radio and the giant computers, to mention only two applications. The number of transistors manufactured is still growing fast from year to year. The figure for 1967 when we first prepared these lectures, was a few times 10^9; that is about one transistor for every man, woman, and child living on Earth. I don't think anyone knows the number of transistors produced nowadays. We are rapidly approaching the stage when a single computer might contain as many as 10^9 transistors (or one of its close relations, namely field-effect transistors or charge-coupled devices).

I'm afraid the discussion of the transistor will turn out to be an anti-climax. It will seem to be too simple. If you find it too simple, remember that we have trodden a long tortuous path to get there. It might be worthwhile to recapitulate the main hurdles we have scaled:

(i) Postulation of Schrödinger's equation.

(ii) The solution of Schrödinger's equation for a rigid, three-dimensional potential well.

(iii) Postulation of Pauli's principle and the introduction of spin.

(iv) Formulation of free electron theory where electrons fill up the energy levels in a potential well up to the Fermi energy.

(v) Derivation of the band structure by a combination of physical insight and Schrödinger's equation.

(vi) Introduction of the concept of holes.

(vii) Demonstration of the fact that electrons at the bottom of the conduction band (and holes at the top of the valence band) may be regarded as free provided that an effective mass is assigned to them.

(viii) Determination of the Fermi level and carrier densities in extrinsic semiconductors.

(ix) The description of a p–n junction in terms of opposing currents.

(x) Explanation of minority-carrier injection.

So if you want to go through a logical chain of reasoning and want to explain the operation of the transistor from first principles, those above are the main steps in the argument.

Now let us see the transistor itself. It consists of two junctions with one semiconductor region common to both. This is called the *base*, and the other two regions are the *emitter* and the *collector* as shown in Fig. 9.10 for a p–n–p transistor. There are also n–p–n transistors; the ensuing explanation could be made to apply to them by judicious changing of words.

Consider first the emitter–base p–n junction. It is forward biased (*p*ositive on *p*-side for those who like mnemonics). This means that large numbers of carriers flow, holes into the base, electrons into the emitter. Now the holes arriving into the base region will immediately start the

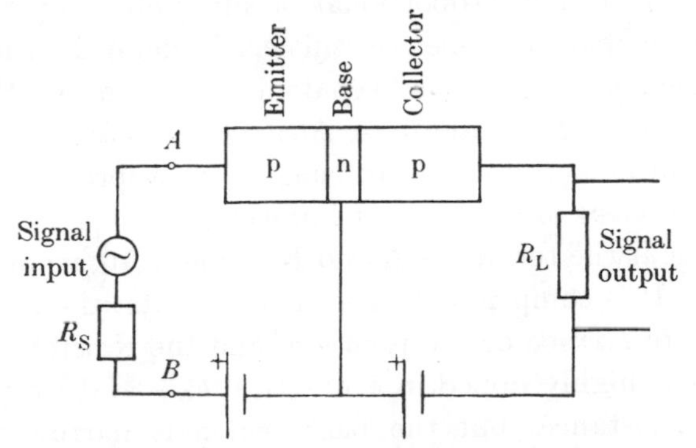

Fig. 9.10. The p–n–p transistor as an amplifier.

process of recombination with electrons. But, as explained before, time and space are needed to annihilate the injected minority-carriers. Hence for a narrow base region ($\ll 1$ mm) the hole current leaving the base region will be almost identical to the hole current entering from the emitter. Now what happens to the holes when they arrive to the collector region? They see a negative voltage (the base–collector junction is reverse biased) and carry on happily towards the load. Thus practically the same current that left the generator finds its way to the load.†

So far there is nothing spectacular; the current gain of the device

† The above argument is not quite correct because the emitter current is not carried solely by holes. There is also an electron flow from the base into the emitter. However, transistors are designed in such a way that the conductivity of the base is well below the conductivity of the emitter (a typical figure may be a factor of a hundred); thus the minority-carrier flow from the base to the emitter is usually negligible.

is somewhat below unity. Why is this an amplifier? It is an amplifier because the voltage gets amplified by a large factor. This is because the input circuit is a low-impedance circuit; a low voltage is thus sufficient to cause a certain current. This current reappears in the high impedance output circuit and is made to flow across a large load resistance, resulting in a high output voltage. Hence the transistor in the common base circuit is a voltage amplifier.

We should, however, know a little more about this amplifier. Can we express its properties in terms of the usual circuit parameters: impedances, current sources, and voltage sources? How should we attempt the solution of such a problem? Everything is determined in principle. If the bias voltages are fixed and an a.c. voltage is applied to the input of the transistor in Fig. 9.10, then the output current is calculable. Is this enough? Not quite. We have to express the frequency dependence in the form of rational fractions (this is because impedances are either proportional or inversely proportional to frequency) and then an equivalent circuit can be defined. It is a formidable job; it can be done and it has been done but, of course, the calculation is far too lengthy to include here. Although we cannot solve the complete problem it is quite easy to suggest an approximate equivalent circuit on the basis of our present knowledge.

Looking in at the terminals A and B of Fig. 9.10, what is the impedance we see? It is comprised of three components: the resistance of the emitter, the resistance of the junction, and the resistance of the base. The emitter is highly doped in a practical case and we may therefore neglect its resistance, but the base region is narrow and of lower conductivity and so we must consider its resistance. Hence we are left with r_e (called misleadingly the *emitter resistance*; it is in fact the resistance of the junction) and r_b (base resistance), forming the input circuit shown in Fig. 9.11a.

What is the resistance of the output circuit? We must be careful here. The question is how will the a.c. collector current vary as a function of the a.c. collector voltage? According to our model the collector current is quite independent of the collector voltage. It is equal to αi_e where i_e is the emitter current and α is a factor very close to unity. Hence our first equivalent output circuit must simply consist of the current generator shown in Fig. 9.11b. In practice the impedance turns out to be less than infinite (a few hundred thousand ohms is a typical figure); so we should modify the equivalent circuit as shown in Fig. 9.11c.

Having got the input and output circuits we can join them together to get the equivalent circuit of the common base transistor† (Fig. 9.11d).

We have not included any reactances. Can we say anything about them? Yes, we can. We have already worked out the junction capacity of

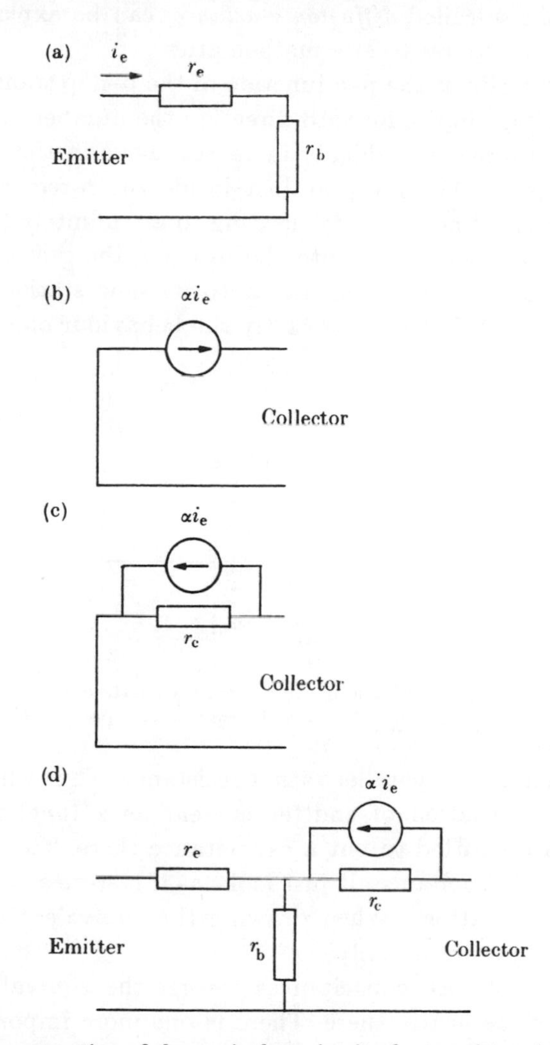

FIG. 9.11. The construction of the equivalent circuit of a transistor. (a) The emitter-base junction. (b) In first approximation the collector current depends only on the emitter current. (c) In a more accurate representation there is a collector resistance r_c as well in the collector circuit. (d) The complete low-frequency equivalent circuit.

† This exceedingly simple construction cannot be done in general but is permissible in the present case when $r_c \gg r_b$.

a reverse biased junction. That capacity should certainly appear in the output circuit in parallel with r_c. There are also some other reactances as a consequence of the detailed mechanism of current flow across the transistor. We can get the numerical values of these reactances if we have the complete solution. But luckily the most important of these reactances, the so-called *diffusion reactance*, can be explained qualitatively without recourse to any mathematics.

Let us look again at the p–n junction of the p–n–p transistor. Applying a step voltage in the forward direction the number of holes able to cross into the n-region suddenly increases. Thus in the first moment, when the injected holes appear just inside the n-region, there is an infinite gradient of hole density, leading to an infinitely large diffusion current. As the holes diffuse into the n-region the gradient decreases, and finally the current settles down to its new stationary value as shown in Fig. 9.12. But this is exactly the behaviour one would expect

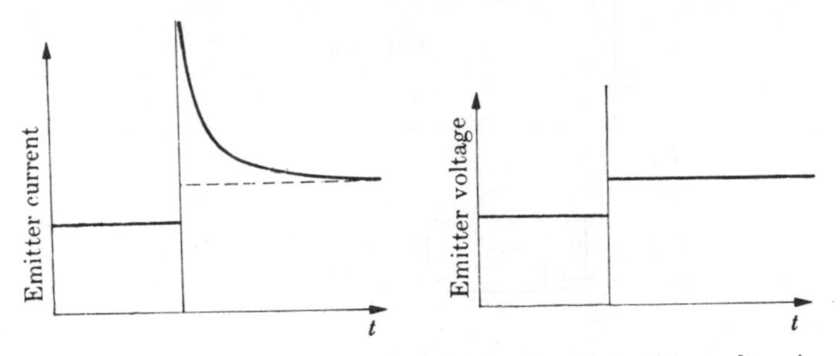

FIG. 9.12. The emitter current as a function of time when the emitter voltage is suddenly increased. It looks like the current response of a parallel RC circuit.

from a capacitance in parallel with a resistance. Thus when we wish to represent the variation of emitter current as a function of emitter voltage we are entitled to put a capacitance there. This is not a real, honest-to-god, capacitance; it just looks as if it were a capacitance but that's all that matters. When drawing the equivalent circuit we are interested in appearance only!

Including now both capacitances we get the equivalent circuit of Fig. 9.13. We are nearly there. There is one more important effect to consider: the frequency-dependence of α. It is clear that the collector current is in phase with the emitter current when the transit time of the carriers across the base region is negligible, but α becomes complex (and its absolute value decreases) when this transit time is comparable with the period of the a.c. signal. We cannot go into the derivation

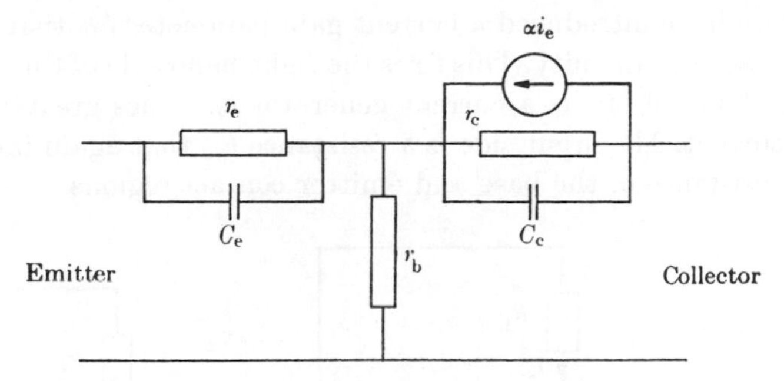

FIG. 9.13. A good approximation to the high-frequency equivalent circuit of a transistor.

here, but α may be given by the simple formula†

$$\alpha = \frac{\alpha_0}{1+j\dfrac{\omega}{\omega_\alpha}}, \tag{9.20}$$

where ω_α is called the *alpha cut-off frequency.* The corresponding equivalent circuit is obtained by replacing α in Fig. 9.13 by that given in eqn. (9.20). And that is the end as far as we are concerned. Our final equivalent circuit represents fairly well the frequency-dependence of a commercially available transistor.

We have seen that the operation of the transistor can be easily understood by considering the current flow through it. The frequency-dependence is more complicated, but still we have been able to point out how the various reactances arise.

It has been convenient to describe the common base transistor configuration, but of course the most commonly used arrangement is the common emitter, shown in Fig. 9.14a. Again most of the current i_e from the forward-biased emitter–base junction gets to the collector so we can write‡

$$i_c = \alpha i_e,$$

as before, and

$$i_c = i_e - i_b$$

$$= \frac{\alpha i_b}{1-\alpha} = h_{fe}\, i_b,$$

† Not to depart from the usual notations we are using j here as honest engineers do but had we done the analysis with our chosen $\exp(-i\omega t)$ time dependence we would have come up with $-i$ instead of j.

‡ The full expression for i_e should contain a term dependent on the emitter-to-collector voltage. This is usually small. Look it up in a circuitry book if you are interested in the finer details.

where we have introduced a current gain parameter h_{fe} that is usually much greater than unity. This fixes the right-hand side of the equivalent circuit of Fig. 9.14c as a current generator h_{fe} times greater than the input current. The input side is a resistance h_{ie} that again includes the series resistance of the base and emitter contact regions.

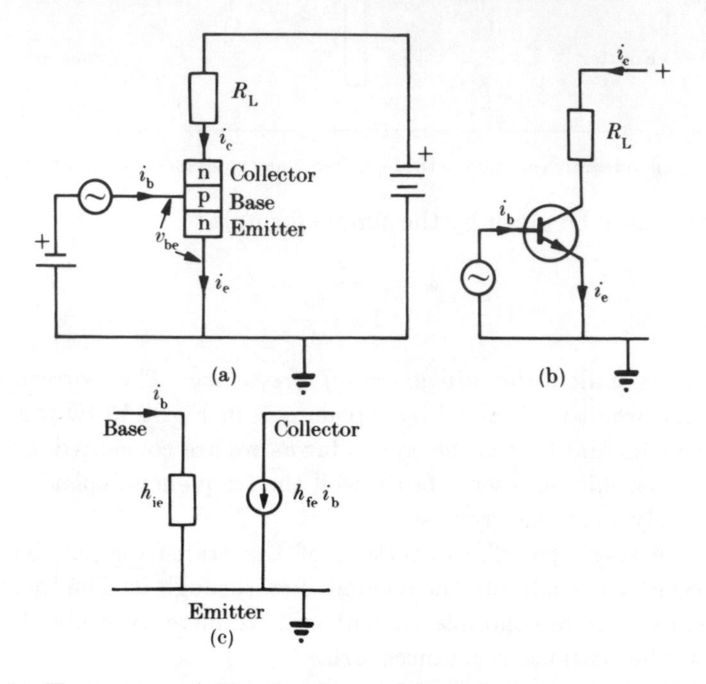

FIG. 9.14. The common emitter transistor. (a) General circuit arrangement, showing current and voltage nomenclature. (b) Circuit diagram. (c) Equivalent circuit of the transistor.

Note that the major part of transistor amplifier design is based on the simple equivalent circuit of Fig. 9.14c. At high frequencies, of course, the capacitances discussed have to be added.

I have so far talked about the applications of transistors as amplifiers, that is, analogue devices. Historically, these applications came first because at the time of the invention of the transistor there was already a mass market in existence eager to snap up transistor amplifiers—particularly for portable devices. The real impact of the transistor came however not in the entertainment business but in computers. Admittedly, computers did exist before the advent of the transistor, but they were bulky, clumsy, and slow. The computers you know and respect, from giant ones down to pocket calculators, depend on the good services of transistors. One could easily write a thousand pages about the circuits used in various computers—the trouble is that by the time the thousandth page is jotted down, the first one is out of date. The rate of technical change in this field is simply breathtaking, much higher than ever before in any branch of technology. Fortunately, the principles are not difficult. For building a logic circuit all we need is a device with two stable states, and that can be easily provided by a transistor, e.g. in a form (Fig. 9.15) quite similar to its use as an amplifier. When the base current, $I_B = 0$ (we use capital letters to describe the d.c. current),

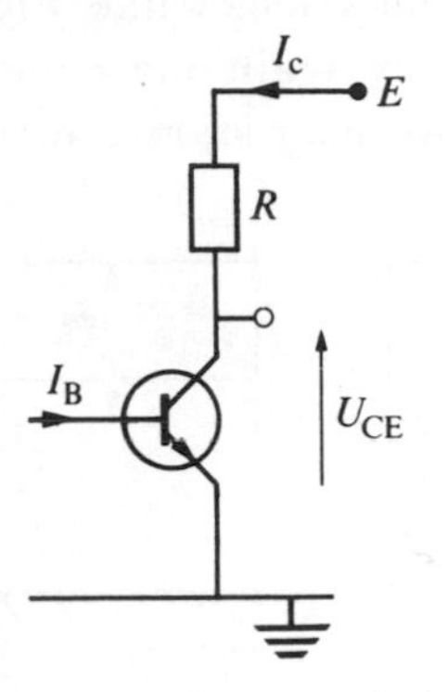

Fig. 9.15. A transistor as a logic element.

no collector current flows, $I_C = 0$, and consequently, $U_{CE} = E$. If a base current is impressed upon the circuit then a collector current flows and U_{CE} is close to zero. Hence we have a 'high' and a 'low' output voltage which may be identified with a logical '1' or '0' (or the other way round). I shall not go into any more details, but I would just like to mention some of the acronyms in present-day use for which transistors are responsible. They include DTL (diode–transistor logic), TTL (transistor–transistor logic), and ECL (emitter coupled logic).

9.7. Metal–semiconductor junctions

Junctions between metals and semiconductors had been used in radio engineering for many years before the distinction between p- and n-type semiconductors was appreciated. Your grandfathers probably played about with 'cat's whiskers' in their early 'crystal sets', as radios were then called, stressing the importance of the piece of coal or whatever was used as the semiconductor detector.

The behaviour of metal–semiconductor junctions is more varied to describe than that of p–n junctions. We find that there is different behaviour on the one hand with p- and n-type semiconductors, and on the other when the metal work function is greater or less than that of the semiconductor.

We shall first consider the case of an n-type semiconductor in contact with a metal whose work function is greater than that of the semiconductor. The semiconductor work function is defined as the energy difference between an electron at the Fermi energy and the vacuum level. The fact that there are usually no electrons at the Fermi energy need not bother us—we do not have to explain definitions. So the band picture of the two substances looks like Fig. 9.16a. When they are joined together we may apply again our general theorem and make the Fermi levels equal. Thus we may start the construction of Fig. 9.16b

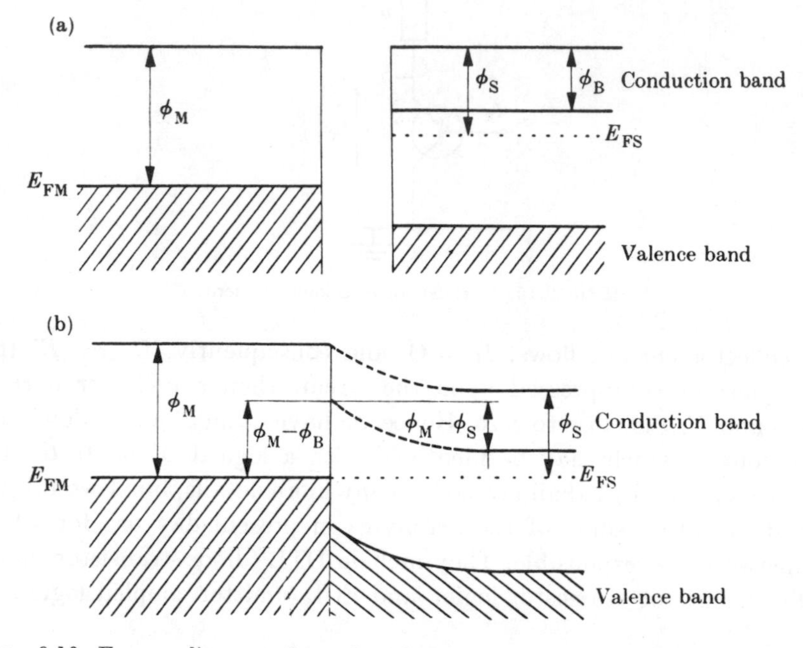

FIG. 9.16. Energy diagrams for a junction between a metal and an n-type semiconductor ($\phi_M > \phi_S$), (a) before contact, (b) after contact the Fermi levels agree ($E_M = E_{FS}$).

by drawing a horizontal line for the Fermi energy, and a vertical one for the junction. We leave the metal side unchanged because we shall assume that 'band-bending' cannot occur in a metal.† We are really saying here that all the potential drop will take place in the semiconductor, which, in view of the much smaller number of carriers there,

† In a metal the charge inequality is confined to the surface.

is a reasonable assumption.† Away from the junction we draw the valence band edge, the conduction band edge, and the vacuum level in the same position (relative to the Fermi level) as for the bulk material, in Fig. 9.16a. These are shown as solid lines. Now with an infinitely small gap the vacuum levels are equal; thus we may join them with a dotted line in Fig. 9.16b; the conduction and valence band edge must also be continued parallel to the vacuum level.

What can we say about the charges? We may argue in the same fashion as for a metal–metal junction. In the first instance, when the metal and the semiconductor are brought together, the electrons from the conduction band cross over into the metal in search of lower energy. Hence a certain region in the vicinity of the junction will be practically depleted of mobile carriers. So we may talk again about a depletion region and about the accompanying potential variation, which is, incidentally, the same thing as the 'band-bending' obtained from the band picture.

So the two pictures are complementary to a certain extent. In the first one the 'band-bending' is a consequence of the matching of the Fermi levels and vacuum levels, and the charge imbalance follows from there. In the second picture electrons leave the semiconductor, causing a charge imbalance and hence a variation in the potential energy. Whichever way we look at it, the outcome is a potential barrier between the metal and the semiconductor. Note that the barrier is higher from the metal side.

In equilibrium the number of electrons crossing over the barrier from the metal to the semiconductor is equal to the number crossing over the barrier from the semiconductor side. We may say that the current I_0 flows in both directions.

Let us apply now a voltage; according to the polarity, the electrons' potential energy on the semiconductor side will go up or down. For a forward bias it goes up, which means that we have to draw the band edges higher up. But the vacuum level at the junction stays where it was. So the effect of the higher band edges is a smaller curvature in the vicinity of the junction and a reduced potential barrier, as shown in Fig. 9.17. Now all electrons having energies above $\phi_M - \phi_S - eU_1$ may cross into the metal. By analogy with the case of the p–n junction it follows that the number of carriers (capable of crossing from the semiconductor into the metal) has increased by a factor $\exp eU_1/kT$,

† We met a very similar case before when discussing p–n junctions. If one of the materials is highly doped, all the potential drop takes place in the other material.

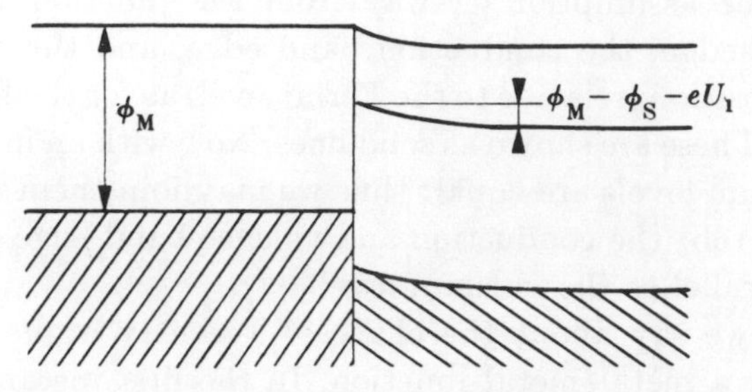

FIG. 9.17. The junction of Fig. 9.16 under forward bias. The potential barrier for electrons on the semiconductor side is reduced by eU_1.

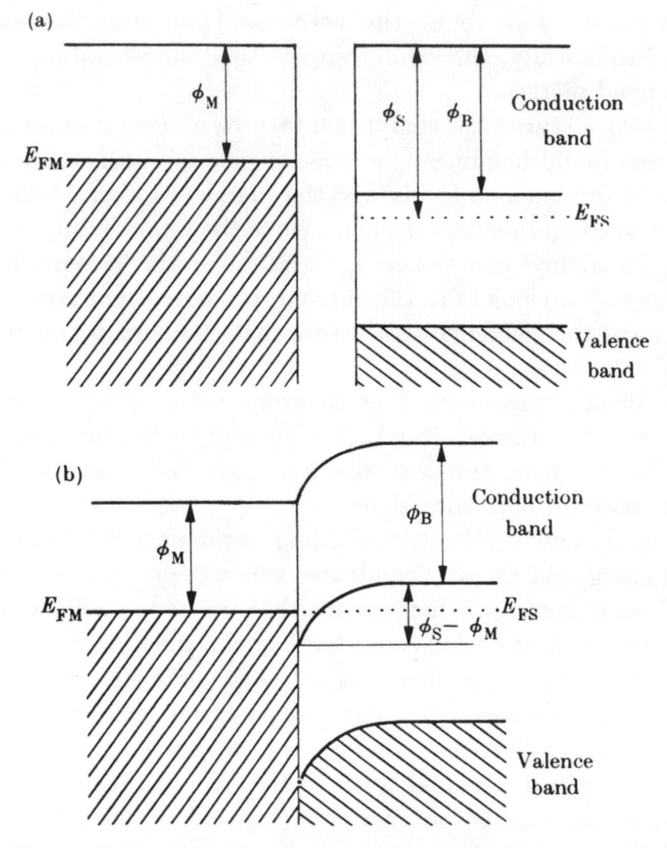

FIG. 9.18. Energy diagrams for a junction between a metal and an n-type semiconductor ($\phi_M < \phi_S$), (a) before contact, (b) after contact the Fermi levels agree ($E_{FM} = E_{FS}$).

and hence the current has increased by the same factor. Since the current from the metal to the semiconductor has not changed, the total current is

$$I = I_0(\exp eU_1/kT - 1), \tag{9.21}$$

that is, a junction of this type is a rectifier.

What happens when the work function of the metal is smaller than that of the n-type semiconductor? The situation before and after contact is illustrated in Fig. 9.18. Now to achieve equilibrium electrons had to move from the metal to the semiconductor, establishing there an accumulation region. There is no potential barrier now from whichever side we look at the junction. As a consequence the current flow does not appreciably depend on the polarity of the voltage. This junction is *not* a rectifier.

9.8. The role of surface states; real metal–semiconductor junctions

The theory of metal–semiconductor junctions as presented above is a nice, logical, consistent theory that follows from the physical picture we have developed so far. It has, however, one major disadvantage; it is not in agreement with experimental results, which seem to suggest that all metal–semiconductor junctions are rectifiers independently of the relative magnitudes of the work functions. This does not necessarily mean that the theory is wrong. The discrepancy may be caused by the physical realization of the junction. Instead of two clean surfaces lining up, there might in practice be some oxide layers and the crystal structure might be imperfect. This may be one of the reasons why 'real' junctions behave differently from 'theoretical' junctions. The other reason could be that the theory, as it stands, is inadequate and to get better agreement with experiments we must take into account some hitherto neglected circumstance.

Have we taken into account anywhere that our solids are of finite dimensions? Yes, we have; we determined the number of allowed states from the boundary conditions. True, but that's not the only place where the finiteness of the sample comes in. Remember, in all our models leading to the band picture we have taken the crystal as perfectly periodic and we have taken the *potential* as perfectly periodic. This is surely violated at the surface. The last step in the potential curve should be different from the others, that is, the potential profile in the solid should rather be chosen in the form displayed in Fig. 9.19.

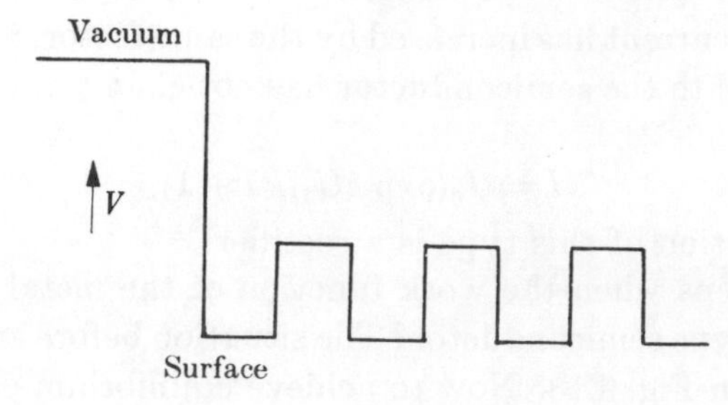

FIG. 9.19. The potential energy distribution near the surface of a crystal.

It was shown some years ago that the assumption of such a surface barrier would lead to the appearance of some additional discrete energy levels in the forbidden gap, which are generally referred to as surface states.

Assuming now an n-type semiconductor some of these surface states may be occupied by electrons that would otherwise be free to roam around. Some of the donor atoms will therefore have uncompensated positive charges leading to 'band-bending' as shown in Fig. 9.20. Thus the potential barrier is already there before we even think of making a metal contact.

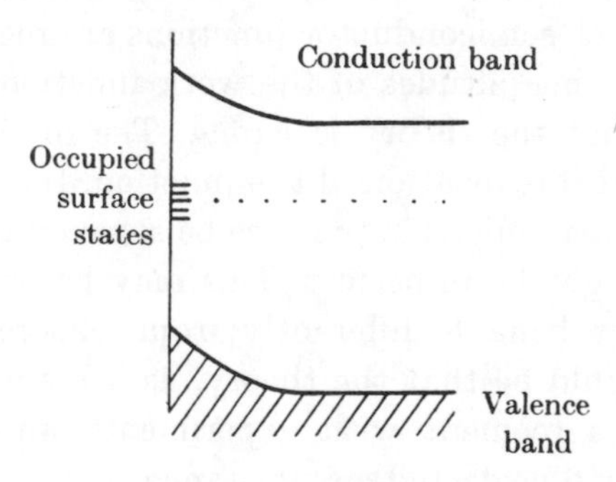

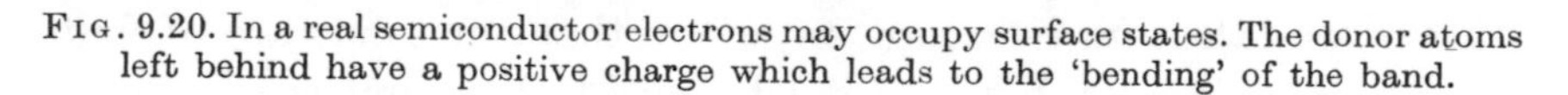

FIG. 9.20. In a real semiconductor electrons may occupy surface states. The donor atoms left behind have a positive charge which leads to the 'bending' of the band.

What happens when we do make contact between the semiconductor and the metal? Let us choose the case when the metal has the lower work function when according to our previous theory the junction is not rectifying. Then, as we have agreed before (and it is still valid) electrons must flow from the metal to the semiconductor until equilibrium is established. But if there is a sufficient number of empty surface states still available then the electrons will occupy those without much effecting the height of the potential barrier. So the potential barrier stays and the junction is rectifying.

It would be difficult in a practical case to ascertain the share of these 'theoretical' surface states (called also Tamm states) in determining the behaviour of the junction, because surface imperfections are also there and those can equally well trap electrons. It seems, however, quite certain that it is the surface effects that make all real metal–semiconductor junctions behave in a similar manner.

Finally, I would like to mention ohmic contacts, that is, contacts which do not care which way the voltage is applied. To make such a contact is not easy; it is more an art than a science. It's an important art though, since all semiconductor devices have to be connected to the outside world.

The two most often used recipes are: (i) to make the contact with alloys containing metals (e.g. In, Au, Sn) that diffuse into the surface forming a gradual junction; or (ii) to make a heavily doped semiconductor region (usually called n^+ or p^+) with about 10^{24} carriers per cubic metre in between the metal and semiconductor to be connected.

9.9. Metal–insulator–semiconductor junctions

Let us now make life a little more complicated by adding one more component and look at metal–insulator–semiconductor junctions. What happens as we join the three materials together? Nothing. If the insulator is thick enough to prevent tunnelling (the situation that occurs in all practical devices of interest), the metal and the semiconductor are just unaware of each other's existence.

What does the energy diagram look like? For simplicity we shall assume that the Fermi levels of all three materials coincide before we join them together. The energy diagram then takes the form shown in Fig. 9.21 where the semiconductor is taken as n-type.

Are there any surface states at the semiconductor–insulator interface? In practice there are, but their influence is less important than for metal–semiconductor junctions so we shall disregard them for the time being.

Let us now apply a positive voltage to the metal as shown in Fig. 9.22a. Will a current flow? No, there can be no current through the insulator. The electrons will nevertheless respond to the arising electric field by moving towards the insulator. That is as far as they can go, so

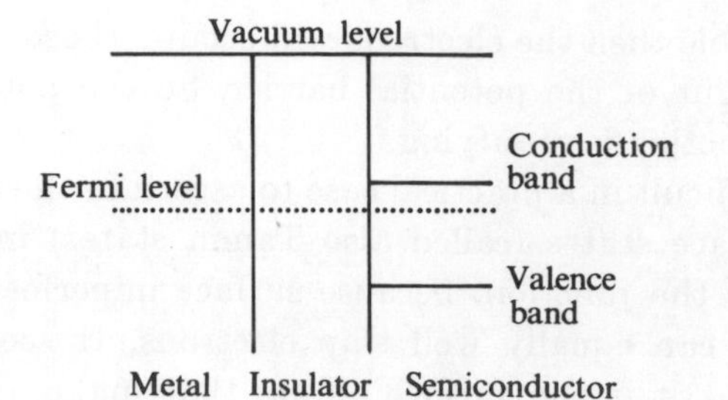

FIG. 9.21. Energy diagram for a metal-insulator-n-type semiconductor junction at thermal equilibrium.

they will accumulate in front of the insulator. Their distribution will be something like that shown in Fig. 9.22b, where N_{en} is the equilibrium concentration in the bulk semiconductor and x is the distance away from the insulator. The shape of the curve may be obtained from the same considerations as in a p–n junction. The diffusion current (due to the gradient of the electron distribution) flowing to the right must be equal to the conduction current (due to the applied field) flowing to the left. The corresponding energy diagram is shown in Fig. 9.22c, where the Fermi level in the semiconductor is taken as the reference level. Looking at the energy diagram we may now argue backwards and say that eqn. (8.17) must still be roughly valid so the electron density is approximately an exponential function of the distance of the Fermi level from the bottom of the conduction band. Hence the electron density is increasing towards the insulator.

Next, let us apply a negative voltage to the metal (Fig. 9.23a). The electrons will be repelled, creating a depletion region as in a reverse biased p–n junction. In fact, we could determine the width of the depletion region (see Example 9.3) by a method entirely analogous to that

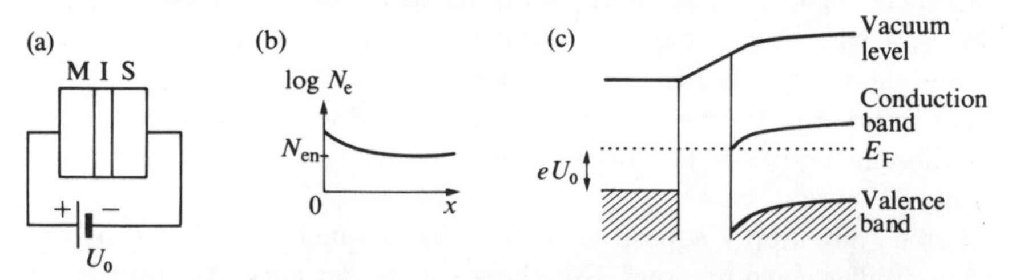

FIG. 9.22. A metal-insulator-n-type semiconductor junction under forward bias. (a) Schematic representation. (b) Variation of electron density in the semiconductor as a function of distance. (c) Energy diagram.

developed in Section 9.2. Alternatively, we can argue that the electron distribution will be of the shape shown in Fig. 9.23b and we may talk again about the balance of diffusion and conduction currents. Finally, the band bending picture is shown in Fig. 9.23c, from which we can also conclude that the electron density is decreasing towards the insulator. What will happen as we apply higher and higher reverse bias? The obvious answer is that the depletion region will widen. What else could one expect? It is difficult to believe at first hearing but the fact

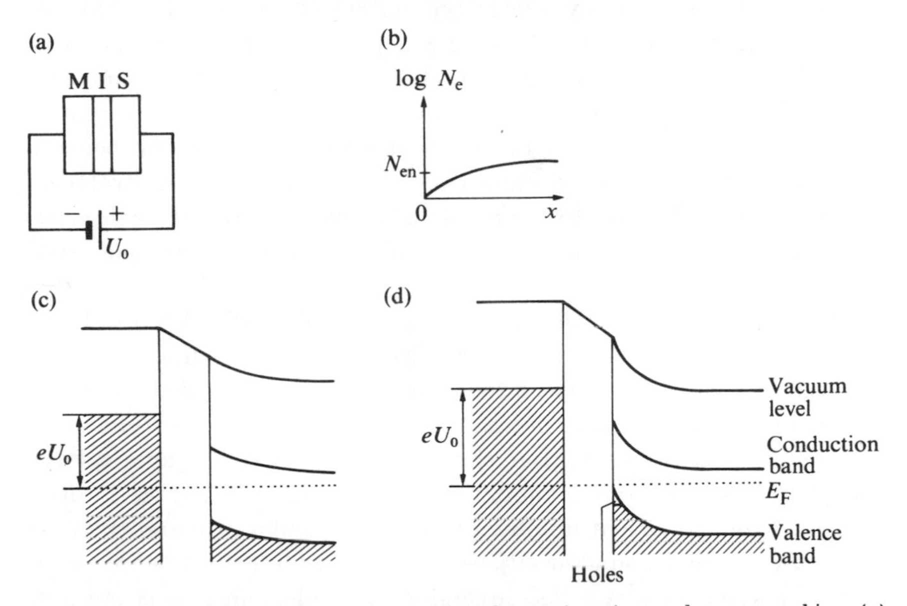

FIG. 9.23. A metal-insulator-n-type semiconductor junction under reverse bias. (a) Schematic representation. (b) Variation of electron density in the semiconductor as a function of distance. (c) Energy diagram at moderate voltage. (d) Energy diagram at a voltage high enough for producing holes.

(fortunate as it happens) is that holes will appear. Can we explain this phenomenon by any of our models? If we consider only ionized donor atoms and mobile electrons, as in the model developed in Section 9.2 we haven't got the slightest chance of creating holes. On the other hand if we adopt the notion that the density of a carrier at any point is determined by the distance in energy from the Fermi level to the edge of the particular band, then holes have acquired the right to appear. All we need to do is to apply a sufficiently large reverse bias (Fig. 9.23d) which will bring the Fermi level right down, close to the top of the valence band. Thus according to this model holes may become the majority carriers near to the surface of an n-type semiconductor. Odd isn't it? The problem still remains though that the holes must come from somewhere. The only process known to produce holes in an n-type semiconductor is thermal generation of electron–hole pairs. But aren't the rates of generation and recombination equal? Wouldn't the holes generated thermally immediately disappear by recombination? This is true indeed under thermal equilibrium conditions, but our junction is not necessarily in thermal equilibrium.

Let us look again at the whole process, considering time relationships as well. At $t = 0$ we apply a negative voltage to the junction. Most of the electrons clear out by t_1 leaving a depletion region of the order of 1 μm behind. What happens now to thermally generated electron–hole pairs? The electrons move away from the insulator and the holes move towards the insulator. Not much recombination will occur because both the electron and hole densities are small and they are separated in space. What will happen to the holes? They will congregate in the vicinity of the insulator where they can find a nice comfortable potential minimum.

The conditions of equilibrium are rather complicated. At the end, say by t_2, the hole difiusion current away from the insulator must be equal to the hole conduction current towards the insulator, and the rates of generation and recombination must balance each other. It is then quite reasonable to conclude that if the applied negative voltage is large enough, i.e. the potential minimum at the insulator surface is deep enough, then a sufficient number of holes can congregate and the part of the n-type semiconductor adjacent to the insulator will behave as if it were p-type. This is called inversion. The phenomenon is not restricted of course to n-type semiconductors. Similar inversion occurs in a metal–insulator–p-type semiconductor junction.

Our conclusion so far is that under equilibrium conditions inversion may occur. Whether equilibrium is reached or not depends on the time constants t_1 and t_2. How long is t_1? As far as I know no one has measured it, but it can't take long for electrons to clear out of a 1 μm part of the material. If we take a snail moving with a velocity of 1 m/hour it will need about 3 ms to cover 1 μm. Thus electrons, which may be reasonably expected to move faster than snails, would need very little time indeed to rearrange themselves and create a depletion region. On the other hand in sufficiently pure materials the thermal generation time constant might be as long as a few seconds. Thus if all the operations we perform in a metal–insulator–semiconductor junction are short in comparison with the generation time of electron-hole pairs then the minority carriers will not have the time to appear, a mode of operation called the *deep depletion mode*.

Inversion and deep depletion are some further representations of the multifarious phenomena of semiconductor physics. They are certainly interesting but are they useful? Can a device through which no current flows be of any use at all in electronics? The secret is that current can flow *along* the insulator surface. The emerging devices are very important indeed. Under acronyms like MOSFETs and CCDs they are the flagbearers of the microelectronics revolution. I shall talk about them a little later.

9.10. The tunnel diode

So far we have considered impurity semiconductors with very low impurity contents, typically less than one part per million. We have characterized the impurity type and density by working out where the Fermi level is, and have found that in all cases it is well within the energy gap. This has meant, among other things, that the sums are much simpler, for we are able to approximate to the Fermi function. However, when the impurity level becomes very high (typically about $10^{24}/m^3$ or about 0·01%) the Fermi level moves right up into the conduction band (or down into the valence band for a p-type impurity). The semiconductor is then said to be 'degenerate'. The tunnel diode is the only device we shall discuss in which degenerate semiconductors are used on both the p- and n-sides of the junction.

The energy diagram at thermal equilibrium is given in Fig. 9.24a, where, for simplicity, we take the difference between the Fermi level and the band edge as the same on both sides. It is interesting to see that the 'built-in' potential is larger than the energy gap; thus the number of electrons crossing over the potential barrier at thermal equilibrium must be small. Hence I_0 in the rectifier equation is small and the rectifying characteristic is rather elongated, as shown in Fig. 9.25a.

Looking carefully at the diagram you may realize that another mechanism of electron flow may also be effective. Remember, tunnel diodes are highly doped, and high doping means a narrow transition region. Thus, electrons as well as moving *over* the potential barrier may also tunnel *through* the potential barrier, and if one puts in the figures it turns out that the tunnelling current is the larger of the two. Hence we may imagine thermal equilibrium as the state in which the tunnelling currents are equal and in opposite directions.

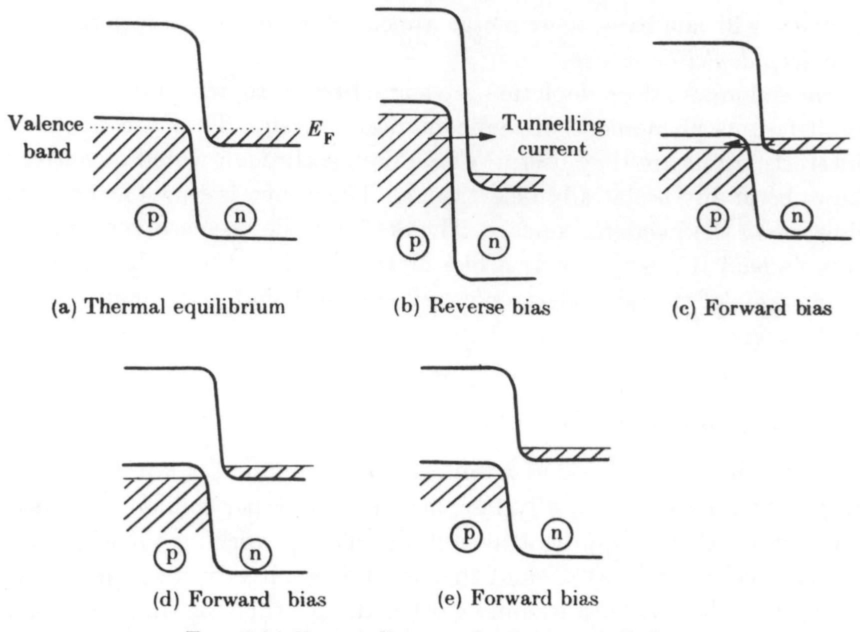

FIG. 9.24. Energy diagrams for the tunnel diode.

What happens now if we apply a reverse bias? As may be seen in Fig. 9.24b the number of electrons tunnelling from left to right is increased because the electrons on the p-side face a large number of empty states on the n-side. We could work out this current by considering a rectangular potential barrier (the one we so skilfully solved when first confronted with Schrödinger's equation) and taking account of the occupancy of states on both sides. It is one of those fairly lengthy and tedious calculations that are usually left as an exercise for the student. So I would rather leave the work to you and go on to discuss qualitatively the case of forward bias (Fig. 9.24c). The situation is essentially the same as before, but now the electrons tunnel from right to left. Increasing the applied voltage the number of states available on the p-side increases and so the current increases too. Maximum current flows when electrons on the n-side have access to all the empty states on the p-side, that is, when the Fermi level on the n-side coincides with the valence band edge on the p-side (Fig. 9.24d).

Increasing the bias further there will be an increasing number of electrons finding themselves opposite the forbidden gap. They cannot tunnel because they have no energy levels to tunnel into. Hence the current must decrease, reaching zero when the top of the valence band on the p-side coincides with the bottom of the conduction band on the n-side (Fig. 9.24e). Therefore the plot of current against voltage must look like that shown in Fig. 9.25b. But this is not the total current; it is the current due to tunnelling alone. We can get the total current by simply taking the algebraic sum of the currents plotted in Fig. 9.25a and b, which is a permissible procedure since the two mechanisms are fairly independent of each other. Performing the addition we get the

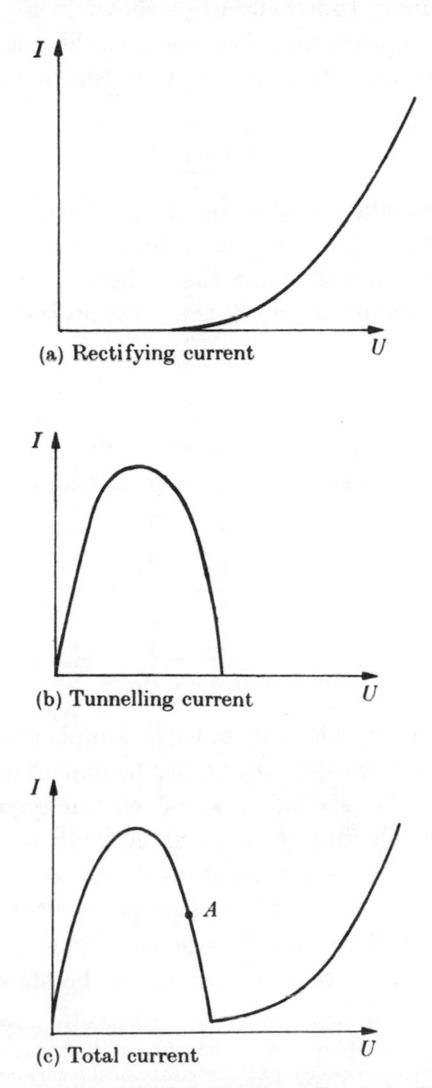

FIG. 9.25. The current in a tunnel diode is the sum of the tunnelling current and of the usual rectifying current.

I–U characteristics (Fig. 9.25c) that we would be able to measure on a real tunnel diode.†

You know now everything about the tunnel diode with the exception of the reason why it can perform some useful function. The answer follows from the *I–U* characteristics. There is a region where the slope is negative which is usually referred to as a negative resistance. In case you have not heard this curious phrase before I shall briefly explain it.

Consider an ordinary tuned circuit as shown in Fig. 9.26. If we start it oscillating in some way, and then leave it, the oscillations will decay exponentially, their amplitude falling with time according to

$$\exp\left(-\frac{R}{2L}t\right). \tag{9.22}$$

Physically, the resistance R absorbs the oscillating energy and gets a little hotter. If we now put a negative resistance in series with R, odd things happen. In the particular case when the negative resistance $(-R_1)$ is equal in magnitude to R, the total resistance becomes

$$R-R_1 = 0 \tag{9.23}$$

and the exponential becomes unity. This means that an oscillation once started in the circuit will continue with no decay, the negative resistance replenishing all the energy dissipated as heat in the real resistance.

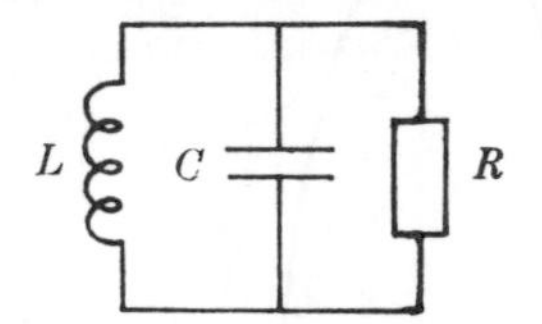

FIG. 9.26. A tuned circuit.

If we could get energy like this out of a simple circuit isolated from the rest of the world except for the R, L, and C we have drawn, it would contravene the second law of thermodynamics and make perpetual motion fairly straightforward. As this does not happen, we can conclude that a 'negative resistance' has to be an *active* circuit device that is connected to a power supply other than the oscillating signal with which it interacts. This is very true of the tunnel diode, since to act as a negative resistance it has to be biased with a battery

† In fact, there is some deviation from this characteristic owing to the inevitable presence of some energy levels in the forbidden gap. There is thus some additional tunnelling, which becomes noticeable in the vicinity of the current minimum. But even including this effect the ratio of current maximum to current minimum may be as high as 15 in a practical case.

to the point A in Fig. 9.25c. The power to overcome the circuit losses comes from this battery.

If the magnitude of the negative resistance in Fig. 9.25c is greater than the loss resistance R, the initiatory signal will not only persist; it will grow. Its magnitude will, of course, be limited by the fact that the tunnel diode can be a negative resistance for only a finite voltage swing (about 0·2 V). Thus, given a negative resistance circuit engineers can make oscillators and amplifiers. The particular advantage of tunnel diodes is that as the junctions are thin, the carrier transit times are shorter than in a transistor and very high-frequency operation (up to about 10^{11} Hz) is possible. Their limitation is that with their inherently low voltage operation, they are very low-power devices.

9.11. The backward diode

This is essentially the same thing as the tunnel diode only the doping is a little lighter. It is called a *backward diode* because everything is the other way round. It has low impedance in the reverse direction and high impedance in the forward direction, as shown in Fig. 9.27.

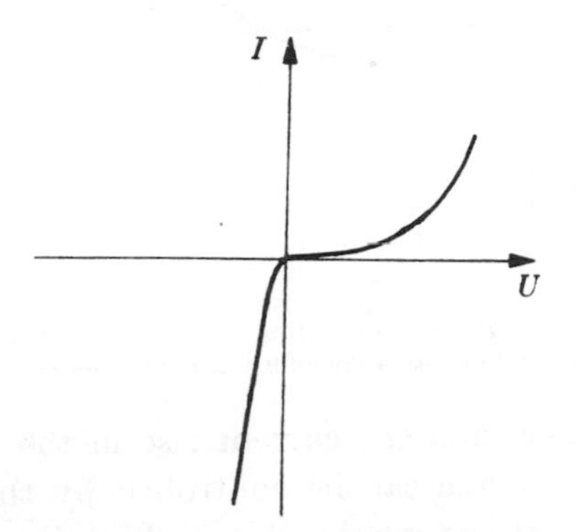

FIG. 9.27. The current voltage characteristics of a backward diode.

The secret of the device is that the doping is just that much lighter (than that of the tunnel diode) as to line up the band edges (the top of the valence band on the p-side to coincide with the bottom of the conduction band on the n-side) at zero bias. Hence for a forward bias there is no tunnelling, just the 'normal' flow, which is very small. In the reverse direction, however, a large tunnelling current may flow.

The backward diode is a very efficient rectifier (of the order of one to a thousand) for low voltages. For higher voltages, of course, the 'forward' current may become significant.

9.12. The Zener diode and the avalanche diode

You should not dwell too heavily on the memory of the backward diode; it is rather exceptional. I'm pleased to say that from now on *forward* means forward and *reverse* means reverse.

We shall now consider what happens at higher voltages. In the forward direction the current goes on increasing, and eventually the diode will be destroyed when more energy is put in then can be conducted away. This is a fascinating topic for those engineers whose job is to make high-power rectifiers but it is of limited scientific interest for the rest of us.

There is considerably more interest in the reverse direction. It is an experimental fact that breakdown occurs very sharply at a certain reverse voltage as shown in Fig. 9.28. Since the 'knee' of this breakdown

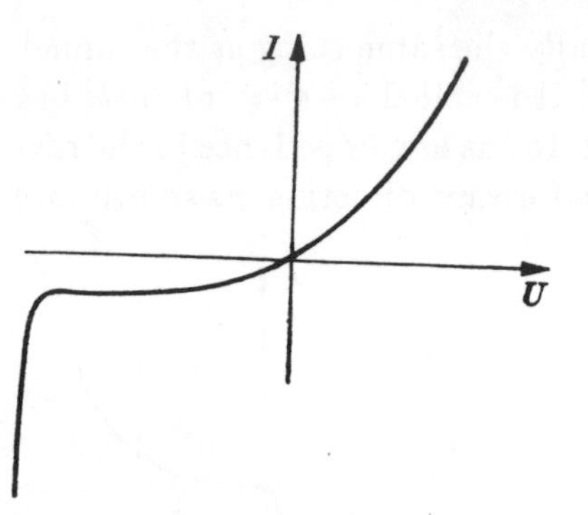

FIG. 9.28. The current voltage characteristics of a p–n junction showing the sudden increase in current at a specific value of reverse voltage.

curve is much sharper than the current rise in the forward direction, and since the knee voltage can be controlled by the impurity levels, this effect has applications whenever a sudden increase in current is required at a certain voltage. The diode can therefore be used as a voltage stabilizer or a switch. In the latter application it has the further advantage that the breakdown is not only sharp but occurs very fast as well.

The breakdown may occur by two distinct mechanisms: (i) Zener breakdown, (ii) avalanche breakdown.

(i) The mechanism suggested by Zener (1934) may be explained as follows.† At low reverse bias there is only the flow of minority

† There were, of course, no p–n junctions at the time. The mechanism was suggested for bulk breakdown to which, incidentally, it doesn't apply. The explanation turned out to be applicable to breakdown in p–n junctions.

electrons from the p-side to the n-side. As the reverse bias is increased, at a certain voltage the bands begin to overlap, and tunnelling current may appear. The tunnelling current does appear if the doping is large enough and the junction is narrow enough. But Zener diodes (in contrast to tunnel diodes and backward diodes) are designed in such a way that practically no tunnelling occurs when the bands just overlap; the potential barrier is too wide (Fig. 9.29a). However, as the reverse bias is increased the

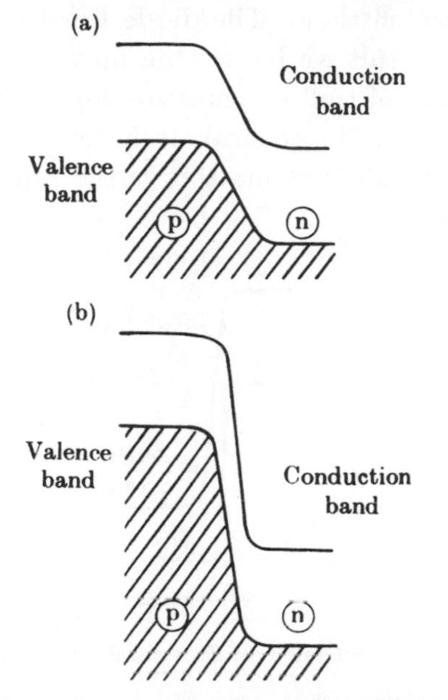

FIG. 9.29. A heavily doped p–n junction (a) in thermal equilibrium, (b) at reverse bias. The width of the potential barrier decreases as the reverse bias is increased.

width of the barrier decreases (Fig. 9.29b) leading—above a certain voltage—to a very rapid rise in current.

(ii) Avalanche diodes differ from Zener diodes by having somewhat smaller impurity density. The depletion layer is then wider and the Zener breakdown would occur at a considerably higher voltage. However, before the tunnelling current has a chance to become appreciable another mechanism takes over, which—very aptly—is designated by the word *avalanche*.

You know that electrons in a solid are accelerated by the applied electric field. They give up the kinetic energy acquired when they collide with lattice atoms. At a sufficiently high electric field an electron may take up enough energy to ionize a lattice atom, that is to create an electron–hole pair. The newly created electrons and holes may in turn liberate further electron–hole pairs, initiating an avalanche.

Note that the two mechanisms are quite different, as may be clearly seen in Fig. 9.30. In both cases the electron moves from the valence band of the p-type material into the conduction band of the n-type material, but for Zener breakdown it moves horizontally, whereas for avalanche breakdown it must move vertically. But although the mechanisms are different, nevertheless, in a practical case it is difficult to distinguish between them. The diode breaks down and that is the only experimental result we have. One may attempt to draw the distinction on the basis of the temperature-dependence of the two breakdown mechanisms but, in general, it is not worth the effort. For a practical application all that matters is the rapid increase in current, whatever its cause.

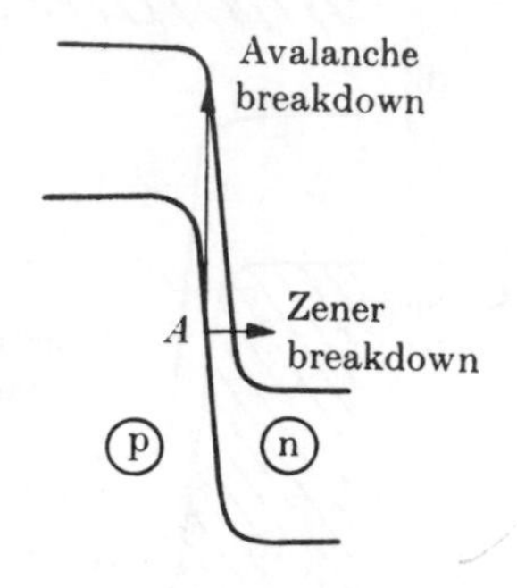

FIG. 9.30. The mechanism of Zener and avalanche breakdown.

Avalanche diodes may also be used for generating microwaves. The principles of operation (as for most microwave oscillators) are fairly complicated. The essential thing is that both during the avalanche process and the subsequent drift of the created carriers, the current and the electric field are not in phase with each other. By judicious choice of the geometry one may get 180° phase difference between voltage and current, at least for a certain frequency range. But this is nothing more than a frequency-dependent negative resistance. Putting the diode in a cavity, the oscillator is ready.

9.13. Varactor diodes

As I have mentioned above, and have shown mathematically in eqn. (9.19), the capitance of a reverse-biased p–n junction is voltage-dependent. In other words the capacitance is *variable*, and that is what the name 'varactor' seems to stand for.

Varactor diodes are p–n junctions designed for variable-capacitance operation. Is a variable capacitance good for anything? Yes, it is the basis of the so-called 'parametric amplifier'. How does a parametric amplifier work? This is really a circuit problem but I had better explain its operation briefly.

Imagine just an ordinary resonant circuit oscillating at a certain frequency. The charge on the capacitor then varies sinusoidally as shown in Fig. 9.31a.

Suppose the plates of the capacitor are pulled apart when Q reaches its maximum and are pushed back to the initial separation when Q is zero. This is shown in Fig. 9.31b, where d is the distance between the plates. When Q is finite and the plates are pulled apart, one is doing work against coulombic attraction. Thus energy is pumped into the resonant circuit at the times t_1, t_3, t_5, etc. When Q is zero no energy need be expended to push the plates back. The energy of the resonant circuit is therefore monotonically increasing.

To see more clearly what happens let us try to plot the voltage against time. From t_0 to t_1 it varies sinusoidally. At t_1 the separation between the plates is suddenly increased, that is, the capacitance

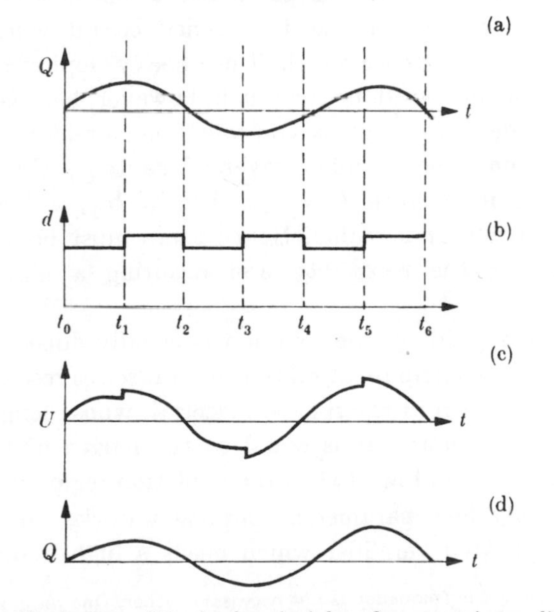

FIG. 9.31. Illustration of the basic principles of parametric amplification.

decreased. The charge on the plates could not change instantaneously; so the reduced capacitance must lead to increased voltage ($Q = CU$ must stay constant). The voltage across the capacitor therefore jumps abruptly at t_1, t_3, t_5, etc., and it is unaffected at t_2, t_4, t_6, etc., as shown in Fig. 9.31c.

We may argue in a similar manner that the charge must also increase. When the plate is pushed back at t_2 the voltage is not affected but the capacitance goes back to its original value. Hence when the capacitor is charged again Q will come to a higher peak value, as shown in Fig. 9.31d.

Note that the important thing is to vary the capacitance in the resonant circuit at twice the resonant frequency. Amplification is achieved then at the expense of the energy available to vary the capacitance. It may be shown (both theoretically and experimentally) that the variation of the capacitance need not be abrupt. Any reasonable variation of the capacitance at twice the rate of the signal frequency would do.†

It is interesting to note that the possibility of parametric amplification has been known for over fifty years but it has become practical only after the advent of the p–n junction.

Now with what sort of properties would we like to endow our p–n junction to be suitable for this particular application? Well, it will be the integral part of some sort of tuned circuit where losses are generally unwelcome. Hence we shall use heavy doping to reduce the resistance. We should not dope too much, however, because that would lead to narrow depletion regions and low Zener breakdown. Since the varactor diode must operate under reverse bias (to get the capacitance), its useful range is between $U = 0$ and $U = U_B$; a low-breakdown voltage is obviously undesirable. Hence there must be a compromise between reducing the resistivity and ensuring a high breakdown voltage.

In practice the p-side of the junction is usually doped very heavily so that it does not contribute at all to the total series resistance. All the depletion layer is then in the n-type material, whose length is limited to the possible minimum. It is equal to the length of the depletion region just below breakdown (when the depletion region is the longest).

That is roughly how parametric amplifiers work. But is it worth making a complicated amplifier which needs a high-frequency pump

† The magic factor 2 in frequency is not necessary either. One may 'pump' at other frequencies too but this more general problem is really beyond the scope of this book.

oscillator when a 'simple' transistor will amplify just as well? The limitation of a transistor is that it will be a source of noise as well as gain. All amplifiers introduce additional noise† due to the random part of their electronic motion, so the emitter and collector currents in a transistor are fairly copious noise sources. As there is almost no standing current in a varactor, it introduces very little noise; so parametric amplifiers are worth their complications in very sensitive receivers, e.g. for satellite communication links, radio astronomy, and radar.

9.14. Photo-diodes and lamps

Semiconductor devices have vastly changed the electronics industry in the past decade but have so far had little impact on lighting methods. We still rely heavily on glowing pieces of wire producing reddish light with an efficiency of less than 5%; large awkwardly shaped tubes containing poisonous chemicals seem to be the only improvement. Many people think that the next decade will see a change in lighting comparable to that which occurred in circuitry in the past decade. For the present we shall limit ourselves to a brief comment on how light and semiconductors interact by describing some devices that detect light, one that generates electricity, and one that generates light.

The simplest light-detection method is photo-conduction. As described before photons incident upon a piece of semiconductor create extra carriers. The increase in conductivity may be related to the intensity of the input light. This is how the CdS cells that are used extensively in exposure meters and in automatic shuttering devices in cameras work.

Let us return now to our favourite subject, the p–n junction, which may also be used as a light detector. We choose reverse bias because for sensitive detection we require a large fractional change—it is very noticeable if one microamp current doubles but it is quite difficult to see a one microamp change in one milliamp.‡ If the photons shine on the p-side they create electrons that are minority carriers, and these will be accelerated across the reverse biased junction. This is the basis of a sensitive photo-detector that is made by producing a shallow layer of

† Noise due to electric currents was mentioned briefly in a previous footnote (p. 4). A fuller discussion is beyond our present scope, but if you wish for further reading in this interesting topic see F. N. H. Robinson, *Noise and fluctuations in electronic devices and circuits* (O.U.P., 1974).

‡ This is not just a question of instrumentation. It is fundamental, since any background current will have a noise current component proportional to its square root.

p-type material on an n-type substrate so that the junction is very close to the illuminated surface (Fig. 9.32).

Another interesting effect occurs when there is no bias voltage at all, so that the junction is in its equilibrium state in the dark. The incident light creates extra electron–hole pairs just as before. The extra electrons created in the p-type material diffuse to the junction and slide down the slope of the 'built-in' potential curve into the n-type material. Similarly,

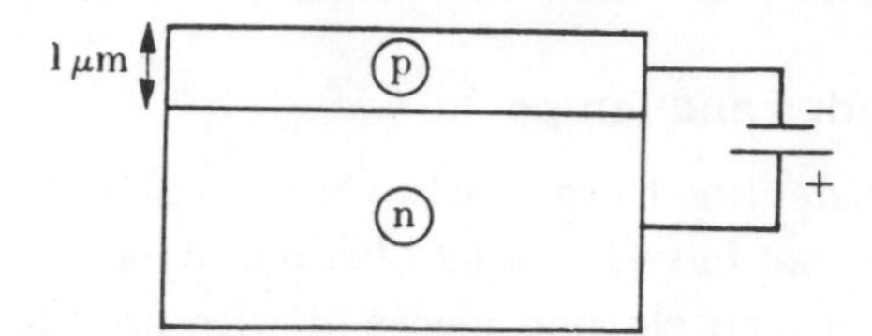

FIG. 9.32. The p–n junction as a light detector.

the extra holes created in the n-type material 'slide up' into the p-type material. Hence the p-type material becomes positively charged and the n-type material negatively charged, setting up a voltage in the forward direction. This is a voltage measurable on open circuit appropriately called the *photo-voltage*.

If an external circuit with a load is connected to the p–n junction (Fig. 9.33) there will be a current flowing across the load. Thus the p–n junction may also be used for electric power generation, or more

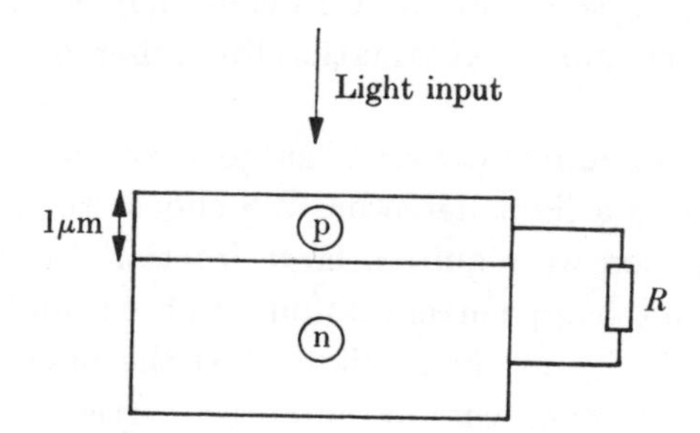

FIG. 9.33. The p–n junction as a generator of electric power.

correctly for the conversion of light into d.c. power. If the input light happens to come from the sun, the p–n junction is called a solar battery. This is useful in a spaceship where there is plenty of sunshine but a scarcity of other sources of energy.

Many processes have a converse, and we may quite legitimately ask whether a junction in non-equilibrium will produce light. The answer is yes; we can indeed get light out of a forward-biased p–n junction. At forward bias lots of minority carriers are injected that will duly recombine and, if we are lucky, they will radiate their energy in the form of electromagnetic radiation.

In order to have reasonable efficiency we must choose direct-gap materials in which band-to-band transitions are more probable. A further requirement is that it should be possible to make both n- and p-types of the materials (a condition quite important for the manufacture of p–n junctions), which unfortunately rules out CdS with its useful energy gap in the visible region. These semiconductor lamps are still in the research stage and an ideal material to match the peak response of the human eye has not yet been found. The most successful material so far has been GaAs, with best efficiency of about 10% radiating just below the visible region. Red lamps have been made with GaAs–GaP alloys and heavily doped GaP,† having an efficiency of about 1–5%.

9.15. Infrared detectors

Semiconductors may also be employed as detectors of infrared radiation: the only difference between visible and infrared radiation is the longer wavelength of the latter. Hence semiconductors with smaller energy gaps must be used. These devices are efficient up to a wavelength of about 10 μm (corresponding to an energy gap of 0·08 eV). There are some semiconductors with even smaller energy gaps, and in principle the same technique of exciting band-to-band transitions could be used for longer wavelengths. In practice, the range of wavelengths from 10 to 100 μm is, however, covered by impurity semiconductors. The increased conductivity is then obtained by exciting an electron from a donor level into the conduction band (or from the valence band into an acceptor level).

Infrared radiation between 100 μm and 1 mm is usually detected with the aid of the so called 'free carrier absorption'. This is concerned with the excitation of electrons from lower to higher energy levels in the conduction band. The number of electrons available for conduction does not change here but the mobility does change owing to the perturbed energy distribution of the electrons. The change (not necessarily increase) in conductivity may then be related to the strength of the

† GaP is an indirect gap material. The radiative transition that leads to red light emission is between *two* impurity levels (Zn and O doping). It apparently radiates like a direct-gap semiconductor.

incident infrared radiation. Since electrons may be excited to higher energy levels by lattice vibrations as well (thus masking the effect of the input electromagnetic wave), the crystal is usually cooled to liquid helium temperatures.

9.16. Field-effect transistors

We have seen a number of two-terminal devices; let us now discuss a device that has a control electrode as well, like the vacuum triode or the transistor. It was originally proposed by Shockley in 1952, and received the rather inappropriate name of *field-effect transistor*, abbreviated as FET. It consists simply of a piece of semiconductor—let us suppose n-type—to which two ohmic contacts, called the *source* and the *drain*, are made (Fig. 9.34). As may be seen the drain is positive: thus electrons flow from source to drain. There is also a gate electrode consisting of a heavily doped p-type region (denoted by p^+). Let us assume for the time being that U_{SG}, the voltage between source and gate, is zero. What will be the potential at some point in the n-type material? Since there is an ordinary ohmic potential drop due to the flow of current, the potential grows from zero at the earthed source terminal to U_{DS} at the drain. Hence the p^+n junction is always reverse biased with the reverse bias increasing towards the drain. As a consequence the depletion region has an asymmetrical shape as shown in Fig. 9.34. The drain current must flow in the channel between the depletion regions.

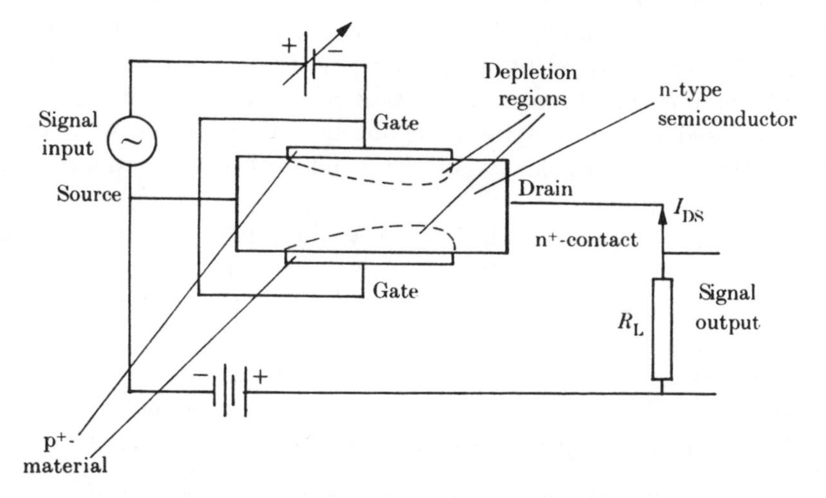

Fig. 9.34. Schematic representation of a field effect transistor (FET). The current between *source* and *drain* is controlled by the voltage on the *gate* electrodes.

If we make the gate negative then the reverse bias, and with it the depletion region, increases, forcing the current to flow through a narrower region, that is through a higher resistance. Consequently, the current decreases. Making the gate more and more negative with respect to the source, there will obviously be a voltage at which the depletion regions join and the drain current decreases to practically zero as shown in Fig. 9.35a. This I_D versus U_{GS} characteristic is strongly reminiscent of that of anode current against grid voltage in a good old triode, the product of a bygone age when electronics was nice and simple.

The physical picture yielding the I_D versus U_{DS} characteristics is a little more complicated. As U_{DS} increases at constant gate voltage there are two effects occurring simultaneously: (i) the drain current increases because U_{DS} has increased, a simple consequence of Ohm's law; (ii) the drain current decreases because increased drain voltage means increased reverse bias and thus a smaller channel for the current to flow. Now will the current increase or decrease? You might be able to convince yourself that when the channel is wide and the increase in U_{DS} means only a relatively small decrease in the width of the channel,

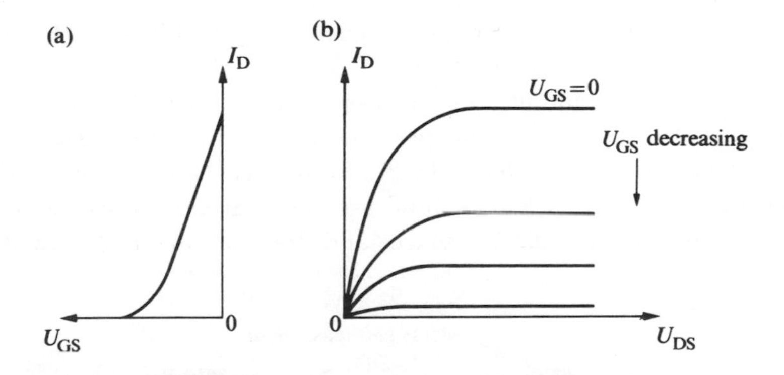

FIG. 9.35. The characteristics of a junction type field-effect transistor.

the second effect is small and the current increases. However as the channel becomes narrower the second effect gradually gains importance and the increase of I_D with U_{DS} slows down, as shown in Fig. 9.35b. At the so-called pinch-off voltage the two effects cancel each other, and they keep their balance for voltages beyond that. The current stays constant; it has reached saturation. The actual value of the saturation current would naturally depend on the gate voltage. At lower gate voltages the saturation current is smaller.

The physical mechanism of current flow in an FET is entirely different from that in a vacuum tube, but the characteristics are similar, and so is the equivalent circuit. A small change in gate voltage, u_{gs} results in a large change in drain current. Denoting the proportionality factor by g_m, called the mutual conductance, a drain current equal to $g_m u_{gs}$ appears. Furthermore, one needs to take into account that the drain current varies with drain voltage as well. Denoting the proportionality constant by r_d (called the drain resistance), we may now construct the equivalent circuit of Fig. 9.36, where i_d, u_{gs}, and u_{ds} are the small a.c. components of drain current, gate voltage, and drain voltage, respectively.

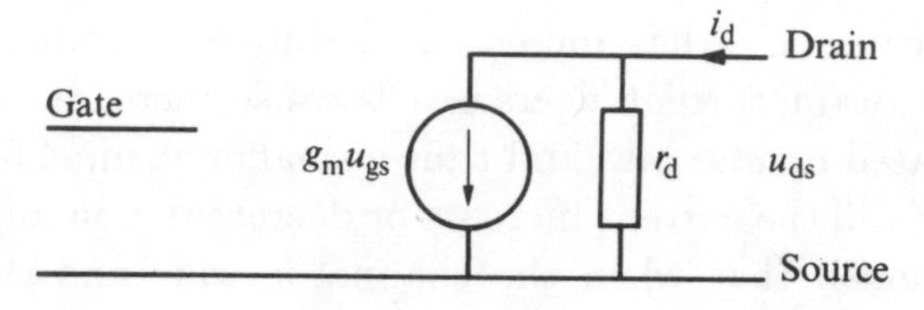

FIG. 9.36. Equivalent circuit of a field-effect transistor.

A modern and more practical variant of this device is the metal–oxide–semiconductor transistor or MOST, also known as metal–oxide–semiconductor field-effect transistor or MOSFET. It is essentially a metal–insulator–semiconductor junction provided with a source and a drain as shown in Fig. 9.37a. To be consistent with our discussion in Section 9.9 we shall assume that the substrate is an n-type semiconductor and the source and drain are made of p^+ material.

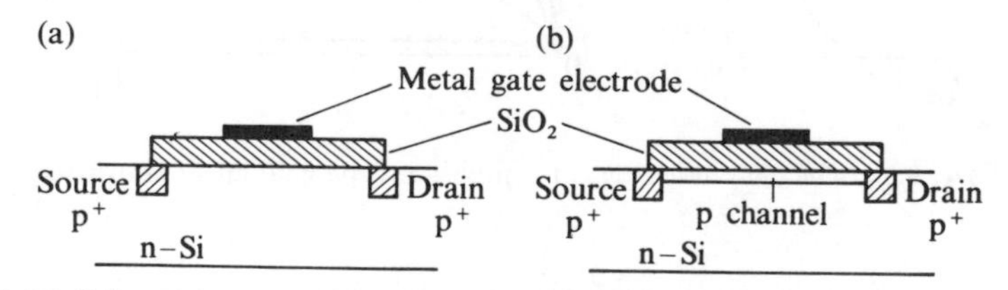

FIG. 9.37. Schematic representation of a MOSFET. (a) Zero gate bias. (b) Forward bias inducing a p-channel.

At zero gate bias no drain current flows because one of the junctions is bound to be reverse biased. What happens as we make the gate negative? Remembering the physical phenomena described in Section 9.9 we may claim that at sufficiently large negative gate voltage inversion will occur, that is the material in the vicinity of the insulator will turn into a p-type semiconductor. Holes may then flow unimpeded from source to drain. The rest of the story is the same as for an ordinary FET (note that since the advent of the MOSFET ordinary FETs got dissatisfied with their simple sounding acronym and opted for the more sonorous title of JUGFET or junction-gate field-effect transistor) and the characteristics are fairly similar, as shown in Fig. 9.38, though in the present case there is no proper current saturation, only a knee in the I_D versus U_{DS} characteristics.

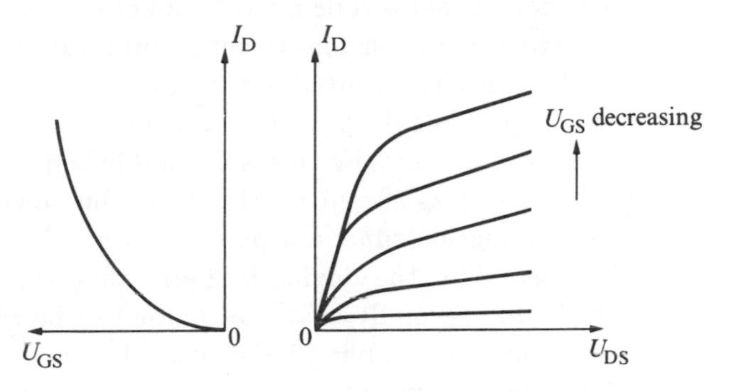

FIG. 9.38. The characteristics of a MOSFET.

The devices in which conduction occurs by inversion are said to operate in the enhancement mode. There is also a *depletion* mode device in which one starts with a p-channel (Fig. 9.37b) and depletes the holes by applying a positive bias to the gate. This is more similar to the traditional FETs.

Naturally, both the enhancement and depletion devices described have their counterparts with n^+ drains and sources and p-type substrate. In principle there is no difference between them. In practice there is some difference, because the surface potential at the Si–SiO_2 interface tends to be positive, thus it is easier to achieve inversion in an n-type material. Sometimes it proves to be advantageous to use both n-channel and p-channel enhancement devices in the same circuit. It is claimed for example that the low power consumption and high stability of these hybrid circuits (usually referred to as CMOS or 'complementary' MOS) made possible the advent of the electronic digital watch.

Otherwise, field-effect transistors can be used for the same purposes as transistors. The tendency in the last few years has been away from transistors and towards MOSFETs because of the very high input impedance and simpler construction. It is early days yet though to claim that MOSFETs have won the race.

9.17. Charge-coupled devices

Charge-coupled devices, abbreviated as CCDs, look very similar to MOSFETs and in today's world looking similar is half the battle. Since companies are rather reluctant to invest into new types of manufacturing processes, a new device that can be made by a known process and is compatible with existing devices is an attractive proposition.

A charge-coupled device is essentially a metal–insulator–semiconductor junction working in the deep depletion mode. As mentioned in Section 9.9, the carriers are not in thermal equilibrium. There is a potential well for holes as was shown in Fig. 9.23d, but owing to the long generation–recombination time in a pure material, it is not occupied by holes. The secret of the device is firstly that the holes are introduced externally and secondly that the charge can be transferred along the insulator surface by applying judiciously chosen voltages to a set of strategically placed electrodes.

Let us look first at three electrodes only, as shown in Fig. 9.39a. There is again an n-type semiconductor upon which an oxide layer is grown, and the metal electrodes are on the top, insulated from each other. We can look at each electrode as part of a metal–insulator–semiconductor junction which can be independently biased. In Fig. 9.39b there are some holes under electrode 1. They had to get there somehow, e.g. they could have got there by injection from a forward biased p–n junction. The essential thing is that they got there and the question is how that positive charge can be transferred from one electrode to the next one.

At $t = t_1$ (see Fig. 9.40) the three electrodes are biased to voltages $-A$, O, O respectively. The corresponding surface potential distribution is shown in Fig. 9.39b. The holes sit in the potential well. At $t = t_2$ we apply a voltage $-A$ to electrode 2, leading to the surface potential distribution of Fig. 9.39c. The holes are still sitting under electrode 1 but suddenly the potential well has become twice as large. Since the holes wish to fill uniformly the space available, some of them will diffuse to electrode 2. At the same time, just to give the holes a gentle nudge, U_1 is slowly returning to zero, so that by t_3 the potential well is entirely under electrode 2. Thus the transfer of charge from electrode 1 to electrode 2 has been completed (Fig. 9.39d).

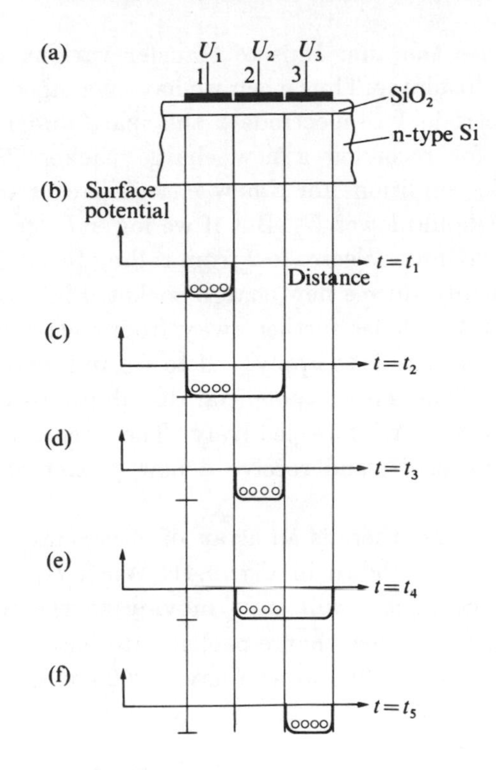

FIG. 9.39. A section of a CCD illustrating the basic principles of charge transfer.

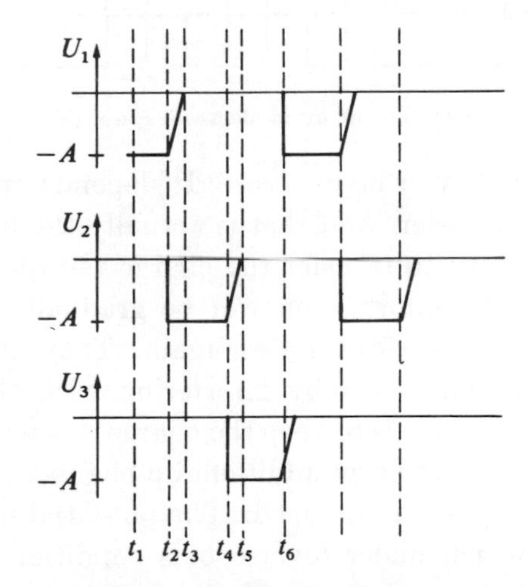

FIG. 9.40. The voltages applied to the three electrodes of Fig. 9.39 as a function of time.

Let me reiterate the aim. It is to transfer various sizes of charge packet along the insulator. Thus when we have managed to transfer the charge from electrode 1 to electrode 2 the space under electrode 1 is again available for receiving a new charge packet. How could we create favourable conditions for a new charge packet to reside under electrode 1? We should lower U_1. But if we lower U_1 to $-A$ what will prevent the charge under electrode 2 from rolling back? Nothing. Thus we cannot as yet introduce a new charge packet. First we should move our original packet of holes further away from electrode 1. Therefore our next move, at $t = t_4$, is to apply $-A$ to U_3 and increase U_2 to zero between t_4 and t_5. The surface potential distributions at t_4 and t_5 are shown in Figs. 9.39e and f respectively. The period ends at t_6. We can now safely lower U_1 and receive a new packet of charge under electrode 1.

In practice of course there is an array of electrodes with each third one joined together as shown in Fig. 9.41. When U_1 is lowered at t_6 our original charge packet will start moving to the next electrode, simultaneously with the new charge packet entering the first electrode. With 3000 electrodes in a line we can have 1000 charge packets stored in the device.

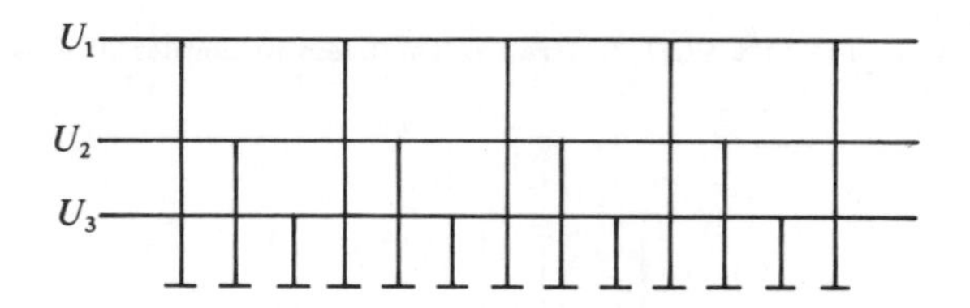

FIG. 9.41. The array of electrodes in a CCD.

How many elements can be in series? It depends on the amount of charge lost at each transfer. And that is actually the limiting factor in speed as well. If we try to transfer the charge too quickly, some of it will get stuck and the information will be gradually corrupted. The troublemakers are the surface states again. They trap and release charge carriers randomly, thereby interfering with the stored information. Thus the best thing is to keep the charge away from the surface. This can be done by inserting an additional p-channel into the junction in much the same way as in Fig. 9.37b. The potential minimum is then in the p-channel, which under reverse bias conditions is entirely depleted of its 'own' carriers and is ready to accept charge packets from the outside. These are called buried-channel devices.

What about other limitations? Well, there is a maximum amount of charge storable, above which the potential minimum disappears. There is a minimum frequency with which the charge can be transferred, below which the information is corrupted by the thermally generated carriers. There is also a minimum size for each cell determined by tunnelling effects (if the cells are too close to each other) and dielectric breakdown (if the insulator is too thin).

What can CCDs be used for? Applications as memory elements are obvious. In fact many people believe that the CCDs have a brilliant future in front of them. They are potentially cheap; a prediction I have come across recently says that eventually they should sell at around 400 a penny, even cheaper than hot cross buns in nursery rhyme times. And the density I have heard mentioned is 4 million bits on a 10 mm by 10 mm chip. Not bad. They have though one disadvantage. The memory is volatile. If power is switched off, the information is lost.

CCDs can perform lots of other tasks as well. I shall just mention two more: delay lines and optical imaging.

Delay lines. The signal to be delayed is sampled (its magnitude is taken and transformed into a short pulse) at discrete intervals, turned into corresponding charge packets, driven step by step through the device and turned back into the original signal at the output.

Optical imaging. Let us imagine a two-dimensional array of charge-coupled devices, say 300 lines, each one containing 300 elements. If a picture is focused upon the surface of the device (which in this case has transparent electrodes) the incident light creates electron–hole pairs proportional to its intensity. The process now has two steps: the 'integrate' period during which U_1 is set to a negative voltage and the holes (in practical devices electron packets are used and everything is the other way round but the principles are the same) are collected in the potential minima, and the 'readout' period during which the information is read out. Light may still be incident upon the device during readout but if the readout period is much shorter than the integration period, the resulting distortions of the video signal are negligible.

9.18. Silicon controlled rectifier

This has four semiconductor layers, as shown in Fig. 9.42. Apart from the ohmic contacts at the end there are three junctions. Suppose that junctions (1) and (3) are forward biased by the external supply, so that (2) must be reverse biased. As the supply voltage increases the current will be limited by junction (2) to a low value, until it gets to the

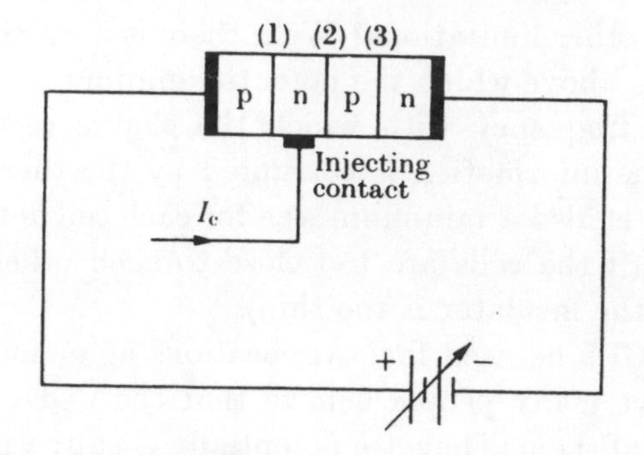

FIG. 9.42. The silicon controlled rectifier (SCR).

reverse avalanche breakdown point. Then its resistance falls very rapidly and the current through the whole device 'switches' to follow a curve approximating to the forward bias junction characteristic starting at this breakdown point, U_s (Fig. 9.43). So far we have described a self-switching arrangement: at a certain applied voltage the device resistance might fall from several megohms to a few ohms. The switch is made even more useful by the additional contact shown in Fig. 9.42, which injects holes into the n-region between junctions (1) and (2) by means of an external positive *control* bias. By injecting extra minority carriers into the junction (2) region the current I_c controls the value of V_s at which the device switches to the *on* position.

This device is used as a switch and in variable power supplies: broadly speaking it is the solid-state version of the gas-filled triode or thyratron. By analogy the name *thyristor* is also used to describe it.

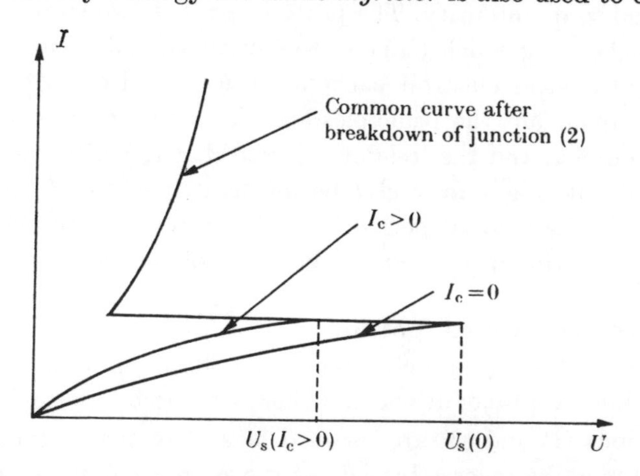

FIG. 9.43. The current voltage characteristics of a SCR. The switching voltage U_s may be controlled by the injected current I_c.

9.19. The Gunn effect

In this and the next two sections I am going to discuss devices that for a change, do not depend on a p–n junction but rather on the bulk properties of semiconductors.

We have shown how desirable it is to have negative resistance and how it can be achieved with a tunnel junction. But an inherent snag with any p–n junction is that it must behave as if there were a capacitor in parallel with the device—we worked out its value in section 9.5. So at high frequencies this capacitor lowers the impedance and causes a falling-off of efficiency. Can we get round this problem by having a negative resistance characteristic, like that of a tunnel diode, in a bulk semiconductor? This is a long-established El Dorado of semiconductor device engineers. Nearly all semiconductors *should* behave like this.

Look again at an E–k curve that we drew earlier (Fig. 7.12a). If this represents the conduction band, the electrons will be clustered about the lowest energy state: $E = 0$, $k = 0$. Now apply a field in the x-direction which accelerates the electrons. So their momentum (which, as we have mentioned before, is proportional to k) will increase as well. This means that our electrons are climbing up the E–k curve. At a certain point the effective mass changes sign as shown in Fig. 7.12c. Now the effective mass is just a concept we introduce to say how electrons are accelerated by a field; so this change of sign means that the electrons go the other way. Current opposing voltage is a negative resistance situation. It seems that there should be a good chance of *any* semiconductor behaving like this but in fact so far this effect has not been discovered. The reason must be that the electrons move for only a short time without collisions. So to get within this time into the negative mass region very high fields are necessary which cause some other trouble, e.g. breakdown or thermal disintegration.

As a matter of fact we do not really need to send our electron into the negative mass region to have a negative differential resistance. If the effective mass of the electron increases fast enough as a function of the electric field, then the reduced mobility (and conductivity) may lead to a reduction of current—and that is a negative *differential* resistance. So there seems no reason why our device could not work in the region where m^* tends rapidly towards infinity. It is a possibility but experiments have so far stubbornly refused to display the effect.

An improvement on the latter idea was put forward by Watkins, Ridley, and Hilsum who suggested that electrons excited into a subsidiary valley of GaAs (see Fig. 8.9) might do the trick. The curvature at the bottom of this valley is smaller; so the electrons acquire the higher effective mass that is our professed aim. In addition there is a higher density of states (it is proportional to $m^{*\frac{3}{2}}$); and furthermore it looks quite plausible that, once an electron is excited into this valley, it would stay there for a reasonable time.

The predicted negative differential resistance was indeed found experimentally a few years later by J. B. Gunn, who gave his name to the device. At low fields most of the conduction-band electrons are in the lower valley. When an electric field is applied the current starts to increase linearly along the line OA in Fig. 9.44. If all electrons had the

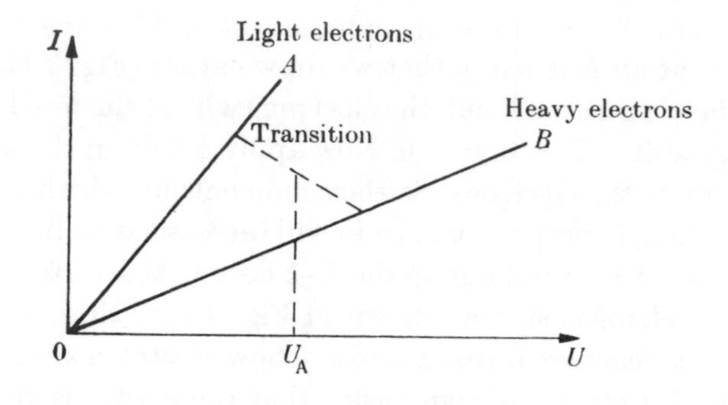

FIG. 9.44. Linear current voltage characteristics for GaAs assuming that only light electrons (OA) or only heavy electrons (OB) are present. The actual characteristics follows the OA line for low voltages and the OB line for high voltages. The transition is shown with dotted lines.

higher effective mass of the upper valley then the corresponding Ohm's law curve would be OB. As the field increases, some electrons (as we mentioned before) gain enough energy (0·36 eV) to get into the higher valley, and eventually most of them end up there. So the actual I–U curve will change from something like OA at low fields to something like OB at high fields. This transition from one to the other can (and in GaAs does) give a negative differential resistance.

Having got the negative resistance all we should have to do is to plug it into a resonant circuit (usually a cavity resonator at high frequencies) and it will oscillate. Unfortunately it is not as simple as that. A bulk negative resistance in a semiconductor is unstable, and is unstable in the sense that a slight perturbation of the existing conditions will grow.

Let us apply a voltage U_A in the negative-resistance region (Fig. 9.44). The corresponding electric field $\mathscr{E}_A = U_A/d$ (d is the length of the sample) is shown in Fig. 9.45, where velocity (proportional to current) is plotted against electric field. Quite obviously, a negative slope on the I–U curve means a negative slope on the v–$\mathscr{E}$ curve.

When everything is homogeneous the voltage, electric field, and

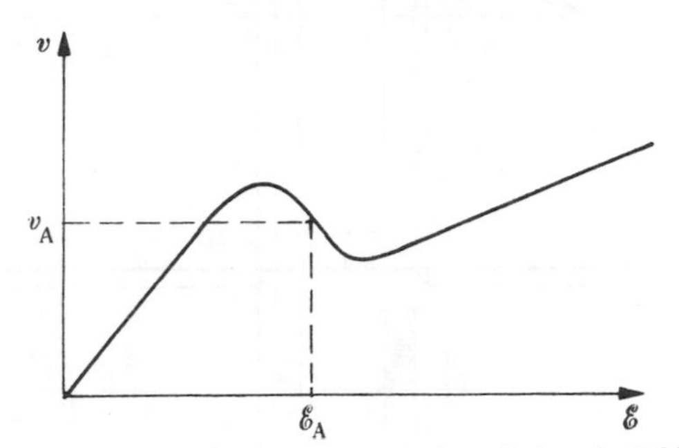

FIG. 9.45. The electron velocity as a function of electric field for GaAs.

carrier density vary along the material as shown by solid lines in Fig. 9.46. Assume now that for some reason the electric field is above its equilibrium value in the x_2–x_5 region (taken here to be quite large for the purpose of illustration). Since the total voltage between 0 and d is constant, an increase of the electric field in the x_2–x_5 region implies a reduction in the regions 0–x_2 and x_5–d, as shown by dotted lines in Fig. 9.46b. As the electric field changes between x_1 and x_3, and x_4 and x_6, there must be an accompanying excess and deficiency of space charge† (dotted lines in Fig. 9.46c). But now the electric field at point x_1 has become lower, and hence the velocity is higher; that is, more carriers flow *into* the x_1–x_3 region. At the same time the electric field at x_3 is higher, and hence the velocity is lower; that is, a smaller number of carriers flow *out* of the x_1–x_3 region. Consequently, the carrier density in the x_1–x_3 region must increase, causing a further increase in the electric field, which causes a further accumulation of space charge, and so on, until the fields on either side are high and low enough respectively to make the resistance positive. The final state is shown in Fig. 9.47; there is a domain (its typical width is about 1 μm in practice) of high field strength as if a piece of high-resistivity material had been inserted.

† Remember Poisson's equation.

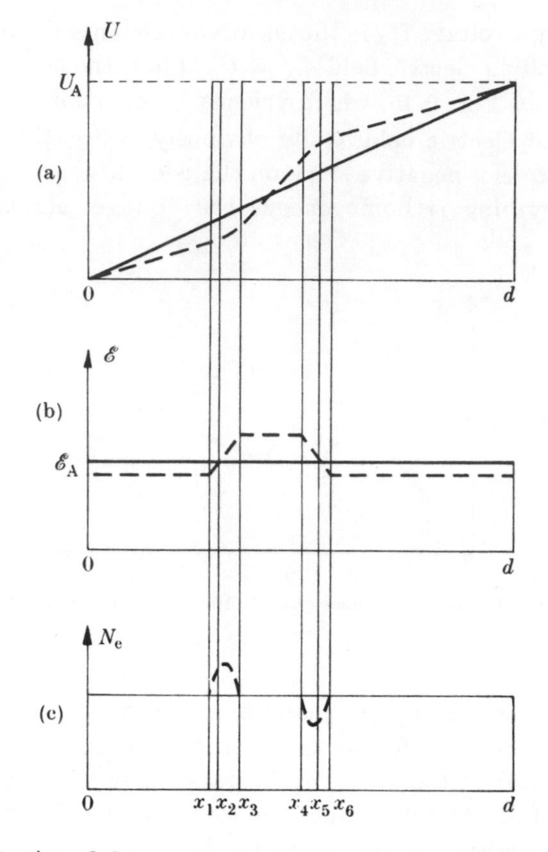

FIG. 9.46. Illustration of the growth of a disturbance when there is a bulk negative resistance.

We are not yet at the end. In order to explain experimental results we must assume that the perturbation occurs at the cathode ($x = 0$), the high-field domain is formed, moves along the material, and is finally extinguished at the anode ($x = d$). Hence the current in the device varies as shown in Fig. 9.48. At $t = t_0$, when the voltage U_A is switched on, the current is I_A. The high-field domain is formed at the cathode between $t = t_0$ and $t = t_1$. This is equivalent to having a high resistance, which means a reduction of current. The current remains constant while the high-field domain moves along the material. At $t = t_2$ (where $t_2 - t_1 = d/v_{\text{domain}}$, and the velocity of the domain is roughly the same as the drift velocity of the carriers) the domain reaches the anode. The high-resistance region disappears and the current climbs back to I_A. By the time t_3 the domain is newly formed at the cathode and everything

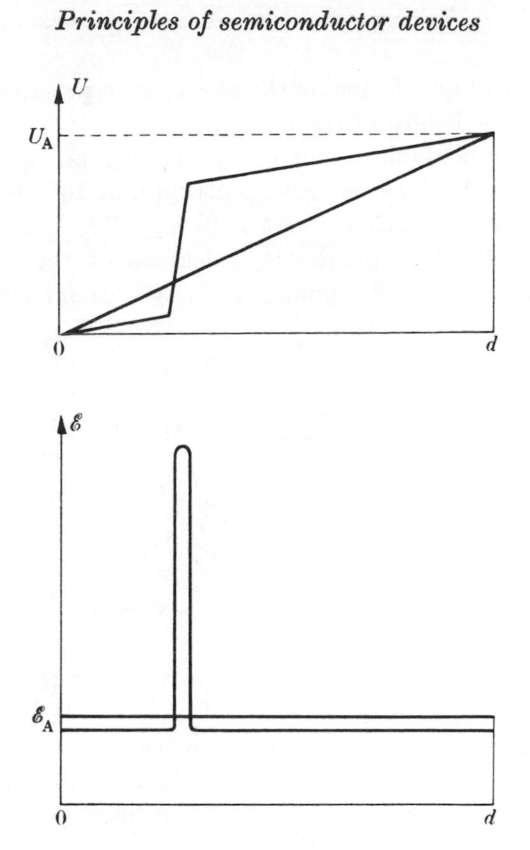

FIG. 9.47. The high field domain fully formed.

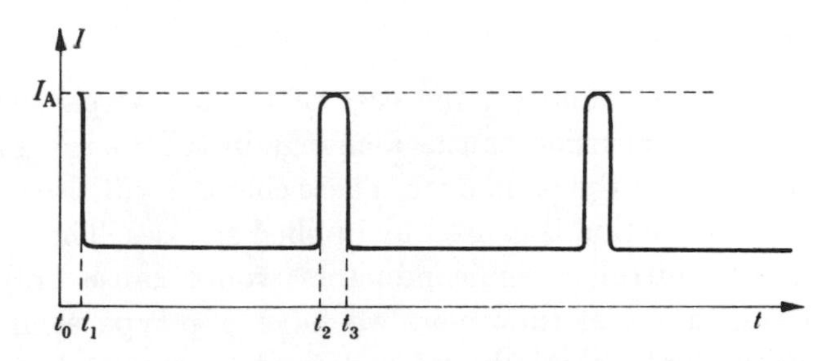

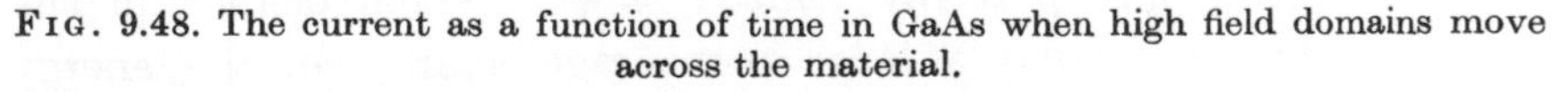
FIG. 9.48. The current as a function of time in GaAs when high field domains move across the material.

repeats itself. We have obtained a periodic current waveform rich in harmonics with a fundamental frequency

$$f = 1/(t_3 - t_1) \cong v_{\text{domain}}/d. \tag{9.24}$$

Thus the Gunn diode has an oscillation frequency governed by the domain transit time. The velocity of the domain is more or less determined by the voltage producing the effect; so in practice the frequency is selected by the length of the device.

A typical Gunn diode is made by growing an epitaxial layer of n-type GaAs, with an electron concentration of 10^{21}–10^{22} m^{-3} on to an n^+-substrate (concentration about 10^{24} m^{-3}). The current flow in the device (Fig. 9.49) is through the thickness of the epitaxial layer. For good quality GaAs the domain velocity is about 10^5 m/s; a 10-μm

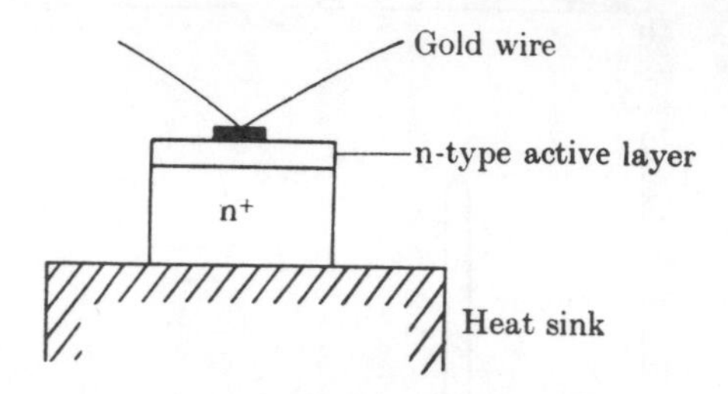

FIG. 9.49. Sketch of a typical Gunn diode.

layer will therefore make an oscillator in the 10^{10} Hz frequency band (the so called X-band of radar).

Note that the transit time mode is not the only mode of operation for this GaAs oscillator. By preventing the formation of domains the bulk negative resistance can be directly utilized.

9.20. Strain gauges

We have noticed before (in the case of thermal expansion) that a change in lattice dimension causes a change in the energy gap as well as in the value of k at the band edge. These changes will also occur if the expansion or contraction is caused by applied stresses. The changes are slight and with intrinsic semiconductors would cause only a small change in resistance. If, however, we have a p-type semiconductor with impurities only partially ionized, a very small change in the energy bands can cause a large percentage change in the energy difference between the impurity level and the band edge. Thus the change in resistance of the material with stress (or strain) is large (Fig. 9.50).

Semiconductor strain gauges are pieces of semiconductor with two ohmic contacts that are of a suitable shape to glue on to the component under test. In general the resistance R can be written

$$R = K(\rho L/A), \tag{9.25}$$

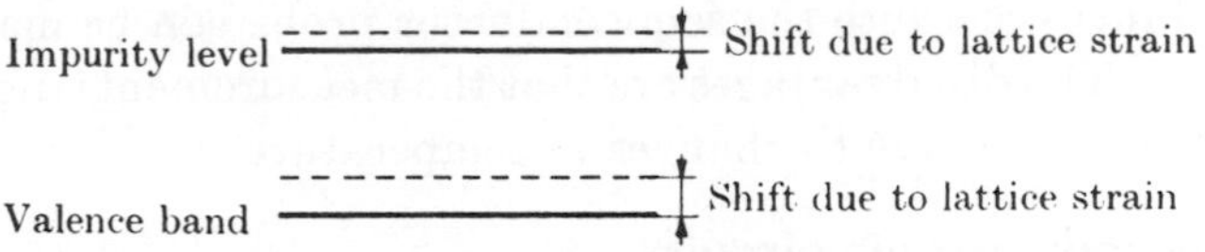

FIG. 9.50. The shift of the energy diagram with strain; this makes semi-conductors suitable materials for strain gauges.

where K is a geometrical constant, L is the length, A the cross-sectional area, and ρ the resistivity of the semiconductor.

Thus

$$\frac{\mathrm{d}R}{R} = \frac{\mathrm{d}L}{L} - \frac{\mathrm{d}A}{A} + \frac{\mathrm{d}\rho}{\rho}, \tag{9.26}$$

which can be rearranged as

$$\frac{\text{Fractional resistance change}}{\text{Strain}} = \frac{\mathrm{d}R/R}{\mathrm{d}L/L} = 1 + 2p + \frac{\mathrm{d}\rho/\rho}{\mathrm{d}L/L}, \tag{9.27}$$

where p is Poisson's ratio. The last term on the right-hand side is called the *gauge factor*,

$$G = \frac{\mathrm{d}\rho/\rho}{\mathrm{d}L/L}. \tag{9.28}$$

For p-type silicon this factor can be between 100 and 200. Of course, for a metal $G \sim 0$ and, since the other two terms on the right-hand side of eqn. (9.27) are of order of unity, the gauge factor gives a measure of the increased sensitivity of strain gauges since semiconductor strain gauges became generally available.

9.21. Measurement of magnetic field by the Hall effect

We can rewrite the Hall effect equation (1.20) in terms of the mobility and of $\mathscr{E}_l$, the applied longitudinal electric field, as

$$\mathscr{E}_{\mathrm{H}} = B\mathscr{E}_l\mu. \tag{9.29}$$

Hence B may be obtained by measuring the transverse electric field, the sensitivity of the measurement being proportional to mobility. One semiconductor is quite outstanding in this respect, n-type indium antimonide. It has an electron mobility of about 8 m^2V^{-1} s^{-1}, an order of magnitude greater than GaAs, and about fifty times greater than Si. In general this is a simpler and more sensitive method of measuring a magnetic field than a magnetic coil fluxmeter, and the method is particularly useful for examining the variation of magnetic field over short distances, because the semiconductor probe can be made exceedingly small. The disadvantages are that the measurement is not absolute, and that it is sensitive to changes in temperature.

9.22. Microelectronic circuits

We shall conclude the discussion of semiconductor devices by saying a few words about the latest techniques for producing them. Since these techniques are suitable for producing very small electronic circuits, they are called *microelectronic* circuits; and because lots of these circuits can be interconnected they are often referred to as *integrated* circuits. It all started with the so-called *planar* technique for making p–n junctions (Fig. 9.51). The crucial property of silicon that makes this possible is its ability to acquire a tenacious 'masking' layer of silicon dioxide. SiO_2 is familiar in an impure state as sand on the beach; it has a ceramic form used for furnace tubes and a crystalline form (quartz) that has good optical properties. It is very hard, chemically resistant, an insulator, and has a high melting point (~1700°C). An oxide layer can be grown by heating the silicon (typically in the form of 25 mm diameter slices, 0·25 mm thick, with an epitaxial layer on the surface) to 1200°C in an oxygen atmosphere. The growth rate is very slow, about 1 μm per hour, and the thickness is thus easily controlled. After the oxide layer is grown, the next stage is to form a pattern by *photo-engraving*. The surface is covered by a thin film of light-sensitive material called *photo-resist*. The pattern is first drawn in black and white, typically 250 times as large as finally required. It is then reduced in size photographically (using a rather sophisticated 'enlarger' backwards) so that the pattern is produced at the required size on a photographic plate. This pattern is repeated using what is called a *step-and-repeat camera*. The photographic plate might contain hundreds or even thousands of the required pattern, repeated every millimeter or so. The plate or *mask* is held close to the oxidized and photoresisted silicon and exposed to ultraviolet light. After development the exposed resist is washed away and that shielded by the pattern on the photographic mask remains.

The one chemical that readily attacks SiO_2 is hydrofluoric acid (HF) but it does not dissolve the photoresist. Hence the 'windows' in the resist can be turned into windows in the SiO_2 by etching with HF. The silicon is now in a form in which it can react within the windows but is elsewhere protected by the SiO_2 layer.

The process that comes next is *diffusion*. The silicon is sealed into a

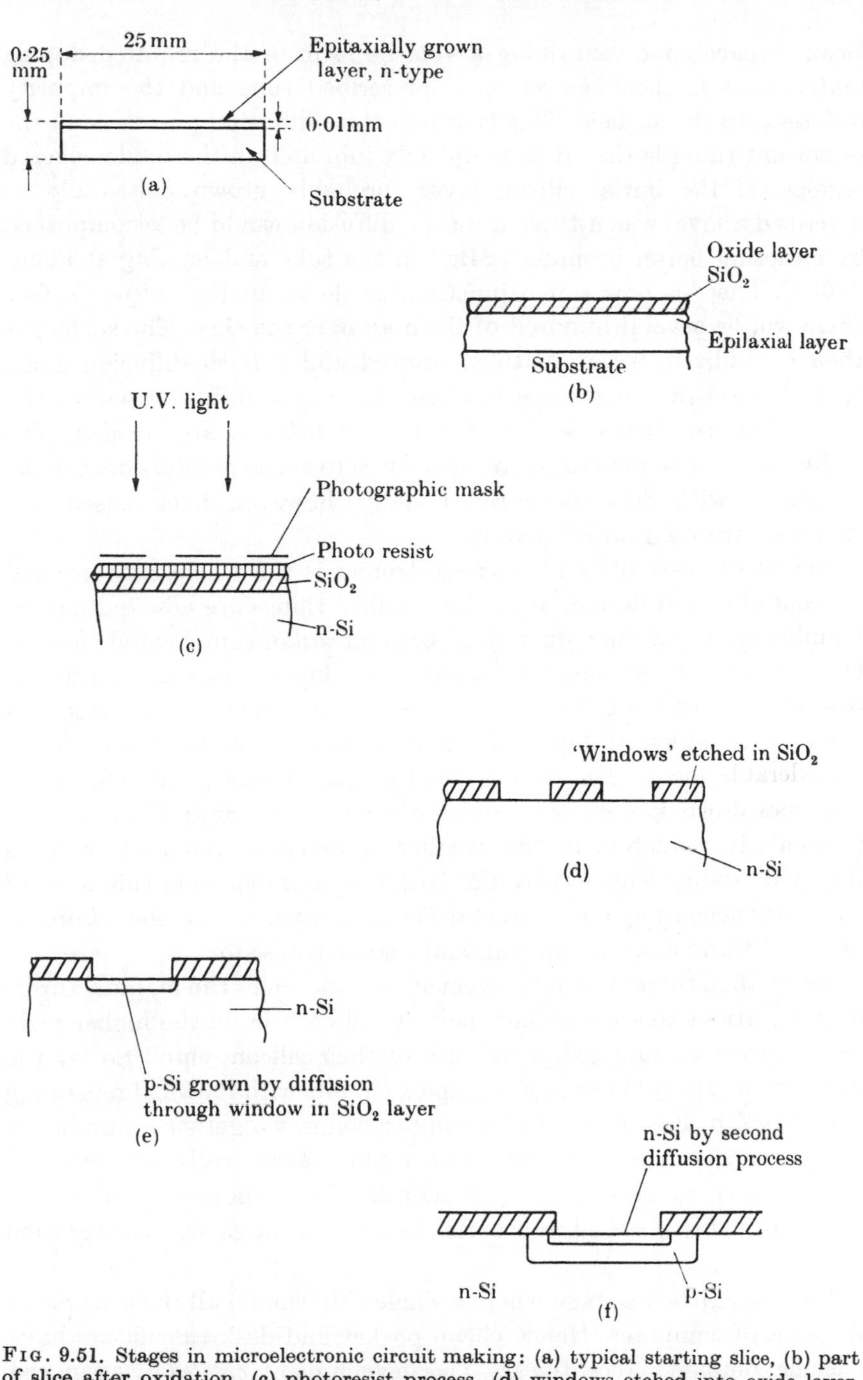

Fig. 9.51. Stages in microelectronic circuit making: (a) typical starting slice, (b) part of slice after oxidation, (c) photoresist process, (d) windows etched into oxide layer, (e) p-type Si diffused into windows, (f) repeat whole process to get an n-type diffusion.

clean furnace tube containing a volatile form of the required doping material. It is then heated for a prescribed time and the impurity diffuses into the surface. This is a solid-state diffusion process, and the important thing is that it is completely inhibited in the oxide-covered regions. If the initial silicon layer (probably grown epitaxially as described above) was n-type, a p-type diffusion would be accomplished by enclosing boron bromide (BBr_3) in the tube and heating at about 1100°C. There is now a p–n junction, as shown in Fig. 9.51e; in fact there will be several hundred of them all over the slice. The surface is then re-oxidized, a new pattern printed and a fresh diffusion done. In Fig. 9.51f the next stage has been an n-type diffusion within the p-diffusion, so that a series of n–p–n transistors are formed. The techniques can be used to isolate areas of Si to act as resistors or to make capacitors with SiO_2 dielectric. Usually, however, back-biased p–n junctions are used for capacitors.

One of the casualties of microelectronics is the whole conventional concept of circuit design. Many 'impossible' things are now quite easy, mainly because of the reduction of size and *cost* of complicated circuits. On the other hand, large capacitors and inductances are quite impossible; circuit tricks have to be used to avoid them at all costs.

Of course, at every stage of making microelectronic circuits there are considerable losses. The overall yield is usually of the order of 10%, and goes down steeply with added complications. Simplifications and particularly reduction in the number of processes required are thus highly desirable. This is why the MOST is so important; this form of transistor needs only one contact diffusion compared with the minimum of two diffusions for a conventional planar transistor.

Every time there is an improvement in techniques the manufacturers do calculations to see whether they should cash in on the higher yield or increase the complexity and size of their silicon 'chip'. So far the decision has always been more complex circuits with the yield remaining about 10% or less. Instead of having to connect together a number of integrated circuit chips up to about 6 mm square, really sophisticated systems with as many as 200,000 identifiable components can now be processed. This is called appropriately enough, large scale integration (LSI).

LSI has reached a stage where a single chip can do all the arithmetic for a small computer. Hence cheap pocket and desk calculators have become common in recent years. This development, *created* by advanced technology rather than public demand, has been made possible by the parallel advances of light-emitting diodes for numerical display. This is called 'compatible technology', i.e. both devices are energized by the same 3–6 V torch batteries.

The final stages of all microelectronic manufacture involve evaporation through photoengraved masks of metallic interconnection and contact patterns, cutting up the slice into the individual chips, and mounting them on suitable 'headers' to be connected to the outside world. There are many fascinating materials problems throughout these techniques, not least in the final packaging process where the conventional fields of engineers, chemists, metallurgists, and physicists meet in an intriguing manner.

The increase in complexity of microelectronics circuits has been a most dramatic feature. So much so that in a world where we have become accustomed to inflation and all our household and professional goods become more expensive each year, we find that the only thing that gets cheaper is electronics. You must have all noticed how the price of pocket calculators has come down, or how the same money buys more powerful calculators as years go by. Will this always be so, will the increased complexity available on a single chip always increase? Looking back to the beginning, the planar process was first developed by the Fairchild Company in the United States in 1960. Initially they made simple transistors. Their Director of Research, Dr. Gordon Moore, noticed that by 1965 the complexity of circuits had doubled each year so that there were then 2^5 components on a single chip. Now in 1978 there should be 2^{18} components if 'Moore's Law' is obeyed. In fact, various projected designs have been published and prototypes made with about 220,000–260,000 components, so Moore's Law is still pretty good. It cannot go on for ever, and many people in the industry think that the end is in sight. As well as the scientific problems of small emitter regions being populated by statistically large fluctuations of impurity density and the small currents being dominated by relatively high noise levels, there are also more financial problems of the large investments needed to produce and market such complex units of electronic equipment. Possibly microelectronics will 'saturate' at around the optical limit of photoengraving, i.e. specific component dimensions of just under 1 μm, in a few years time, although by using electron beam techniques (X-ray lithography is coming as well!) to 'expose' special resists, components can be made down to dimensions of about 10 nm. This is likely to be confined to special devices.

Compatible technology, as mentioned several times before, is a powerful force in electronics. In Chapter 10 you will read about Surface Acoustic Wave devices and in Chapter 11 about Magnetic Bubbles. Both of these ideas have led to a technology where patterns are printed onto small 'chips' of materials using the same photoengraving techniques that we have just decribed. With a chip of piezoelectric ceramic we have a signal processing unit and a chip of garnet gives a storage unit. Thus we have 3 sorts of chips that can be used for practically everything and that can be processed in the same microelectronic factory to produce powerful systems.

I would like to end this Chapter with the social implications of microelectronics. Could one talk about the 'Microelectronics Revolution' and claim at the same time (as I have done in the Introduction) that it is steady progress that counts? Well, it all depends on your definition of a Revolution. If you favour a rather brief time scale, like a day in July 1789 or October 1917, then microelectronics cannot aspire to be treated as a Revolution, but if you think of the effect upon the next few centuries to come then Revolution might be the right word. After all we do talk about the Industrial Revolution meaning the rapid progress in machinery in eighteenth century Britain so Second Industrial Revolution might be the appropriate term for describing the impact of microelectronics.

The First Industrial Revolution assisted our muscles, the Second Industrial Revolution assists our brains and lengthens our arms. The time is not very far when a man in an armchair will be able to look after a factory. Will it lead to unemployment? In spite of many assertions to the contrary I cannot help feeling that it will, and the only solution will be a drastic shortening of the working hours. Is that a good thing? Will people know what to do with their leisure time? I hope so.

Examples.

1. Show that the 'built-in' voltage in a p–n junction is given by

$$U_0 = \frac{kT}{e}\log_e\frac{N_{\text{en}}}{N_{\text{ep}}} = \frac{kT}{e}\log_e\frac{N_{\text{hp}}}{N_{\text{hn}}}$$

where N_{en}, N_{hn} and N_{ep}, and N_{hp} are the carrier densities beyond the transition region in the n and p-type materials respectively.

(Hint: Use eqns. (8.17) and (8.20) and the condition that the Fermi levels must agree.)

2. If both an electric field and concentration gradients are present the resulting current is the sum of the conduction and diffusion currents. In the

one-dimensional case it is given in the form

$$J_e = e\mu_e N_e \mathscr{E} + eD_e \frac{dN_e}{dx},$$

$$J_h = e\mu_h N_h \mathscr{E} - eD_h \frac{dN_h}{dx}.$$

When the p-n junction is in thermal equilibrium there is no current flowing, i.e. $J_e = J_h = 0$. From this condition show that the 'built-in' voltage is

$$U_0 = \frac{D_e}{\mu_e} \log_e \frac{N_{en}}{N_{ep}}.$$

Compare this with the result obtained in example 9.1, and prove the 'Einstein relationship'

$$\frac{D_e}{\mu_e} = \frac{D_h}{\mu_h} = \frac{kT}{e}.$$

Prove that $N_e N_h = N_i^2$ everywhere in the two semiconductors, including the junction region.

3. In a metal–insulator–n-type semiconductor junction the dielectric constants are ϵ_i and ϵ_s for the insulator and the semiconductor respectively. Taking the width of the insulator (sufficient to prevent tunnelling) to be equal to d_i, determine the width of the depletion region as a function of reverse voltage.

4. The doping density across a p–n junction is of the form shown in Fig. 9.52. Both the donor and acceptor densities increase linearly in the range

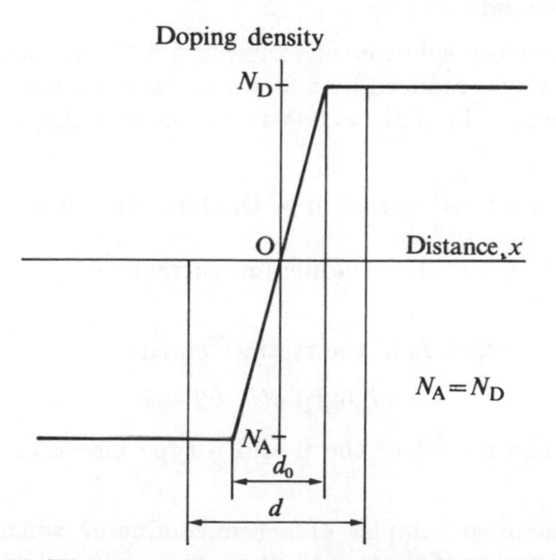

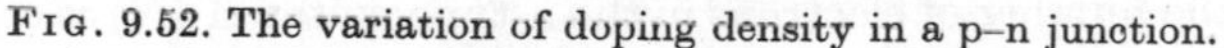
FIG. 9.52. The variation of doping density in a p–n junction.

$|x| < d_0/2$ and are constant for $|x| > d_0/2$. Determine d, the width of the depletion layer, in terms of the parameters given. Assume that the 'built-in' voltage is known.

5. Use the expression given in example 9.1 to evaluate the 'built-in' voltage for junctions in germanium and silicon. At room temperature the following data may be assumed

	μ_h	μ_e	σ_p	σ_n	N_i
	$m^2\ V^{-1}\ s^{-1}$	$m^2\ V^{-1}\ s^{-1}$	$ohm^{-1}\ m^{-1}$	$ohm^{-1}\ m^{-1}$	m^{-3}
Ge	0·17	0·36	10^4	100	$2{\cdot}4\times10^{19}$
Si	0·04	0·18	10^4	100	$1{\cdot}5\times10^{16}$

where σ_p and σ_n are the conductivities of the p- and n-type materials respectively

6. If a forward bias of 0·1 V is applied to the germanium p–n junction given in example 9.5, what will be the density of holes injected into the n-side and the density of electrons injected into the p-side?

7. Calculate the reverse breakdown voltage in an abrupt Ge p-n junction for $N_A = 10^{23}/m^3$, $N_D = 10^{22}/m^3$, $\epsilon = 16$, and breakdown field $\mathscr{E}_{br} = 2.10^7$ V/m.

8. Determine the density distribution of holes injected into an n-type material. Assume that $\partial/\partial t = 0$ (d.c. solution), and neglect the conduction current in comparison with the diffusion current.

(Hint: solve the continuity equation subject to the boundary conditions, $N_h(x = 0)$ = injected hole density, $N_h(x \to \infty)$ = equilibrium hole density in the n-type material.)

9. Determine from the solution of example 9.8 the distance at which the injected hole density is reduced by a factor e. Calculate this distance numerically for germanium where $D_h = 0{\cdot}0044$ m²/sec and the lifetime of holes is 200 μsec.

10. Determine the spatial variation of the hole current injected into the n-type material.

(Hint: Neglect again the conduction current in comparison with the diffusion current.)

11. Express the constant I_0 in the rectifier equation

$$I = I_0(\exp eU_1/kT - 1)$$

in terms of the parameters of the p and n-type materials constituting the junction.

12. Take two identical samples of a semiconductor which are oppositely doped so as the number of electrons in the n-type material (N) is equal to the

number of holes in the p-type material. Denoting the lengths of the samples by L and the cross-sections by A, the number of electrons and holes are:

p-type	number of holes	LAN
	number of electrons	LAN_i^2/N
n-type	number of holes	LAN_i^2/N
	number of electrons	LAN

Thus the total number of carriers in the two samples is

$$2LAN\left(1+\frac{N_i^2}{N^2}\right)$$

Assume now that we join together (disregard the practical difficulties of doing so) the two samples. Some holes will cross into the n-type material and some electrons into the p-type material until finally an equilibrium is established.

Show that the total number of mobile carriers is reduced when the two samples are joined together (that is some electrons and holes must have been lost by recombination).

10. Dielectric materials

Le flux les apporta, le reflux les emporte.

CORNEILLE *Le Cid*

10.1. Introduction

IN discussing properties of metals and semiconductors we have seen that with a little quantum mechanics and a modicum of common sense a reasonable account of experiments involving the transport (the word meaning motion in the official jargon) of electrons emerges. As a dielectric is an insulator, by definition, no transport occurs. We shall see that we can discuss the effects of dielectric polarization adequately in terms of electromagnetic theory. Thus all we need from band theory is an idea of what sort of energy gap defines an insulator.

Suppose we consider a material for which the energy gap is 100 times the thermal energy at 300 K, i.e., 2·5 eV. Remembering that the Fermi level is about halfway across the gap in an intrinsic material, it is easily calculated that the Fermi function is about 10^{-22} at the band edges. With reasonable density of states values this leads to less than 10^6 mobile electrons per cubic metre, which is usually regarded as a value for a good insulator. Thus, because we happen to live at room temperature we can draw the boundary between semiconductors and insulators at an energy gap of about 2·5 eV.

Another possible way of distinguishing between semiconductors and insulators is on the basis of optical properties. Since our eyes can detect electromagnetic radiation between the wavelengths of 400 nm and 700 nm we attribute some special significance to this band. So we may define an insulator as a material in which electron–hole pairs are *not* created by visible light. Since a photon of 400 nm wavelength has an energy of about 3 eV, we may say that an insulator has an energy gap in excees of that value.

10.2. Macroscopic approach

This is really the subject of electromagnetic theory, which most of you already know, so I shall just briefly summarize the results.

A dielectric is characterized by its dielectric constant ϵ, which relates the electric flux density to the electric field by the relationship

$$D = \epsilon \mathscr{E}. \qquad (10.1)$$

In the MKS system ϵ is the product of ϵ_0 (permittivity of free space) and ϵ_r (relative dielectric constant).

The basic experimental evidence (as discovered by Faraday some time ago) comes from the condenser experiment, where by inserting a dielectric between the condenser plates the capacitance increases by a factor ϵ_r. The reason is the appearance of charges on the surface of the dielectric (Fig. 10.1) necessitating the arrival of fresh charges from the battery to keep the voltage constant.

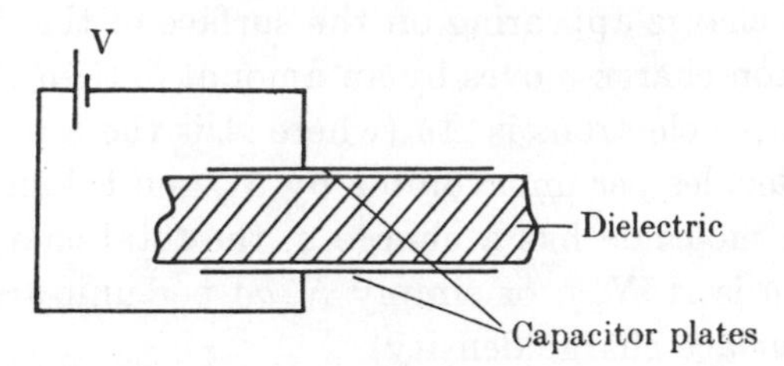

FIG. 10.1. Inserting a dielectric between the plates of a capacitor increases the surface charge.

In vacuum the surface charge density on the condenser plates is

$$Q = \epsilon_0 \frac{V}{d}, \tag{10.2}$$

where d is the distance between the plates. In the presence of the dielectric the surface charge density increases to

$$Q' = \epsilon_0 \epsilon_r \frac{V}{d}. \tag{10.3}$$

Denoting the increase in surface charge density by P, and defining the 'dielectric susceptibility' by

$$\chi = \epsilon_r - 1 \tag{10.4}$$

we may get from eqns. (10.2) and (10.3) the relationships

$$P = D - \epsilon_0 \mathscr{E} \quad \text{and} \quad P = \epsilon_0 \chi \mathscr{E}. \tag{10.5}$$

10.3. Microscopic approach

We shall now try to explain the effect in terms of atomic behaviour, seeing how individual atoms react to an electric field, or even before that recalling what an atom looks like. It has a positively charged nucleus surrounded by an electron cloud. In the absence of an electric field the statistical centres of positive and negative charges coincide. (This is actually true for a class of molecules as well.) When an electric field is applied there is a shift in the charge centres, particularly of the electrons. If this separation is δ and the total charge is q the molecule has an *induced dipole* moment,

$$\mu = q\delta. \tag{10.6}$$

Let us now switch back to the macroscopic description and calculate the amount of charge appearing on the surface of the dielectric. If the centre of electron charge moves by an amount δ, then the total volume occupied by these electrons is $A\delta$ (where A is the area). Denoting the number of molecules per unit volume by N_{m} and taking account of the fact that each molecule has a charge q, the total charge appearing in the volume $A\delta$ is $A\delta N_{\mathrm{m}}q$, or simply $N_{\mathrm{m}}q\delta$ per unit area (this is what we mean by surface charge density).

It is interesting to notice that this polarized surface charge density (which we denoted by P) is exactly equal to the amount of dipole moment per unit volume, which from eqn. (10.6) is also $N_{\mathrm{m}}q\delta$. So we have obtained our first relationship between the microscopic and macroscopic quantities

$$P = N_{\mathrm{m}}\mu. \tag{10.7}$$

For low electric fields we may assume that the dipole moment is proportional to the local electric field, $\mathscr{E}'$:

$$\mu = \alpha\mathscr{E}', \tag{10.8}$$

where α is a constant called the *polarizability*. Notice that the presence of dipoles increases the local field (Fig. 10.2), which will thus always be larger than the applied electric field.

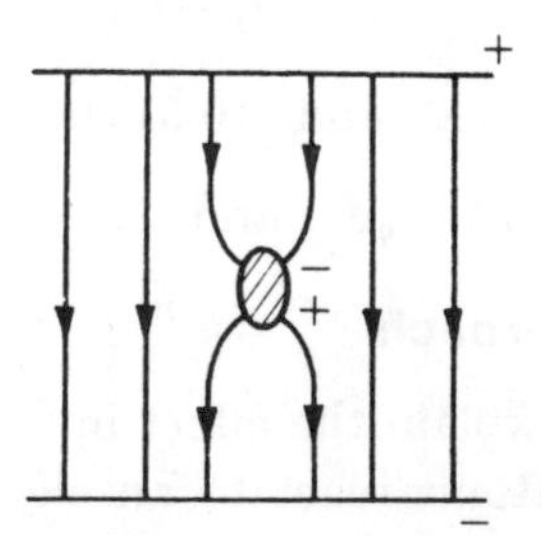

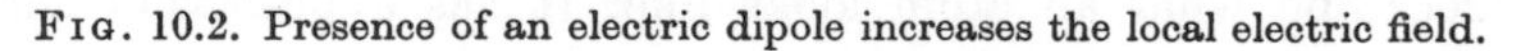

FIG. 10.2. Presence of an electric dipole increases the local electric field.

10.4. Types of polarization

(i) *Electronic.* All materials consist of ions surrounded by electron clouds. As electrons are very light they have a rapid response to field changes; they may even follow the field at optical frequencies.

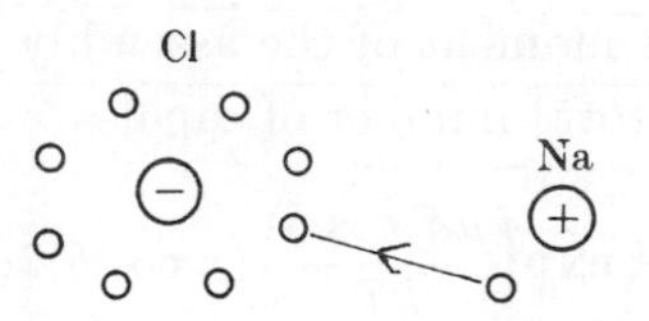

FIG. 10.3. The inter-atomic bond in NaCl is caused by Coulomb attraction. An external electric field will change the separation, thus changing the dipole moment.

(ii) *Molecular.* Bonds between atoms are stretched by applied electric fields when the lattice ions are charged. This is easily visualized with an alkali halide crystal (Fig. 10.3), where small deformations of the ionic bond will occur when a field is applied, increasing the dipole moment of the lattice.

(iii) *Orientational.* This occurs in liquids or gases when whole molecules, having a permanent or induced dipole moment move into line with the applied field. You might wonder why in a weak static field all the molecules do not eventually align just as a weather vane languidly follows the direction of a gentle breeze. If they did, that would be the lowest energy state for the system but we know from Boltzmann statistics that in thermal equilibrium the number of molecules with an energy E is proportional to $\exp(-E/kT)$; so at any finite temperature other orientations will also be present.

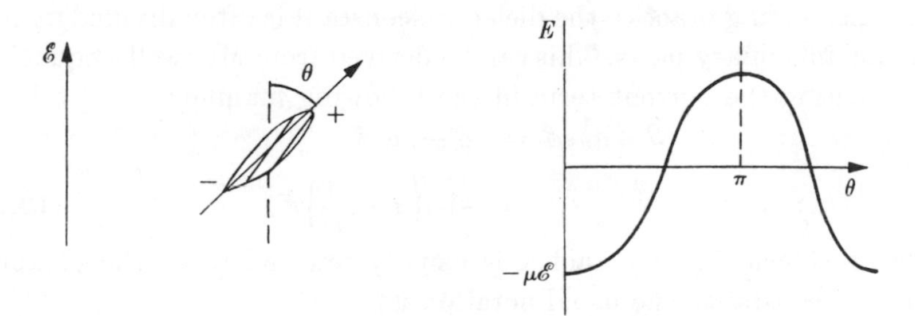

FIG. 10.4. Energy of a dipole in an electric field.

Physically, we may consider the dipole moments as trying to line up but, jostled by their thermal motion, not all of them succeed. Since the energy of a dipole in an electric field, $\mathscr{E}$ is (Fig. 10.4)

$$E = -\mu\mathscr{E}\cos\theta, \tag{10.9}$$

the number of dipoles in a solid angle $\mathrm{d}\Omega$ is

$$A \exp\left(\frac{\mu\mathscr{E}\cos\theta}{kT}\right) 2\pi \sin\theta \,\mathrm{d}\theta, \tag{10.10}$$

where A is a constant. Hence the average dipole moment is given as

$$\langle\mu\rangle = \frac{\text{net moment of the assembly}}{\text{total number of dipoles}}$$

$$= \frac{\int_0^\pi A \exp\left(\frac{\mu\mathscr{E}\cos\theta}{kT}\right)(\mu\cos\theta) 2\pi \sin\theta \,\mathrm{d}\theta}{\int_0^\pi A \exp\left(\frac{\mu\mathscr{E}\cos\theta}{kT}\right) 2\pi \sin\theta \,\mathrm{d}\theta} \tag{10.11}$$

Equation (10.11) turns out to be integrable, yielding

$$\frac{\langle\mu\rangle}{\mu} = L(a) = \coth a - \frac{1}{a}, \quad \text{where} \quad a = \frac{\mu\mathscr{E}}{kT} \tag{10.12}$$

and $L(a)$ is called the Langevin function.

If a is small, which is true under quite wide conditions, eqn. (10.12) may be approximated by

$$\langle\mu\rangle = \frac{\mu^2\mathscr{E}}{3kT}, \tag{10.13}$$

i.e. the polarizability is inversely proportional to the absolute temperature.

10.5. The complex dielectric constant and the refractive index

In engineering practice the dielectric constant is often divided up into real and imaginary parts. This can be derived from Maxwell's equations by rewriting the current term in the following manner:

$$J - \mathrm{i}\omega\epsilon\mathscr{E} = \sigma\mathscr{E} - \mathrm{i}\omega\epsilon\mathscr{E}$$

$$= -\mathrm{i}\omega\left(\epsilon + \mathrm{i}\frac{\sigma}{\omega}\right)\mathscr{E}, \tag{10.14}$$

where the term in the bracket is usually referred to as the complex dielectric constant. The usual notation is†

$$\epsilon \equiv \epsilon'\epsilon_0 \quad \text{and} \quad \frac{\sigma}{\omega} = \epsilon''\epsilon_0, \tag{10.15}$$

with the *loss tangent* defined as $\tan\delta \equiv \epsilon''/\epsilon'$.

† The complex dielectric constant used by electrical engineers is invariably in the form $\epsilon = \epsilon_0(\epsilon' - \mathrm{j}\epsilon'')$. We found a different sign because we had adopted the physicists' time variation, $\exp(-\mathrm{i}\omega t)$.

The refractive index is defined as the ratio of the velocity of light in a vacuum to that in the material,

$$n = \frac{c}{v}$$

$$= \sqrt{\epsilon_r \mu_r} = \sqrt{\epsilon'} \tag{10.16}$$

since $\mu_r = 1$ for most materials that transmit light.

Conventionally, we talk of 'dielectric constant' (or permittivity) for the lower frequencies in the electromagnetic spectrum and of refractive index for light. Equation (10.16) shows that they are the same thing—a measure of the polarizability of a material in an alternating electric field.

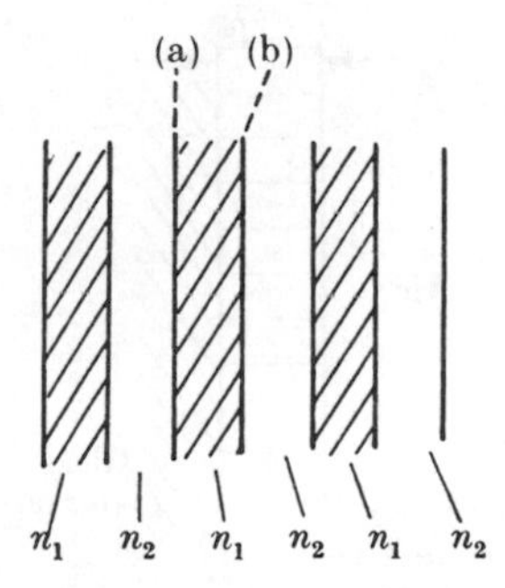

FIG. 10.5. Quarter wavelength layers used to make dielectric mirrors.

A fairly recent and important application of dielectrics to optics has been that of multiply-reflecting thin films. Consider the layered structure represented in Fig. 10.5 with alternate layers of transparent material having refractive indices n_1 and n_2 respectively. At each interface there will be some light reflected, and some transmitted. The *reflection coefficient*, from electromagnetic theory, at an interface like (a) in Fig. 10.5 is

$$r_a = \frac{n_2 - n_1}{n_2 + n_1} \tag{10.17}$$

By symmetry, the reflection coefficient at (b) will be the reverse of this,

$$r_b = \frac{n_1 - n_2}{n_1 + n_2} = -r_a, \tag{10.18}$$

which means that the two reflections have a phase difference of π radians. Now suppose that all the layers are a quarter wavelength thick (their actual thickness will be $n_1(\lambda/4)$ and $n_2(\lambda/4)$ respectively). Then the wave reflected back from (b) will be π radians out of phase with the wave reflected back from (a) because of its extra path length, and another π radians because of the phase difference in eqn. (10.18). So the two reflected waves are 2π radians different; that is, they add up in phase. A large number of these layers, often as many as 17, makes an excellent mirror. In fact, provided good dielectrics (ones with low losses, that is), are used an overall reflection coefficient of 99·5% is possible, whereas the best metallic mirror is about 97–98% reflecting. This great reduction in losses with dielectric mirrors has made their use with low-gain gas lasers almost universal. I shall return to this topic when discussing lasers.

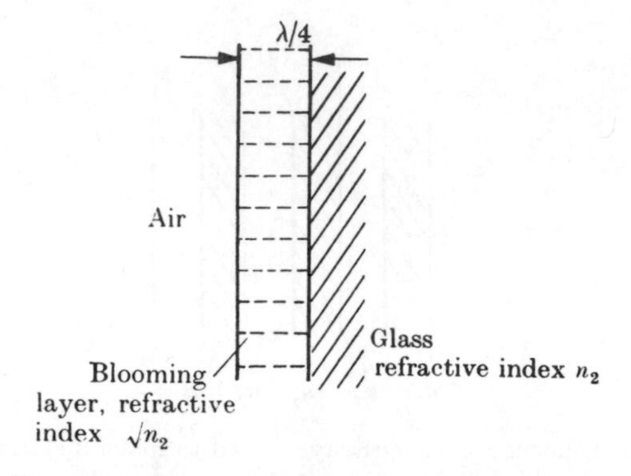

FIG. 10.6. Simple coating for a 'bloomed' lens.

Another application of this principle occurs when the layer thickness is one *half* wavelength. Successive reflections then cancel and we have a reflectionless or 'bloomed' coating, much used for the lenses of microscopes and binoculars. A simpler form of 'blooming' uses only one intermediate layer on the glass surface (Fig. 10.6) chosen so that

$$n_1 = \sqrt{n_2}. \tag{10.19}$$

The layer of the material of refractive index n_1 is this time one quarter wavelength, as can be seen by applying eqn. (10.17)

10.6. Frequency response

Most materials are polarizable in several different ways. As each type has a different frequency of response the dielectric constant will vary with frequency in a complicated manner; e.g. at the highest frequencies (light waves) only the electronic polarization will 'keep up' with the applied field. Thus we may measure the electronic contribution to the dielectric constant by measuring the refractive index at optical frequencies. An important dielectric, water has a dielectric constant of about 80 at radio frequencies, but its refractive index is 1·3, not $(80)^{1/2}$. Hence we may conclude that the electronic contribution is about 1·7 and the rest is probably due to the orientational polarizability of the H_2O molecule.

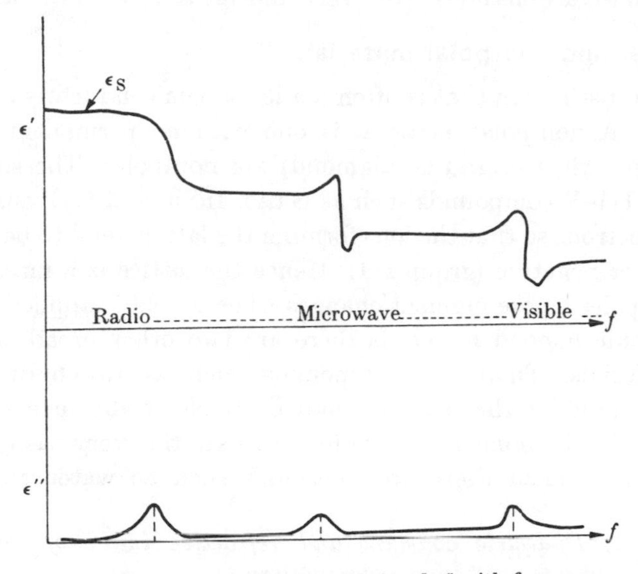

Fig. 10.7. Typical variation of ϵ' and ϵ'' with frequency.

The general behaviour is shown in Fig. 10.7. At every frequency where ϵ' varies rapidly, there tends to be a peak of the ϵ'' curve. In some cases this is analogous to the maximum losses that occur at resonance in a tuned circuit: the molecules have a natural resonant frequency because of their binding in the crystal and they will transfer maximum energy from an electromagnetic wave at this frequency. Another case is the 'viscous lag' occurring between the field and the polarized charge which is described by the Debye equations that we shall presently consider. A consequence of all this is that materials that transmit light often absorb strongly in the ultraviolet and infrared regions, e.g. most forms of glass. Radio reception indoors is comparatively easy because (dry) bricks transmit wireless waves but absorb light; we can listen in privacy. The earth's atmosphere is a most interesting dielectric. Of the fairly complete spectrum radiated by the sun, not many spectral bands reach the earth. Below 10^8 Hz the ionosphere absorbs or reflects; between 10^{10} and 10^{14} Hz there is molecular resonance absorption in H_2O, CO_2, O_2, and N_2; above 10^{15} Hz there is a very high scattering rate by molecules and dust particles. The visible light region (about 10^{14}–10^{15} Hz) has, of course, been of greatest importance to the evolution of life on earth. One wonders what we would all be like if there had been just a little more dust around and we had had to rely on the 10^8–10^{10} Hz atmosphere window for our vision.

10.7. Polar and non-polar materials

This is a distinction that is often made for semiconductors as well as dielectrics. A non-polar material is one with no permanent dipoles. For example, Si, Ge, and C (diamond) are non-polar. The somewhat analogous III–V compounds such as GaAs, InSb, and GaP share their valency electrons so that the ions forming the lattice tend to be positive (group V) or negative (group III). Hence the lattice is a mass of permanent dipoles whose moment changes when a field is applied. As well as these ionic bonded materials there are two other broad classes of polar materials. There are compounds such as the hydrocarbons (C_6H_6 and paraffins) that have permanent dipole arrangements but still have a net dipole moment of zero (one can see this very easily for the benzene ring). Then there are molecules such as water and many

TABLE 10.1. *Dielectric constant and refractive index of polar and non-polar materials*

Material	(Refractive Index)^2	ϵ'	Frequency at which ϵ' measured (Hz)
Non polar:			
C (diamond)	5·66	5·68	10^8
H_2 (liquid)	1·232	1·228	10^7
Weakly polar:			
polythene	2·28	2·30	10^2-10^{10}
paraffin	2·19	2·20	10^3
ptfe (polytetrafluoro-ethylene)	1·89	2·10	10^2-10^9
Polar:			
NaCl (rocksalt)	2·25	5·90	10^3
TiO_2 (rutile)	6·8	94	10^3
SiO_2 (quartz)	2·13	3·85	10^3
Soda glass	2·30	7·60	10^3

transformer oils that have permanent dipole moments and the total dipole moment is determined by their orientational polarizability.

A characteristic of non-polar materials is that, as all the polarization is electronic, the refractive index at optical wavelengths is approximately equal to the square root of the relative dielectric constant at low frequencies. This behaviour is illustrated in Table 10.1.

10.8. The Debye equation

We have seen that frequency variation of relative permittivity is a complicated affair. There is one powerful generalization due to Debye of how materials with orientational polarizability behave in the region where the dielectric polarization is 'relaxing', i.e. the period of the a.c. wave is comparable to the alignment time of the molecule. When the applied frequency is much greater than the reciprocal of the alignment time, we shall call the relative dielectric constant ϵ_∞ (representing atomic and electronic polarization). For much lower frequencies it becomes ϵ_s, the static relative dielectric constant. We need to find an expression of the form

$$\epsilon(\omega) = \epsilon_\infty + f(\omega), \tag{10.20}$$

which for $\omega \to 0$ reduces to

$$f(0) = \epsilon_s - \epsilon_\infty. \tag{10.21}$$

Now suppose that a steady field is applied to align the molecules and then switched off. The polarization and hence the internal field will diminish. Following Debye we shall assume that it decays exponentially with a time constant τ, the characteristic relaxation time of the dipole moment of the molecule

$$P(t) = P_0 \exp(-t/\tau). \tag{10.22}$$

To obtain a frequency response from a time function one must take the Fourier transform, as we know from circuit theory. If $f(\omega)$ is the frequency spectrum due to the time function $P(t)$, then

$$\begin{aligned} f(\omega) &= \int_0^\infty P(t) e^{i\omega t}\, dt \\ &= \int_0^\infty P_0 \exp(i\omega t - t/\tau)\, dt \\ &= \frac{P_0}{-i\omega + 1/\tau} \end{aligned} \tag{10.23}$$

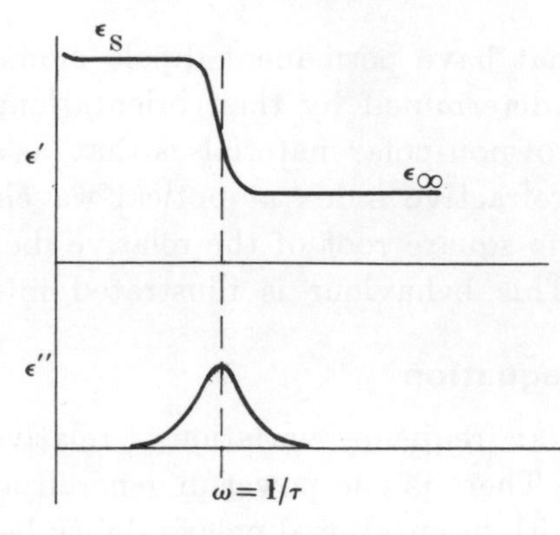

FIG. 10.8. Frequency variation predicted by the Debye equations.

Using the condition (10.21) for the limit when $\omega = 0$, we get

$$\tau P_0 = \epsilon_s - \epsilon_\infty. \tag{10.24}$$

Hence eqn. (10.20) becomes

$$\epsilon(\omega) = \epsilon_\infty + \frac{\epsilon_s - \epsilon_\infty}{-i\omega\tau + 1}, \tag{10.25}$$

which, after the separation of the real and imaginary parts, reduces to

$$\epsilon' = \epsilon_\infty + \frac{\epsilon_s - \epsilon_\infty}{1 + \omega^2\tau^2} \tag{10.26}$$

$$\epsilon'' = \frac{\omega\tau}{1 + \omega^2\tau^2}(\epsilon_s - \epsilon_\infty). \tag{10.27}$$

These equations agree well with experimental results. Their general shape is shown in Fig. 10.8. Notice particularly that ϵ'' has a peak at $\omega\tau = 1$ where the slope of the ϵ' curve is a maximum.

10.9. The effective field

We have remarked that the effective or local field inside a material is increased above its value in free space by the presence of dipoles. Generally it is difficult to calculate this increase but, for a non-polar solid, assumptions can be made that give reasonable agreement with experiment and give some indication of how the problem could be tackled for more complicated materials. Consider the material to which an external field is applied. We claim now that the local electric field at a certain point is the same as that inside a spherical hole. In this approximate picture the effect of all the 'other' dipoles is represented by the charges on the surface of the sphere. Since in this case the surface is not

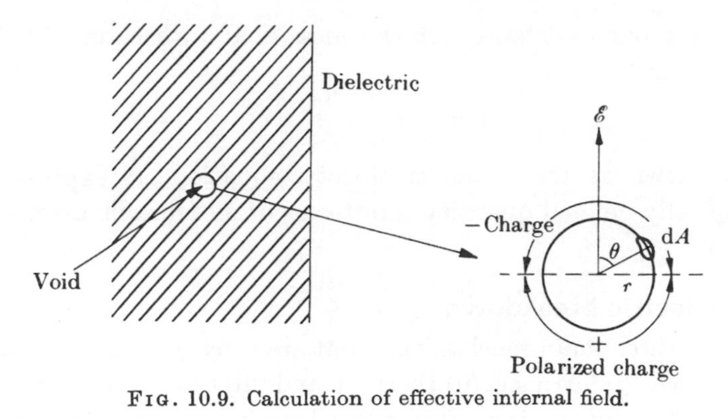

FIG. 10.9. Calculation of effective internal field.

perpendicular to the direction of the polarization vector, the surface charge is given by the scalar product (Fig. 10.9)

$$\mathbf{P}\,.\,\mathrm{d}\mathbf{A} = P\,\mathrm{d}A\cos\theta, \tag{10.28}$$

giving a radial electric field

$$\mathrm{d}\mathscr{E}_r = P\,\mathrm{d}A\cos\theta/4\pi\epsilon_0 r^2 \tag{10.29}$$

in the middle of the sphere. Clearly, when we sum these components the net horizontal field in our drawing will be zero and we have to consider only the vertical field. We get this field, previously called $\mathscr{E}'$, by integrating the vertical component and adding to it the applied field, that is

$$\mathscr{E}' = \oiint_{\text{surface}} \mathrm{d}\mathscr{E}_r\cos\theta + \mathscr{E}, \tag{10.30}$$

whence

$$\begin{aligned}\mathscr{E}' - \mathscr{E} &= \oiint \frac{P\cos^2\theta\,\mathrm{d}A}{4\pi\epsilon_0 r^2}\\ &= \int_0^{\pi} \frac{P\cos^2\theta}{4\pi\epsilon_0 r^2}\, r^2 2\pi\sin\theta\,\mathrm{d}\theta\\ &= \frac{P}{3\epsilon_0}.\end{aligned} \tag{10.31}$$

Substituting for P from eqn. (10.5) and solving for $\mathscr{E}'$ we get

$$\mathscr{E}' = \tfrac{1}{3}(\epsilon'+2)\mathscr{E}. \tag{10.32}$$

This result is clearly acceptable for $\epsilon' = 1$; and it is also consistent with our assumption that $\mathscr{E}'$ is proportional to $\mathscr{E}$.

We can now derive an expression for the polarizability α as well, by combining our expression for the local field with eqns. (10.7) and (10.8), yielding

$$\alpha = \frac{\epsilon' - 1}{\epsilon' + 2} \cdot \frac{3\epsilon_0}{N_m}, \tag{10.33}$$

which is known as the Clausius–Mosotti equation. It expresses the microscopically defined quantity α in terms of measurable macroscopic quantities.

10.10. Dielectric breakdown

There are three main mechanisms that are usually blamed for dielectric breakdown: (i) intrinsic, (ii) thermal, and (iii) discharge breakdown.

Electrical breakdown is a subject to which it is difficult to apply our usual scientific rigour. A well-designed insulator (in the laboratory) breaks down in service if the wind changes direction or if a fog descends. An oil-filled high-voltage condenser will have bad performance, irrespective of the oil used, if there is 0·01 % of water present. The presence of grease, dirt, and moisture is the dominant factor in most insulator design. The flowing shapes of high-voltage transmission line insulators are not entirely due to the fact that the ceramic insulator firms previously made chamber-pots; they also reduce the probability of surface tracking. The onset of dielectric breakdown is an important economic as well as technical limit in capacitor design. Generally one wishes to make capacitors with a maximum amount of stored energy. Since the energy stored per unit volume is $\frac{1}{2}\epsilon\mathscr{E}^2$, the capacitor designers value high breakdown strength even more highly than high dielectric constant.

In general, breakdown is manifested by a sudden increase in current when the voltage exceeds a critical value U_b, as shown in Fig. 10.10. Below U_b, there is a small current due to the few free electrons that must

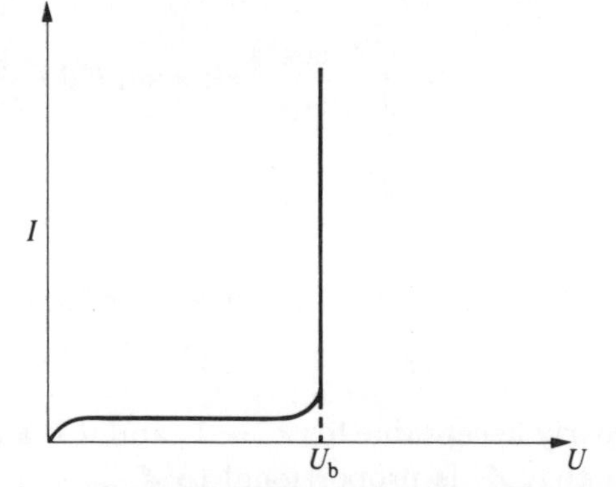

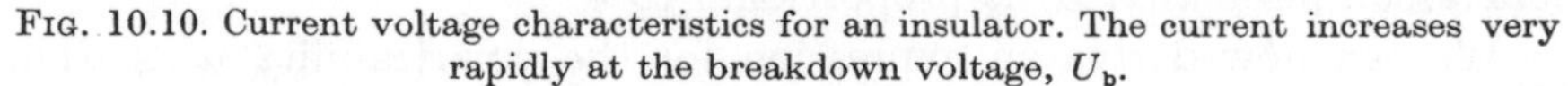
FIG. 10.10. Current voltage characteristics for an insulator. The current increases very rapidly at the breakdown voltage, U_b.

be in the conduction band at finite temperature. When breakdown occurs it does so very quickly, typically in 10^{-8} s in a solid.

(i) *Intrinsic breakdown.* When the few electrons present are sufficiently accelerated (and lattice collisions are unable to absorb the energy) by the electric field they can ionize lattice atoms. The minimum requirement for this is that they give to the bound (valence) electron enough energy to excite it across the energy gap of the material. This is, in fact, the same effect that we mentioned before in connection with avalanche diodes.

(ii) *Thermal breakdown.* This occurs when the operating or test conditions heat the lattice. For example, an a.c. test on a material in the region of its relaxation frequency where ϵ'' is large would cause heating by the lossy dipole interaction rather than by accelerating free electrons. The heated lattice ions could then be more easily ionized by free electrons, and hence the breakdown field could be less than the intrinsic breakdown field measured with d.c. voltages. The typical polymer, polyethylene, has a breakdown field of 3 to 5×10^{8} V/m for very low frequencies, but this falls to about 5×10^{6} V/m around 10^{6} Hz where a molecular relaxation frequency occurs. Ceramics such as steatite and alumina exhibit similar effects.

If it were not for dielectric heating effects breakdown fields would be lower at high frequencies, simply because the free electrons have only half a period to be accelerated in one direction. I mentioned that a typical breakdown time was 10^{-8} s; so we might suppose that at frequencies above 10^{8} Hz breakdown would be somewhat inhibited. This is true, but a fast electron striking a lattice ion still has a greater speed after collision than a slow one, and some of these fast electrons will be further accelerated by the field. Thus quite spectacular breakdown may sometimes occur at microwave frequencies (10^{10} Hz) when high power densities are passed through the ceramic windows of klystrons or magnetrons. Recent work with high-power lasers has shown that dielectric breakdown still occurs at optical frequencies. In fact the maximum power available from a solid laser is about 10^{12} watts from a series of cascaded neodymium glass amplifier lasers. The reason why further amplification is not possible is that the optical field strength disrupts the glass laser material.

(iii) *Discharge breakdown.* In materials such as mica or porous ceramics, where there is occluded gas, the gas often ionizes before the solid breaks down. The gas ions can cause surface damage, which accelerates breakdown. This shows up as intermittent sparking and then breakdown as the test field is increased.

10.11. Piezoelectricity

It is easy to describe the piezoelectric† effect in a few words: a mechanical strain will produce dielectric polarization and, conversely, an applied electric field will cause mechanical strain. Which crystals will exhibit this effect? Experts say that out of the 21 classes of crystals that lack a centre of symmetry, 20 are piezoelectric. Obviously, we cannot go into the details of all these crystals structures here but one can produce a simple argument showing that lack of a centre of symmetry is a necessary condition. Take for example the symmetric cubic structure shown in Fig. 10.11a. If mechanical force is applied (Fig. 10.11b) then the dimensions change but no net electric dipole moment

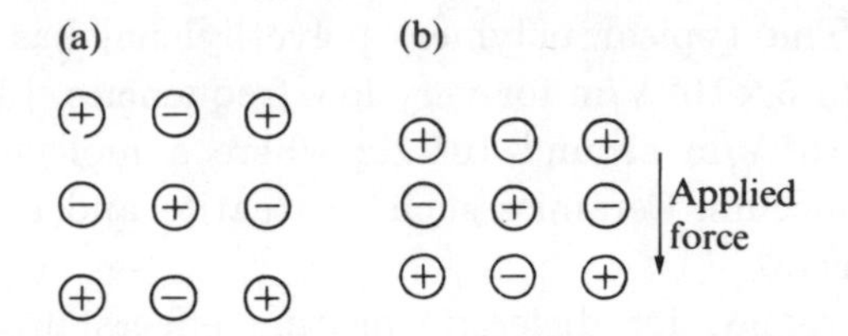

FIG. 10.11. Schematic representation of a symmetric crystal: (a) in the absence of applied force, (b) in the presence of applied force.

is created. If we take however a crystal which clearly lacks a centre of symmetry (Fig. 10.12), and apply a mechanical force, then the centres of positive and negative charge no longer coincide and a dipole moment is

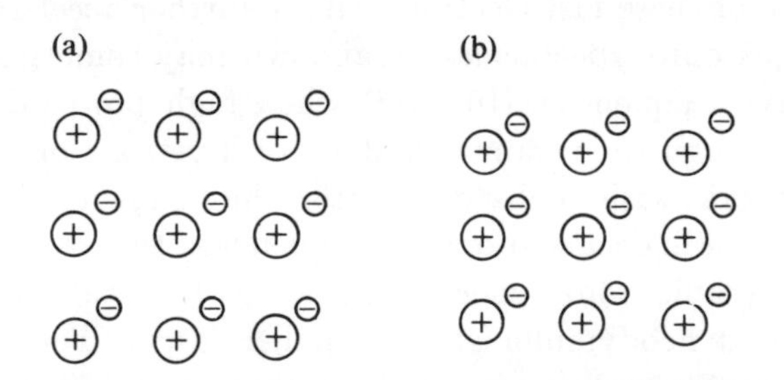

FIG. 10.12. Schematic representation of a crystal that lacks a centre of symmetry; (a) in the absence of applied force, (b) in the presence of applied force.

† Derived from the Greek word meaning 'to press'.

produced. For small deformations and small electric fields the relationships are linear and may be described by the two equations

$$T = cS - e\mathscr{E} \tag{10.34}$$

and

$$D = \epsilon\mathscr{E} + eS, \tag{10.35}$$

where T = stress, S = strain, c = elastic constant, e = piezoelectric constant. With $e = 0$ the above equations reduce to Hooke's law and to the $D = \epsilon\mathscr{E}$ relationship respectively. It may be seen that when $\mathscr{E} = 0$ the electric flux density, D, may be finite in spite of zero electric field. Similarly, an electric field sets up a strain without a mechanical stress being applied.

In general, all the constants in eqns. (10.34) and (10.35) are tensors. In the worst case there are 45 independent coefficients comprising 21 elastic constants, 6 dielectric constants, and 18 piezoelectric constants. In practical cases the situation is not so complicated because we would apply an electric field in some particular direction and would make use of the mechanical displacement in some other specified direction so that you can safely think in terms of those three scalar constants in the above equations.

It follows from the properties of piezoelectrics that they are ideally suited to play the role of electromechanical transducers. Common examples are the microphone, where longitudinal sound vibrations in the air are the mechanical driving force, and the gramophone pick-up, which converts into electrical signals the mechanical wobbles in the groove of a record. For these applications Rochelle salt has been used. More recently ceramics of the barium titanate type, particularly lead titanate, are finding application. They have greater chemical stability than Rochelle salt but suffer from larger temperature variations.

A very important application of piezoelectricity is the quartz (SiO_2) stabilized oscillator, used to keep radio stations on the right wavelength, with an accuracy of about one part in 10^8 or even 10^9. The principle of operation is very simple. A cuboid of quartz (or any other material for that matter) will have a series of mechanical resonant frequencies of vibration whenever its mechanical length L is an odd number of half wavelengths. Thus the lowest mode will be when $L = \lambda/2$. The mechanical disturbance will travel in the crystal with the velocity of sound, which we shall call v_s. Hence the frequency of mechanical oscillation will be $f = v_s/\lambda = v_s/2L$. If the ends of the crystal are metallized, it forms a capacitor that can be put in a resonant electrical circuit having the same resonant frequency, f. The resonant frequency of a valve or transistor oscillator circuit depends a little on things outside the inductor and capacitor of the resonant circuit. Usually these are small effects that can be ignored, but if you want an oscillator that is stable in frequency to one part in 10^8, things like gain variation in the amplifier caused by supply voltage changes or ageing of components become important; on this scale they are virtually uncontrollable. This is where the mechanical oscillation comes in. We have seen it is a function only of the crystal dimension. Provided the electrical frequency is nearly the same, the electrical circuit will set up mechanical as well as electrical oscillations, linked by the piezoelectric behaviour of quartz. The mechanical oscillations will dominate the frequency that the whole system takes up, simply because the amplifier part of the oscillator circuit works over a finite band of frequencies that is greater than the frequency band over which the mechanical oscillations can be driven. A circuit engineer would say that the 'Q' of the mechanical circuit is greater than that of the electrical circuit. The only problem in stabilizing the frequency of the quartz crystal-controlled oscillator is to keep its mechanical dimension, L, constant. This, of course, changes with temperature, so we just have to put the crystal in a thermostatically controlled box. This also allows for slight adjustment to the controlled frequency by changing the thermostat setting by a few degrees. Quartz-controlled master oscillators followed by stages of power amplification are used in all radio and television transmitters, from the most sophisticated, down to the humblest 'ham'.

The reverse effect is used in earphones and in a variety of transducers used to launch vibrations in liquids. These include the 'echo-sounder' used in underwater detection and ultrasonic washing and cleaning plant.

I should like to discuss in a little more detail another application, in which a piezoelectric material, cadmium sulphide (CdS) is used. The basic set-up is shown in Fig. 10.13. An input electric signal is transformed by an electromechanical transducer into acoustic vibrations that are propagated through the crystal and are converted back into electric signals by the second transducer. Assuming for simplicity that the transducers are perfect (convert all the electric energy into acoustic energy) and the acoustic wave suffers no losses, the gain of the device is unity (0 db if measured in decibels). If the crystal is illuminated, that is mobile charge carriers are created, the measured gain is found to decrease to B (Fig. 10.14). If further a variable d.c. voltage is applied

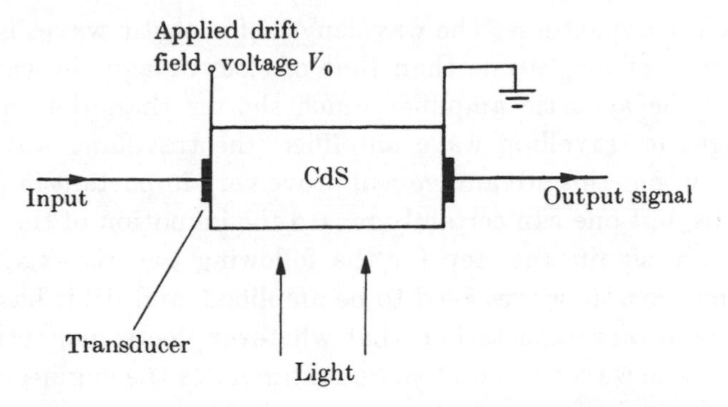

FIG. 10.13. General arrangement of an acoustic amplifier. The applied electric field causes the electrons (photoelectrically generated by the light) to interact with the acoustic wave in the crystal.

across the CdS crystal, the gain varies as shown.

These experimental results may be explained in the following way. The input light creates charge carriers that interact with the acoustic waves via the piezoelectric effect. If the carriers move slower than the acoustic wave there is a transfer of power from the acoustic wave to the charge carriers. If, on the other hand, the charge carriers move faster than the acoustic wave, the power transfer takes place from the carriers to the acoustic wave, or in other words the acoustic wave is amplified.

Hence there is the possibility of building an electric amplifier relying on the good services of the acoustic waves. Since we can make electric amplifiers without using acoustic waves, there is not much point using this acoustic amplifier unless it has some other advantages. The main

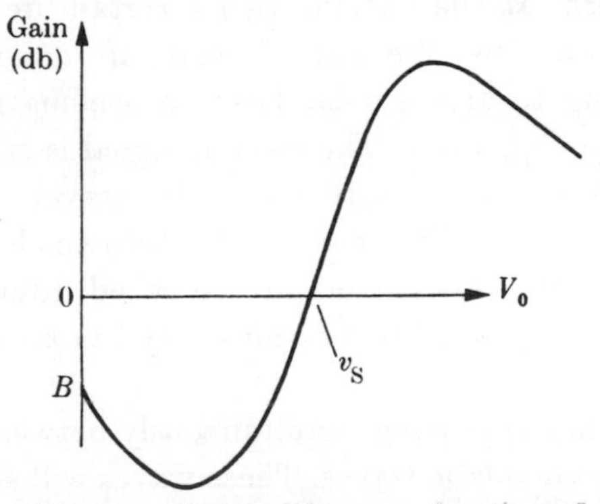

FIG. 10.14. The gain of an acoustic amplifier as a function of the applied voltage. At the voltage where the electron drift velocity is equal to the sound velocity (v_s) the gain changes from negative to positive.

advantage is compactness. The wavelength of acoustic waves is smaller by five orders of magnitude than that of electromagnetic waves, and this makes the acoustic amplifier much shorter than the equivalent electromagnetic travelling wave amplifier (the travelling wave tube). It is unlikely that this advantage will prove very important in practical applications, but one can certainly regard the invention of the acoustic amplifier as a significant step for the following two reasons: (i) it is useful when acoustic waves need to be amplified, and (ii) it has created a feeling (or rather expectation) that whatever electromagnetic waves can do, acoustic waves can do too. So it stimulates the engineers' brains in search of new devices and a host of new devices may one day appear.

Since the first edition of this book was published, a new device of considerable importance has, in fact, evolved from these ideas. It is the Surface Acoustic Wave device (acronym SAW). We are all familiar with the ripples and surface waves that form on the surfaces of disturbed liquids. It was shown long ago by Lord Rayleigh that analogous waves rippled across the surface of solids, travelling with a velocity near to the velocity of sound in the bulk material.

Why do we want to use surface waves? There are two reasons: (1) they are there on the surface so we can easily interfere with them, and (2) the interfering structures may be produced by the same techniques as for integrated circuits. Why do we want to interfere with the waves? Because these devices are used for signal processing which, broadly speaking, requires that the signal which arrives during a certain time interval at the input should be available some time later in some other form at the output.

Let us first discuss the simplest of these devices which will produce the same waveform at the output with a certain delay. The need for such a device is obvious. We want to compare a signal with another one which is going to arrive a bit later. A schematic drawing of the device is shown in Fig. 10.15. The electric signal is transformed by the input transducers into an acoustic one which travels slowly to the output transducer where it is duly reconverted into an electric signal again. The input and output transducers are so called interdigital lines. The principles of this transducer's operations may be explained with the aid of Fig. 10.16.

The electric signal appearing simultaneously between fingers 1 and 2, and 2 and 3 excites acoustic waves. These waves will add up in a certain phase depending on the distance d between the fingers. If d is half a wavelength (wavelength and frequency are now related by the acoustic

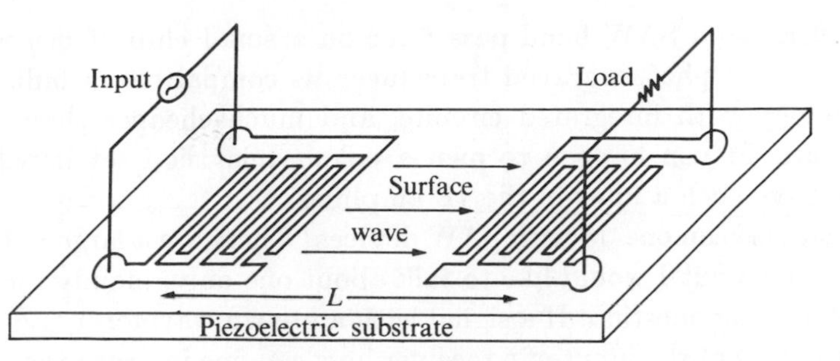

FIG. 10.15. Schematic representation of a surface wave device with two interdigital transducers.

velocity!) and considering that the electric signals are in opposite phase (one is plus-minus, the other one is minus-plus) then the effect from each finger pair adds up and we have maximum transmission. At twice the frequency, where d is a whole wavelength, the contribution from neighbouring finger pairs will be opposite and the transducer will produce no net acoustic wave. This shows that the delay line cannot be a very broad-band device but it shows in addition that by having a frequency-dependent output we might be able to build a filter. The parameters at our disposal are the length of overlap between fingers (l on Fig. 10.16) and the relative position of the fingers, the former controlling the strength of coupling and the latter the relative phase. It turns out that excellent filters can be produced which are sturdy, cheap, and better than anything else available in the MHz region. In the stone-age days of radio, when I was a young man, such a band-pass filter was accomplished by a series of transformers tuned by capacitors. These would appear ridiculously bulky compared to today's microelectronic

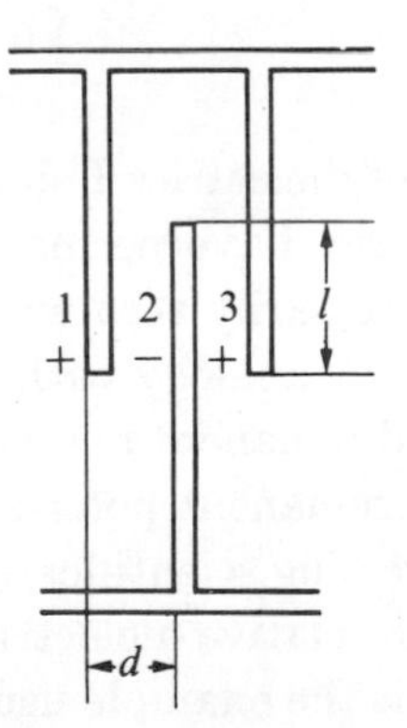

FIG. 10.16. A section of the interdigital transducer.

amplifiers, so a SAW band-pass filter on a small chip of piezoelectric ceramic with photoengraved transducers is compatible in bulk and in technology with integrated circuits, and much cheaper than the old hardware. If you happen to own a colour television set it will most likely have such a filter in the IF amplifier.

What else can one do with SAW devices? There are a large number of applications but I would like to talk about one only, mainly for posing an interesting question. If a signal with a known waveform and a lot of noise arrives at the input of a receiver how can one improve the chances of detecting the signal? The answer is that a device exists which will so transform the signal as to make it *optimally* distinguishable from noise. The device is called, rather inappropriately, a matched filter. It turns out that SAW devices are particularly suitable for their realization. They are vital parts of certain radar systems.

I shall finish this section by mentioning *electrostriction*, a close relative of piezoelectricity. It is also concerned with mechanical deformation caused by an applied electric field but it is not a linear phenomenon. The mechanical deformation is proportional to the square of the electric field and the relationship applies to all crystals whether symmetric or not. It has no inverse effect. The mechanical strain does not produce an electric field via electrostriction. *Biased* electrostriction is however very similar to piezoelectricity. If we apply a large d.c. electric field, $\mathscr{E}_0$, and a small a.c. electric field, $\mathscr{E}_1$, then the relationship is

$$S = \gamma(\mathscr{E}_0+\mathscr{E}_1)^2 \cong \gamma\mathscr{E}_0^2+2\gamma\mathscr{E}_0\mathscr{E}_1,$$

(where γ is a proportionality factor) and we find that the a.c. strain is linearly proportional to the amplitude of the a.c. field.

10.12. Ferroelectrics

There is one more class of dielectrics I should like to mention, which, as well as being piezoelectric, have permanent dipole moments and a polarization that is not necessarily zero when there is no electric field. In fact they get their name by analogy with ferromagnetics, which have a B–H loop, hysteresis, and remanent magnetism. Ferroelectrics have a P–E loop, hysteresis, and remanent polarized charge as shown in Fig. 10.17. They are very interesting scientifically but so far have not found much application. The high relative dielectric constants of the titanates ($BaTiO_3$ is one of them; it is the example usually given of a ferroelectric material) are used in capacitor-making where the essential ferroelectric effects of voltage and temperature changes of capacity are usually an embarrassment. At one time it seemed as if the voltage change of dielectric constant would find application in voltage-tuneable capacitors but varactor diodes won the race. Another potential application (as an externally tuneable phase-shifter) has not materialized either, because of the high hysteresis losses at high frequencies. Both these situations could be changed by improved materials.

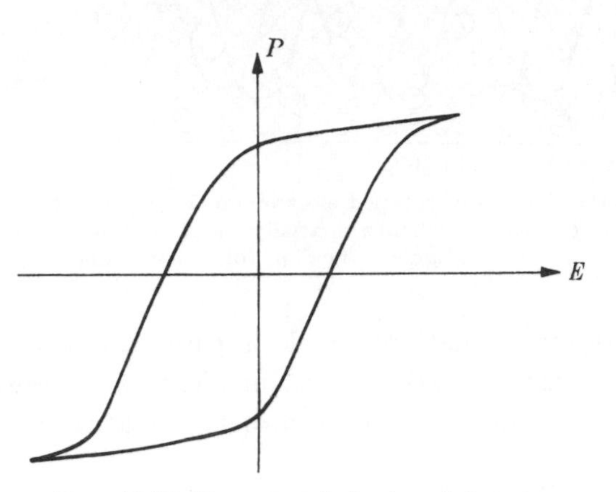

FIG. 10.17. Ferroelectric hysteresis loop.

An application which might be round the corner is concerned with a fairly new subject called volume holography. I shall describe holography later among laser applications; let it suffice for the moment that we want to inscribe an interference pattern in a material in the form of dielectric constant variation. The most extensively used ferroelectric is Fe-doped $LiNbO_3$. The purpose of the Fe doping is to introduce electron traps into the material. Initially, some of the traps are empty, some others are occupied in a fairly uniform manner as shown in Fig. 10.18a. Assume now that a sinusoidal light interference pattern is incident. As a consequence some of the electrons sitting in the traps will be excited into the conduction band, or in other words will be free to move around. Their motion will be governed partly by diffusion and partly by the magnitude of the electric field reigning at that particular point. The larger the electric field (i.e. the light intensity) the farther will the electrons travel. Hence when the light is removed and the electrons get trapped again their spatial distribution will correspond to the interference pattern slightly shifted as shown in Fig. 10.18b. A non-uniform charge distribution implies an electric field distribution

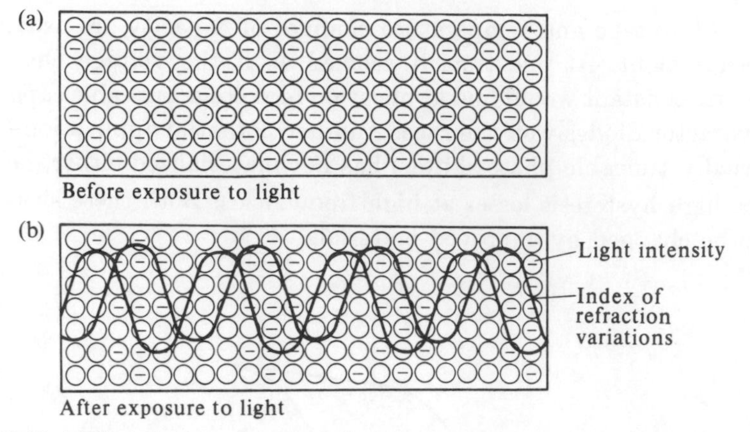

FIG. 10.18. (a) Distribution of trapped electrons in Fe-doped $LiNbO_3$. (b) Redistribution of the trapped electrons when a spatially varying electromagnetic wave (to be specific, light) is incident for a short while.

(remember Poisson's equation) which, in turn, causes a small change in the dielectric constant. Hence the incident interference pattern is stored in the material as variation in dielectric constant.

10.13 Optical fibres

I have tried to show that dielectric properties have importance in optics as well as at the more conventional electrical engineering frequencies. That there are no sacred boundaries in the electromagnetic spectrum is shown very clearly by a recent development in communication engineering. This involves the transmission (guiding) of electromagnetic waves for a long distance. In principle it can be done at any frequency but only in the visible region do *dielectric waveguides* have distinct advantages. The particular configuration used is a fibre of rather small diameter (say 5–50 μm) made of glass or silica. Whether this transmission line is practical or not will clearly depend on the attenuation. Have we got the formula for the attenuation of a dielectric waveguide? No, we have not performed that specific calculation but we do have a formula for the propagation coefficient of a plane wave in a lossy medium and that gives a sufficiently good approximation.

Recall eqn. (1.38)

$$k = (\omega^2\mu\epsilon + \mathrm{i}\omega\mu\sigma)^{\frac{1}{2}}, \tag{10.36}$$

and assuming this time that

$$\omega\epsilon \gg \sigma, \tag{10.37}$$

we get the attenuation coefficient

$$\alpha = k_{\text{imag}} = \tfrac{1}{2}\frac{\omega\sqrt{\epsilon'}}{c}\frac{\sigma}{\omega\epsilon} = \tfrac{1}{2}\frac{\omega\sqrt{\epsilon'}}{c}\tan\delta. \tag{10.38}$$

The usual measure is the attenuation in decibels for a length of one kilometre, which may be expressed from eqn. (10.38) as follows

$$A = 20\log_{10}\exp(1000\,\alpha) = 8680\,\alpha = 4340\,\frac{\omega\sqrt{\epsilon'}}{c}\tan\delta\ \text{db/km}. \tag{10.39}$$

For optical communications to become feasible, A should not exceed 20 db/km, first pointed out by Kao and Hockham in 1966. Taking an operational frequency of $f = 3.10^{14}$ Hz, a typical dielectric constant $\epsilon' = 2{\cdot}25$, and the best material available at the time with $\tan\delta \approx 10^{-7}$, we get

$$A \approx 4.10^{3}\ \text{db/km}, \tag{10.40}$$

a far cry from 20. No doubt materials can be improved, but an improvement in $\tan\delta$ of more than 2 orders of magnitude looked at the time somewhat beyond the realm of practical possibilities. Nevertheless, the work began and Fig. 10.19 shows the improvement achieved. The

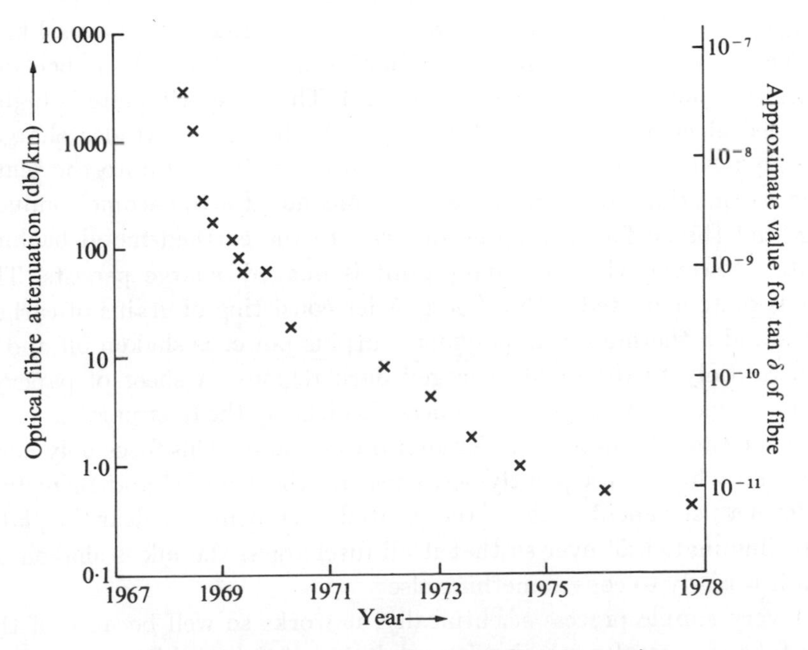

FIG. 10.19. Improvement in optical fibre attenuation.

critical 20 db was reached at the end of 1969, and by 1978 the figure was down below 0·5 db/km, an amazing performance. It is quite likely that at the right frequency using the right materials the attenuation may be further decreased by two or three orders of magnitude. Optical communications are here to stay. Give another decade or so and all major cities will be linked by optical fibres. This is obviously what Puck had in mind when he promised to 'put a girdle about the earth'.

If you ever encounter a problem that appears to be too daunting, remember the story of the optical fibres. It is an excellent illustration of the American maxim (born in the optimism of the post-war years) that the impossible takes a little longer.

10.14 The Xerox process

This great development of the past decade or so enables the production of high quality reproductions of documents quickly and easily. This has in turn made decision making more democratic, bureaucrats more powerful, and caused vast forests of trees to disappear to provide the extra paper consumed. Scientifically the principles are simple. The heart of the Xerox machine is a plate made of a thin layer of amorphous semiconductor on a metal plate. The semi-conductor is a compound of As, Se, and Te. It is almost an insulator so that it behaves like a dielectric, but it is also photo-conductive, i.e. in the light it becomes more conducting (remember section 8.6). The dielectric plate is highly charged electrostatically by brushing it with wire electrodes charged to about 30 kV. The document to be copied is imaged onto the plate. The regions that are white cause the semiconductor to become conducting, and the surface charge leaks away to the earthed metal backing plate. However where the dark print is imaged, charge persists. The whole plate is dusted with a fine powder consisting of grains of carbon silica and a thermosetting polymer. Surplus power is shaken off and it adheres only to the highly charged dark regions. A sheet of paper is then pressed onto the plate by rollers. It picks up the dust particles, and is then treated by passing under an infra-red lamp. This fuses polymeric particles which subsequently set, encasing the black C and SiO_2 dust to form a permanent image of the printed document. To clear the plate, it is illuminated all over so that it all discharges, the ink is shaken off and it is ready to copy something else.

A very simple process scientifically, it works so well because of the careful and very clever technological design of the machine.

Examples.

1. Sketch qualitatively how you would expect the permittivity and loss tangent to vary with frequency in those parts of the spectrum that illustrate the essential properties, limitations, and applications of the following materials; window glass, water, transformer oil, polythene, and alumina.

2. What is the atomic polarizability of argon? Its susceptibility at 273 K and 1 atmosphere is $4{\cdot}35\ 10^{-4}$.

3. A long narrow rod has an atomic density $5.10^{28}/\text{m}^3$. Each atom has a polarizability of 10^{-40} farad m^2. Find the internal electric field when an axial field of 1 V/m is applied.

4. The tables† show measured values of dielectric loss for thoria (ThO_2) containing a small quantity of calcium. For this material the static and high frequency permittivities have been found from other measurements to be

$$\epsilon_s = 19{\cdot}2\epsilon_0, \qquad \epsilon_\infty = 16{\cdot}2\epsilon_0.$$

Assume that oriental polarization is responsible for the variation of $\tan\delta$. Use the Debye equations to show that by expressing the characteristic relaxation time as

$$\tau = \tau_0 \exp(H/kT)$$

(where τ_0 and H are constants) both of the experimental curves can be approximated. Find τ_0 and H.

If a steady electric field is applied to thoria at 500 K, then suddenly removed, indicate how the electric flux density will change with time.

$f = 695$ Hz		$f = 6950$ Hz	
T K	$\tan\delta$	T K	$\tan\delta$
555	0·023	631	0·026
543	0·042	621	0·036
532	0·070	612	0·043
524	0·086	604	0·055
516	0·092	590	0·073
509	0·086	581	0·086
503	0·081	568	0·086
494	0·063	543	0·055
485	0·042	518	0·025
475	0·029	498	0·010

5. Discuss the limitations of simple theories of breakdown in several non-

† Data taken from Ph.D. thesis of J. Wachtman, University of Maryland, 1962, quoted in *Physics of Solids*, Wert and Thomson, (McGraw-Hill), 1964.

laboratory situations of practical importance.

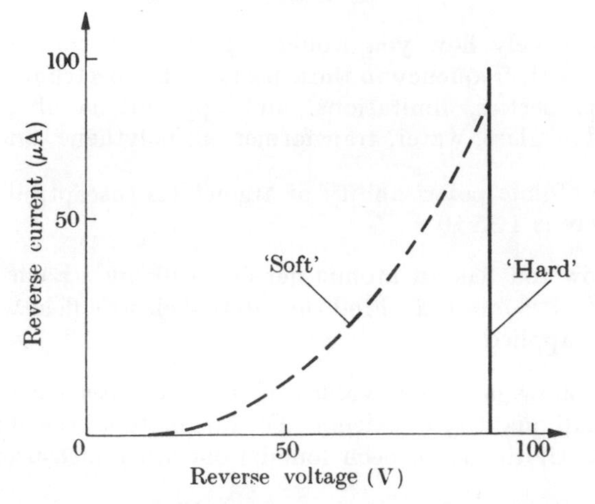

6. A capacitor is to be made from a dielectric having a breakdown field strength $\mathscr{E}_b$ and a relative permittivity ϵ_r. The electrodes are metal plates fixed to the sides of a slab (thickness 0·5 mm) of the dielectric. Owing to a slight distortion of one of the plates, one third of its area is separated from the dielectric by an air-filled gap of thickness 1 μm. The remaining two-thirds of the plate and the whole of the second plate are in intimate contact with the dielectric. The breakdown field $\mathscr{E}_b$ is 2·0 MV/m for the dielectric and 3·0 MV/m for air. ϵ_r is 1000. Discuss the effect of the gap (compared with a gap-free capacitor), on (a) the capacitance and (b) the breakdown voltage.

7. The figure shows two types of breakdown that can occur in the reverse characteristic of a p-n junction diode. The 'hard' characteristic is the desired avalanche breakdown discussed in Chapter 9. The 'soft' characteristic is a fault that sometimes develops with disastrous effect on the rectification efficiency. It has been suggested that this is due to precipitates of metals such as copper or iron in the silicon, leading to local breakdown in high field regions. (Goetzberger and Shockley (1960). 'Metal precipitates in p-n junctions', *J. appl. Phys.*, **31**, 1821.) Discuss briefly and qualitatively this phenomenon in terms of the simple theories of breakdown given in this chapter.

8. Derive an expression for the gain of the piezoelectric ultrasonic amplifier (Hutson, McFee, and White (1961), *Phys. Rev. Lett*, **7**, 237).

(Hint: In the one-dimensional case we can work in terms of scalar quantities. Our variables are: ε, D, T, S, J, N_e. The equations available are: eqns. (10.34) and (10.35) for the relationship between the mechanical and electrical quantities, the equation for the electron current including both a conduction and a diffusion term (given in example 9.2), the continuity equation for electrons, and one of Maxwell's equations relating D to N_e. Altogether there are five equations and six variables. The missing equation is the one relating strain to stress for an acoustic wave. It is of the form

$$\frac{\partial^2 T}{\partial z^2} = \rho_m \frac{\partial^2 S}{\partial t^2},$$

where ρ_m is the density of the piezoelectric material.

The gain may be derived in a manner analogous to that adopted in Chapter 1 for the derivation of the dispersion relations for electromagnetic and plasma waves. The steps are as follows:

(i) Assume that the a.c. quantities are small in comparison with the d.c. quantities (e.g. the a.c. electric field is much smaller than the applied d.c. electric field) and neglect the products of a.c. quantities.

(ii) Assume that the a.c. quantities vary as $\exp(-i(\omega t - kz))$ and reduce the linear differential equation system to a set of linear equations.

(iii) Derive the dispersion equation from the condition that the linear equation system must have a non-trivial solution.

(iv) Substitute $k = \omega/v_s + \delta$ (where $v_s = (c/\rho_m)^{\frac{1}{2}}$ is the velocity of sound in the medium) into the dispersion equation and neglect the higher powers of δ.

(v) Calculate the imaginary part of δ which will determine (by its sign) the growing or attenuating character of the wave. Show that there is gain for $v_0 > v_s$ where v_0 is the average velocity of the electrons.)

9. Using the data for KCl given in question 4 of Chapter 5 estimate a frequency at which the relative permittivity of KCl will relax to a lower value.

11. Magnetic materials

Quod superest, agere incipiam quo foedere fiat
naturae, lapis hic ut ferrum ducere possit,
quem Magneta vocant partio de nomine Grai,
Magnetum quia fit patriis in finibus ortus
hunc homines lapidem mirantur; ...
....
Hoc genus in rebus firmandumst multa prius quam
ipsius rei rationem reddere possis,
et nimium longis ambagibus est adeundum;

LUCRETIUS *De Rerum Natura*

To pass on, I will begin to discuss by what law of nature it comes about that iron can be attracted by that stone which the Greeks call magnet from the name of its home, because it is found within the national boundaries of the Magnetes. This stone astonishes men ....
In matters of this sort many principles have to be established before you can give a reason for the thing itself, and you must approach by exceedingly long and roundabout ways:

11.1. Introduction

THERE are some curious paradoxes in the story of magnetism that make the topic of considerable interest. On the one hand the lodestone was one of the earliest known applications of science to industry—the compass for shipping; and ferromagnetism is of even more crucial importance to industrial society today than it was to early navigators. On the other hand the origin of magnetism eluded explanation for a long time and the theory is still not able to account for all the experimental observations.

It is supposed that the Chinese used the compass around 2500 B.C. This may not be true, but it is quite certain that the power of lodestone to attract iron was known to Thales of Miletos in the sixth century B.C. The date is put back another two hundred years by William Gilbert (the man of science in the court of Queen Elizabeth the First), who wrote in 1600 that 'by good luck the smelters of iron or diggers of metal discovered magnetite as early as 800 B.C.' There is little doubt about the technological importance of ferromagnetism today. In the United Kingdom as much as 3×10^{10} watts of electricity are generated at times;

electrical power in this quantity would be hopelessly impractical without large quantities of expertly controlled ferromagnetic materials. Evidence for the statement that the theory is not fully understood may be obtained from any honest man who has done some work on the theory of magnetism.

11.2. Macroscopic approach

By analogy with our treatment of dielectrics I shall summarize here briefly the main concepts of magnetism used in electromagnetic theory. As you know, the presence of a magnetic material will enhance the magnetic flux. Thus the relationship

$$\mathbf{B} = \mu_0 \mathbf{H}, \tag{11.1}$$

valid in a vacuum, is modified to

$$\mathbf{B} = \mu_0(\mathbf{H}+\mathbf{M}) \tag{11.2}$$

in a magnetic material, where $\mathbf{M}$ is the magnetic dipole moment per unit volume, or shortly, magnetization. It is related to the magnetic field by the relationship

$$\mathbf{M} = \chi_{\mathrm{m}}\mathbf{H},$$

where χ_{m} is the magnetic susceptibility. Substituting the above equation into eqn. (11.2) we get

$$\begin{aligned}\mathbf{B} &= \mu_0(1+\chi_{\mathrm{m}})\mathbf{H} \\ &= \mu_0\mu_{\mathrm{r}}\mathbf{H},\end{aligned} \tag{11.3}$$

where μ_{r} is called the *relative permeability.*

11.3. Microscopic theory (phenomenological)

Our aim here is to express the macroscopic quantity, M, in terms of the properties of the material at atomic level. Is there any mechanism at atomic level that could cause magnetism? Reverting for the moment to the classical picture, we can say yes. If we imagine the atoms as systems of electrons orbiting round protons they can certainly give rise to magnetism. We know this from electromagnetic theory, which maintains that an electric current, I, going round in a plane will produce a magnetic moment†

$$\mu_{\mathrm{m}} = IS, \tag{11.4}$$

† It is an unfortunate fact that the usual notation is μ both for the permeability and for the magnetic moment. I hope that, by using the subscripts 0 and r for permeability and m for magnetic moment, the two things will not be confused.

where S is the area of the current loop. If the current is caused by a single electron rotating with an angular frequency ω_0, then the current is $e\omega_0/2\pi$ and the magnetic moment becomes

$$\mu_m = \frac{e\omega_0}{2} r^2, \tag{11.5}$$

where r is the radius of the circle. Introducing now the angular momentum

$$\Pi = mr^2\omega_0 \tag{11.6}$$

we may rewrite eqn. (11.5) in the form

$$\mu_m = \frac{e}{2m} \Pi. \tag{11.7}$$

Remember that the charge of the electron is negative; the magnetic moment is thus in a direction opposite to the angular momentum.

We now ask what happens when an applied magnetic field is present. Consider a magnetic dipole that happens to be at an angle θ to the direction of the magnetic field (Fig. 11.1). The magnetic field produces a

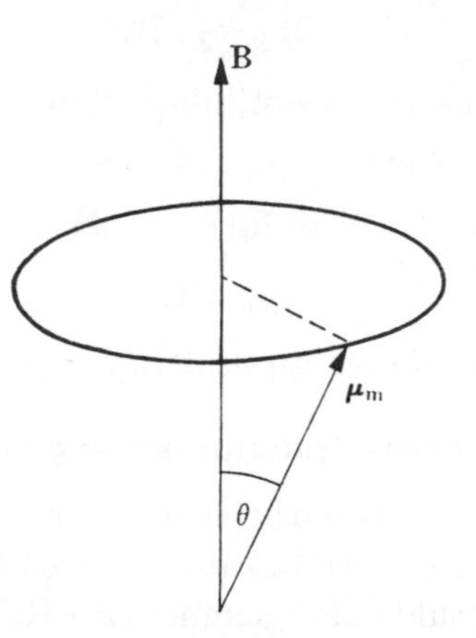

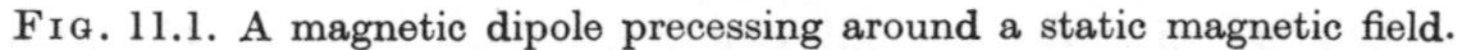

FIG. 11.1. A magnetic dipole precessing around a static magnetic field.

torque $\boldsymbol{\mu}_m \times \mathbf{B}$ on the magnetic dipole. Since the torque is perpendicular both to $\boldsymbol{\mu}_m$ and $\mathbf{B}$, the change in the angular momentum will also be perpendicular, causing the magnetic dipole to precess around the magnetic field. From kinetics we can easily show that the frequency of precession is

$$\omega_L = \frac{eB}{2m}, \tag{11.8}$$

which is usually called the *Larmor frequency*. If the magnetic dipole precesses, some electric charge must go round. So we could use eqn. (11.5) to calculate the magnetic moment due to the precessing charge. Replacing ω_0 by ω_L we get

$$(\mu_m)_{ind} = Br^2e^2/4m, \tag{11.9}$$

where r is now the radius of the precessing orbit. The sign of this induced magnetic moment can be deduced by remembering Lenz's law. It must oppose the magnetic field responsible for its existence.

We are now in a position to obtain M from the preceding microscopic considerations. If there are N_a atoms per unit volume and each atom contains Z electrons, the total induced magnetic dipole moment is

$$M = N_a Z(\mu_m)_{ind}. \tag{11.10}$$

Hence the magnetic susceptibility is

$$\chi_m = \frac{M}{H} = -\frac{N_a Z e^2 r^2 \mu_0}{4m}. \tag{11.11}$$

χ_m given by the above equation is a small number, rarely exceeding 10^{-5}, but the remarkable thing is that it is negative. This is in marked contrast with the analogous case of electric dipoles, which invariably give a positive contribution.† The reason for this is that the electric dipoles line up whereas the magnetic dipoles precess in a field. Magnetic dipoles can line up as well. The angle of precession will stay constant in the absence of losses but not otherwise. In the presence of some loss mechanism the angle of precession gradually becomes smaller and smaller as the magnetic dipoles lose energy; in other words, the magnetic dipoles do line up. They will not align completely because they occasionally receive some energy from thermal vibrations that frustrates their attempt to line up. This is exactly the same argument we used for dielectrics and we can therefore use the same mathematical solution. Replacing the electric energy in eqn. (10.12) by the magnetic energy, we get the average magnetic moment in the form

$$\langle\mu_m\rangle = \mu_m L(a), \qquad a = \frac{\mu_m \mu_0}{kT} H. \tag{11.12}$$

Denoting by N_m the number of magnetic dipoles per unit volume we get for the total magnetic moment

$$M = N_m \langle\mu_m\rangle. \tag{11.13}$$

† The electric susceptibility can also be negative but that is caused by free charges and not by electric dipoles.

At normal temperatures and reasonable magnetic fields, $a \ll 1$ and eqn. (11.13) may be expanded to give

$$M = N_{\mathrm{m}}\mu_{\mathrm{m}}^2\mu_0 H/3kT, \tag{11.14}$$

leading to

$$\chi_{\mathrm{m}} = N_{\mathrm{m}}\mu_{\mathrm{m}}^2\mu_0/3kT, \tag{11.15}$$

which is definitely positive and varies inversely with temperature.

At the other extreme of very low temperatures all the magnetic dipoles line up; this can be seen mathematically from the fact that the function $L(a)$ (plotted in Fig. 11.2) tends to unity for large values of a.

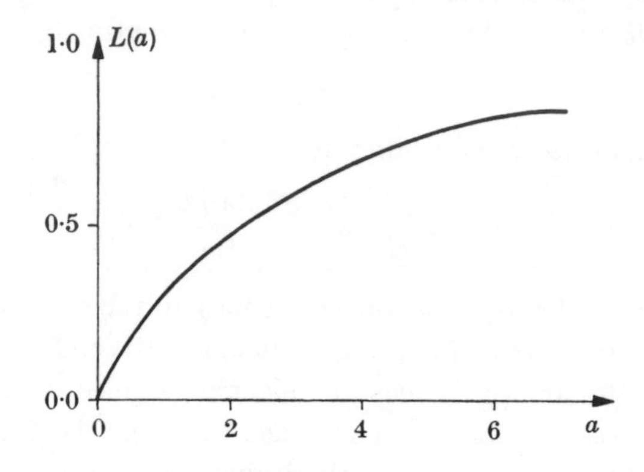

FIG. 11.2. The Langevin function, $L(a)$.

The total magnetic moment is then

$$M_{\mathrm{s}} = N_{\mathrm{m}}\mu_{\mathrm{m}}, \tag{11.16}$$

which is called the *saturation magnetization* because this is the maximum contribution the material can provide.

We have now briefly discussed two distinct cases: (i) when the induced magnetic moment opposes the magnetic field, called *diamagnetism*; and (ii) when the aligned magnetic moments strengthen the magnetic field, called *paramagnetism*. Both phenomena give rise to small magnetic effects that are of little use when the aim is the production of high magnetic fluxes. What about our most important magnetic material, iron? Can we explain its properties with the aid of our model? Not in its present state. We have to modify our model by introducing the concept of the internal field. This is really the same sort of thing that we did with dielectrics. We said then that the local electric field differs from the applied electric field because of the presence of the electric dipoles in the material. We may argue now that in a magnetic material the local magnetic field is the sum of the applied magnetic field and the internal magnetic field, and we may assume (as Pierre Weiss did in 1907) that this internal field is proportional to the magnetization, that is

$$H_{\text{int}} = \lambda M, \tag{11.17}$$

where λ is called the *Weiss constant*.

Using this newly introduced concept of the internal field, we may replace H in eqn. (11.12) by $H+\lambda M$ to obtain for the magnetization:

$$\frac{M}{N_{\text{m}}\mu_{\text{m}}} = L\left\{\frac{\mu_{\text{m}}\mu_0}{kT}(H+\lambda M)\right\}. \tag{11.18}$$

Thus for any given value of H we need to solve eqn. (11.18) to get the corresponding magnetization. It is interesting to note that eqn. (11.18) still has solutions when $H = 0$. To prove this, let us introduce the notations

$$b = \mu_{\text{m}}\mu_0\lambda M/kT, \qquad \theta = N_{\text{m}}\mu_{\text{m}}^2\mu_0\lambda/3k, \tag{11.19}$$

and

$$\frac{M}{N_{\text{m}}\mu_{\text{m}}} = \frac{T\mu_{\text{m}}\mu_0\lambda M/kT}{3N_{\text{m}}\mu_{\text{m}}^2\mu_0\lambda/3k} = \frac{Tb}{3\theta} \tag{11.20}$$

We can now rewrite eqn. (11.18), for the case $H = 0$, in the form

$$\frac{T}{3\theta}b = L(b). \tag{11.21}$$

Plotting both sides of the above equation (Fig. 11.3) it becomes

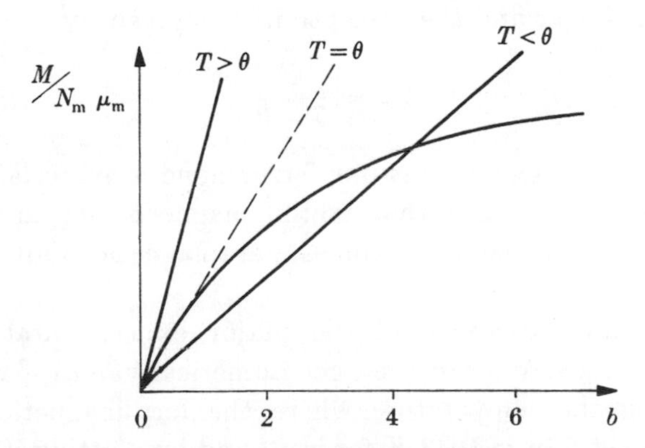

Fig. 11.3. Graphical solution of eqn. (11.21). There is no solution, that is the curves do not intersect each other, for $T > \theta$.

apparent that there is a solution when $T < \theta$ but no solution when $T > \theta$.

What does it mean if there is a solution? It means that M may be finite for $H = 0$; that is, the material can be magnetized in the absence of an external magnetic field. This is a remarkable conclusion. We have managed to explain with a relatively simple model the properties of permanent magnets.

Note that there is a solution only below a certain temperature. Thus, if our theory is correct, permanent magnets should lose their magnetism above this temperature. Is this borne out by experiment? Yes, it is a well-known experimental fact (discovered by Gilbert) that permanent magnets cease to function above a certain temperature. What happens when $T > \theta$? There is no magnetization for $H = 0$ but we do get some magnetization for finite H.

The mathematical solution may be obtained from eqn. (11.18), noting that the argument of the Langevin function is small, and we may use again the approximation $L(a) \cong a/3$, leading to

$$\frac{M}{N_{\mathrm{m}}\mu_{\mathrm{m}}} = \frac{\mu_m\mu_0}{3kT}(H+\lambda M), \tag{11.22}$$

which may be solved for M to give

$$M = \frac{N_{\mathrm{m}}\mu_{\mathrm{m}}^2\mu_0/3k}{T-N_{\mathrm{m}}\mu_{\mathrm{m}}^2\mu_0\lambda/3k}H = \frac{C}{T-\theta}H, \tag{11.23}$$

where

$$C = N_{\mathrm{m}}\mu_{\mathrm{m}}^2\mu_0/3k \tag{11.24}$$

is called the *Curie constant* and θ the *Curie temperature*. The M–H relationship is linear and the susceptibility is given by

$$\chi_{\mathrm{m}} = \frac{C}{T-\theta} \tag{11.25}$$

Thus we may conclude that our ferromagnetic material (the name given to materials like iron that exhibit magnetization in the absence of applied magnetic fields) becomes paramagnetic above the Curie temperature.

We have now explained all the major experimental results on magnetic materials. We can even get numerical values if we want to. By measuring the temperature where the ferromagnetic properties disappear we get θ (it is 1043 K for iron) and by plotting χ_{m} above the Curie temperature as a function of $1/(T-\theta)$ we get C (it is about unity for iron). With the aid of eqns. (11.19) and (11.24) we can now express the unknowns μ_m and λ as follows:

$$\lambda = \frac{\theta}{C} \quad \text{and} \quad \mu_m = \left(\frac{3kC}{N_m \mu_0}\right)^{\frac{1}{2}}. \tag{11.26}$$

Taking $N_m = 8\ 10^{28}/\text{m}^3$, and the above-mentioned experimental results, we get for iron

$$\lambda \cong 1000 \quad \text{and} \quad \mu_m \cong 2\ 10^{-23}\text{A m}^2. \tag{11.27}$$

The value for the magnetic dipole moment of an atom seems reasonable. It would be produced by an electron going round a circle of 0·1 nm radius about 10^{15} times per second. One can imagine that, but it is much harder to swallow a numerical value of 1000 for λ. It means that the internal field is 1000 times as large as the magnetization. When all the magnetic dipoles line up, M comes to about 10^6 A/m, leading to a value for the internal flux density

$$B_{\text{int}} = \mu_0 \lambda M \cong 10^3 \ \text{weber/m}^2, \tag{11.28}$$

which is about two orders of magnitude higher than the highest flux density ever produced. Where does such an enormous field come from? It is a mysterious problem, and we shall leave it at that for the time being.

11.4. Domains and the hysteresis curve

We have managed to explain the spontaneous magnetization of iron but as a matter of fact freshly smelted iron does not act as a magnet. How is this possible? If below the Curie temperature all the magnetic moments line up spontaneously, how can the outcome be a material exhibiting no external magnetic field? Weiss, with remarkable foresight, postulated the existence of a domain structure. The magnetic moments do line up within a domain but the magnetizations of the various domains are randomly oriented relative to each other, leading to zero net magnetism.

The three most important questions we need to answer are as follows:

(i) Why does a domain structure exist at all?

(ii) How thick are the domain walls?

(iii) How will the domain structure disappear as the magnetic field increases?

It is relatively easy to answer the first question. The domain structure comes about because it is energetically unfavourable for all the magnetic moments to line up in one direction. If it were so then, as shown in Fig. 11.4a, there would be large magnetic fields and consequently a large amount of magnetic energy outside the material. This magnetic energy would be reduced if the material would break up into domains as shown in Fig. 11.4b to e. But why would this process ever stop? Shouldn't the material break up into as many domains as it possibly could, down to a single atom? The reason why this would not happen is because domains must have boundaries and, as everyone knows, it is an expensive business to maintain borders of any kind.

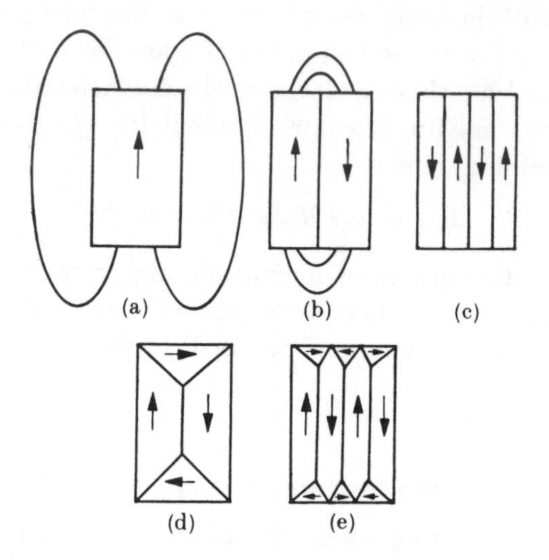

FIG. 11.4. The formation of domains (from C. Kittel, Introduction to Solid State Physics, John Wiley and Sons, New York).

Customs officials must be paid, not mentioning the cost of guard towers and barbed wires with which some borders are amply decorated. Thus some compromise is necessary. The more domains there are, the smaller will be the magnetic energy outside but the more energy will be needed to maintain the boundary walls. When putting up one more wall needs as much energy as the achieved reduction of energy outside, an equilibrium is reached, and the energy of the system is minimized.

We have now managed to provide a reasonable answer to question (i). It is much more difficult to describe the detailed properties of domains and their dependence on applied magnetic fields. We must approach the problem, as Lucretius said, 'by exceedingly long and roundabout ways'.

However much I dislike talking about crystal structure there is no escape now, because magnetic properties do depend on crystallographic directions. I am not suggesting that magnets are ever made of single-crystal materials, but in order to interpret some of the properties of ordinary polycrystalline magnets, we have to know the magnetic properties of the single crystals.

In Fig. 11.5 the magnetization curve (B against H) of iron is plotted for three different directions in the crystal. It may be seen that magnetization is relatively easy in the AB direction and harder in the AC and AG directions, or in other words, it is easier to magnetize iron along a cube edge than along a face or a body diagonal. This does not mean of course that all magnetic materials follow the same pattern. In nickel,

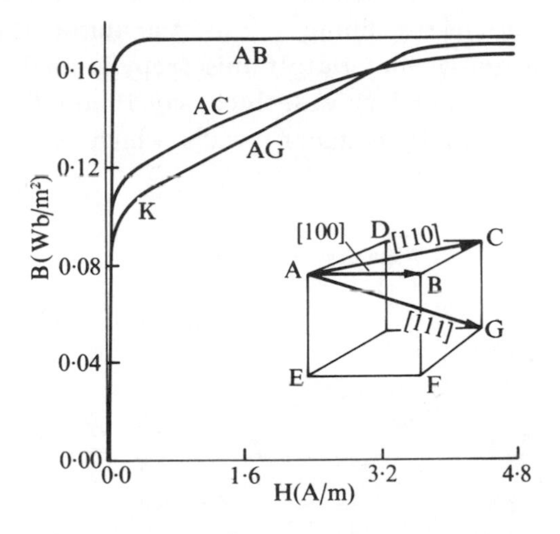

FIG. 11.5. Magnetization curve of single crystal iron in three different crystallographical directions.

another cubic crystal, the directions of easy and difficult magnetization are the other way round. What matters is that in most materials magnetization depends on crystallographic directions. The phenomenon is referred to as anisotropy, and the internal forces which bring about this property are called anisotropy forces.

Let us now see what happens at the boundary of two domains, and choose for simplicity two adjacent domains with opposite magnetizations as shown in Fig. 11.6. Note that the magnitude of the magnetic moments is unchanged during the transition but they rotate from an 'up' position into a 'down' position. Why is the transition gradual?

Because the forces responsible for lining up the magnetic moments (let's call them for the time being 'lining up' forces) try to keep them parallel. If we wanted a sudden change in the direction of the magnetic moments, we should have to do a lot of work against the 'lining up' forces and consequently there would be a lot of 'lining up' energy present. The anisotropy forces would act the opposite way. If 'up' is an easy direction (the large majority of domains may be expected to line up in an easy direction) then 'down' must also be an easy direction. Thus most of the directions in between must be looked upon unfavourably by the anisotropy forces. If we want to rotate the magnetic moments, a lot of work needs to be done against the anisotropy forces, resulting in large anisotropy energy. According to the foregoing argument the thickness of the boundary walls will be determined by the relative magnitudes of the 'lining up' and the anisotropy forces in the particular ferromagnetic material. If anisotropy is small, the transition will be slow and the boundary wall thick, say 10 μm. Conversely, large anisotropy forces lead to boundary walls which may be as thin as 0·3 μm.

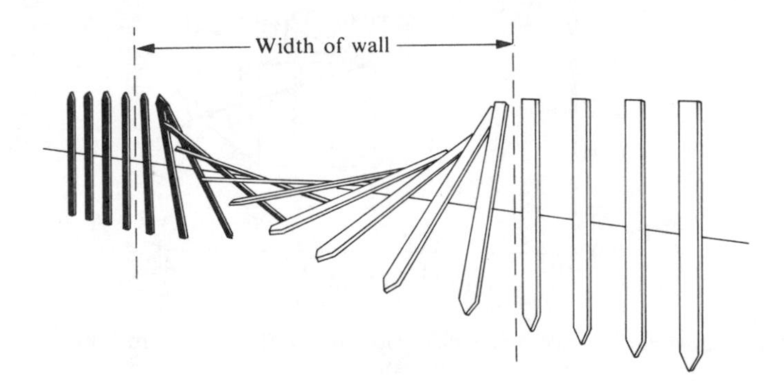

FIG. 11.6. Rotation of magnetic moments at a domain wall.

We are now ready to explain the magnetization curves of Fig. 11.5. When the piece of single-crystal iron is unmagnetized, we may assume that there are lots of domains and the magnetization in each domain is in one of the six easy directions. Applying now a magnetic field in the AB direction we find that as the magnetic field is increased the domains which lay originally in the AB direction will increase at the expense of the other domains, until the whole material contains only one single domain. Since the domain walls move easily the magnetic field required to reach saturation is small.

What happens if we apply to the same single-crystal iron a magnetic field in the AG direction? First the domain walls will move just as in the previous case until only three easy directions are left, namely AB, AD, and AE, i.e. those with components in the AG direction. This may be achieved with very little magnetic field, but from then on (K in Fig. 11.5) the going gets hard. In order to increase the magnetization further the magnetic moments need to change direction, which can only happen if the internal anisotropy forces are successfully overcome. This requires more effort, hence the slope of the magnetization curve changes and saturation will only be achieved at greater magnetic fields.

Is this explanation still correct for polycrystalline materials? Well, a polycrystalline material contains lots of single crystal grains, and the above argument applies to each of the single crystals; thus the magnetization curve of a polycrystalline material should look quite similar to that of a single-crystal material in a difficult direction. As you know from secondary school this is not the case. Fig. 11.5 does not tell the whole story. The magnetization curve of a typical ferromagnetic material exhibits *hysteresis* as shown in Fig. 11.7. Starting with a completely demagnetized material we move up the curve along 2, 3, 4, 5 as the magnetic field is increased. Reducing then the magnetic field we get back to point 6 which is identical with point 4, but further decrease takes place along a different curve. At 7 there is no applied magnetic field but B is finite. Its value, $B = B_r$, is the so called *remanent* flux density. Reducing further the magnetic field B takes the values along 8, 9, 10. Returning from 10 we find that 11 is identical with 9 and then proceed further along 12 and 13 to reach finally 4. Note that the value of H at 13 is called the *coercive force* or *field*, denoted by H_c. It represents the amount of magnetic field that needs be applied for the flux density to vanish.

The loop 4, 7, 8, 9, 12, 13, 4 is referred to as the hysteresis loop. It clearly indicates that the magnetization of iron is an irreversible phenomenon. The paths 4, 5 and 9, 10 suggest that rotation from easy into difficult directions is reversible, thus the causes of irreversibility should be sought in domain movement. Because of the presence of all sorts of defects in a real material, the domain walls move in little jerks causing the magnetization to increase in a discontinuous manner (region 2, 3 magnified in Fig. 11.7). The walls get stuck once in a while and then suddenly surge forward, setting up in the process some eddy currents and sound waves which consume energy. If energy is consumed the process cannot be reversible and that is the reason for the existence of the hysteresis loop.

Is it possible to describe more accurately the movement of domains? One can go indeed a little further by taking into account the effect of *magnetostriction*, which, as you may guess, is the magnetic counterpart of electrostriction. Strictly speaking, one should distinguish between magnetostriction and piezomagnetism, the magnetic counterpart of piezoelectricity. But biased magnetostriction (see discussion on biased electrostriction in Section 10.11) is phenomenologically equivalent to

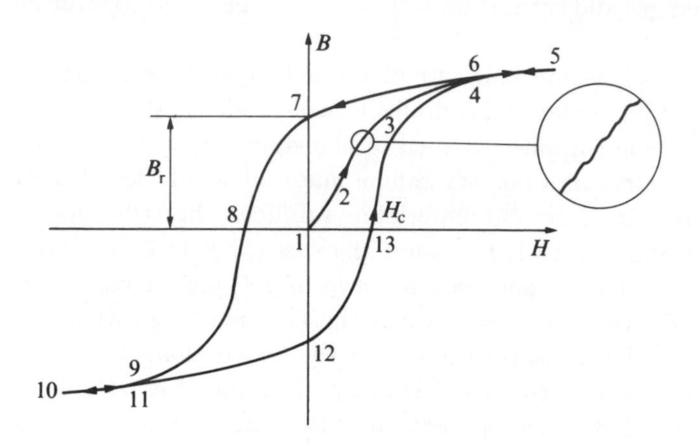

FIG. 11.7. The magnetization curve of a typical ferromagnetic material.

piezomagnetism, and piezomagnetism has not been much investigated anyway; thus most authors just talk about magnetostriction. Disregarding the problem of nomenclature, the relevant fact is that by applying a magnetic field the dimensions of the material change†, and conversely, strain in the material leads to changes in magnetization and may also effect the directions of easy magnetization. Now if the material exhibits a large anisotropy and stresses are present as well, then there will be everywhere *local* easy directions resisting the movement of domain walls. The stresses may be caused by the usual defects in crystals and particularly by impurities. In addition, a cluster of non-magnetic impurity atoms might be surrounded by domains (see Fig. 11.8), a stable configuration which cannot be easily changed.

Does all this matter very much? In practical terms this is probably the most important materials science problem that we have touched

† It is, incidentally, the cause of the humming noise of transformers.

upon. The area of the *B–H* loop defines the hysteresis loss, which is a major source of loss in electrical machines. Most of the time something like a million megawatts of electricity is being generated around the world, all by generators with hysteresis losses of order 0·5–1·0 per cent.

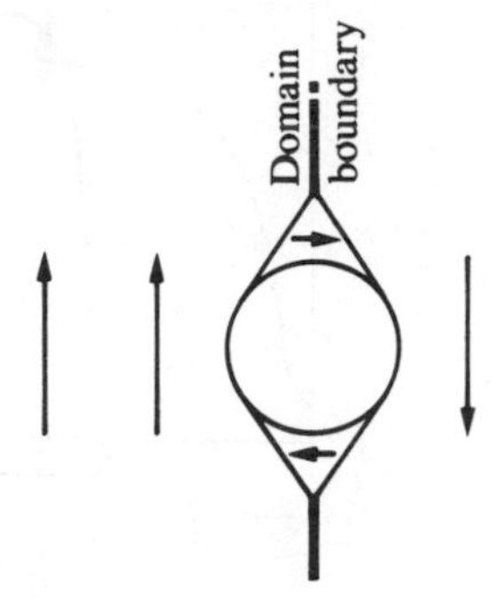

FIG. 11.8. Non-magnetic impurity surrounded by a domain.

Then a large fraction of this electricity goes into motors and transformers with more iron losses. If all inventors were paid a 1 per cent royalty on what they saved the community, then a good way to become a millionaire would be to make a minute improvement to magnetic materials. Is there any good scientific way to set about this? Not really. We know that anisotropy, magnetostriction, and local stresses are bad but we cannot start from first principles and suggest alloys which will have the required properties. The considerable advances that have been made in 'soft' magnetic materials, as the narrow *B–H* loop types are called, have been largely achieved by extensive and expensive trial and error. To Gilbert's seventeenth-century crack about 'good luck' we must add *diligence* for the modern smelters of iron.

Iron containing silicon is used in most electrical machinery. An alloy with about 2% silicon, a pinch of sulphur, and critical cold rolling and annealing processes is used for much rotating machinery. Silicon increases the resistivity, which is a good thing because it reduces eddy-current losses. Iron with a higher silicon content is even better and can be used in transformer laminations, but it is mechanically brittle and therefore no good for rotating machinery. Where small quantities of very low-loss material are required and expense is not important, as for induction loading of submarine cables and radiofrequency transformers, Permalloy (78·5% Ni, 21·5% Fe) is often used. A further improvement is achieved in the material called 'Supermalloy' which contains a little molybdenum and manganese as well. It is very easily magnetizable in small fields (Fig. 11.9a) and has no magnetostriction.

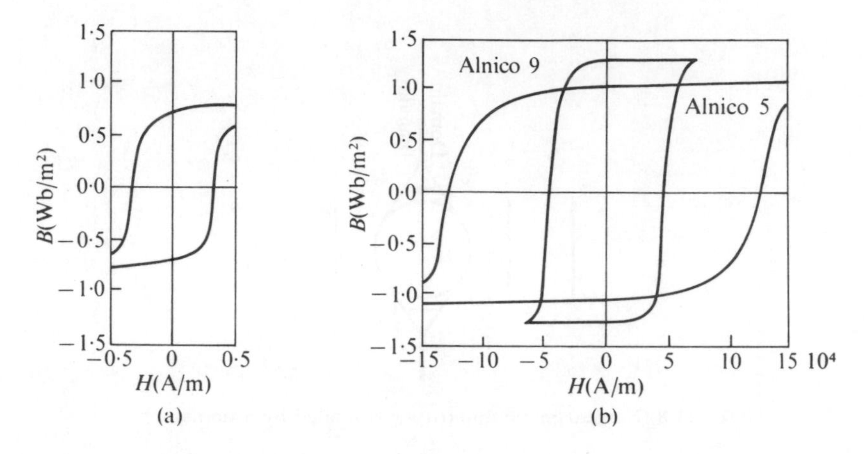

FIG. 11.9. Hysteresis loops of (a) Supermalloy and (b) Alnico 5 and 9. Note the factor 10^5 between the horizontal scales of (a) and (b).

We may now mention a fairly new and rather obvious trick. If anisotropy is bad, and anisotropy is due to crystal structure, then we should get rid of the crystal structure. What we obviously need is an amorphous material. The problems are purely technological and a solution has been recently found in the form of squirting a stream of molten metal on a rotating drum followed usually by a stress relief anneal at about 300°C. The resulting magnetic material will have the form of long thin ribbons typically about 40 μm thick and 1 mm wide. Their main advantage is that they can be produced easily and relatively cheaply with magnetic properties nearly as good as those of commercial alloys, which require careful melting and elaborate sequences of rolling and annealing. The presently available amorphous materials have not quite reached the quality of Permalloy but can already be used in many electronic applications.

The situation is somewhat different in power applications such as transformers. There the traditional materials are cheaper, but amorphous materials may still represent the better choice on account of lower losses; their higher cost may be offset in the long term by lower power consumption (not mentioning possible future legislation in some countries requiring higher efficiency in electrical equipment).

Permanent magnets. What kind of materials are good for permanent magnets? Well, if we want large flux density produced, we need a large value of B_r. What else? We need a large H_c. Why? A rough answer is that the high value of B_r needs to be protected. If for some reason we are not at the $H = 0$ point, we don't want to lose much flux, therefore the B–H curve should be as flat as possible.

A more rigorous argument in favour of large H_c can be produced by taking account of the so-called demagnetization effect, but in order to explain that, I shall have to make a little digression and go back to electromagnetic theory. First of all note that in a ring magnet (Fig. 11.10a) $B = B_r =$ constant everywhere in the material to a very good approximation. Of course such a permanent magnet is of not much interest because we cannot make any use of the magnetic flux. It may though be made available by cutting a narrow gap in the ring as shown in Fig. 11.10b. What will be the values of B and H in the gap? One may argue from geometry that the magnetic lines will not spread out (this is why we chose a narrow gap so as to make the calculations simpler) and the flux density in the gap will be the same as in the magnetic material.

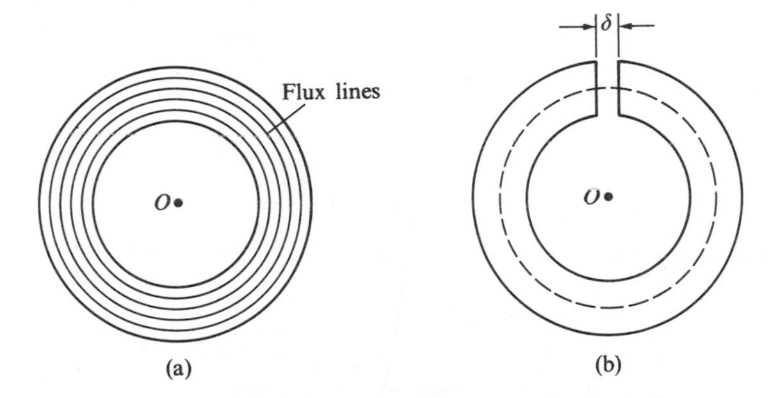

FIG. 11.10. (a) Magnetic field lines inside a permanent magnet. (b) The same magnet with a narrow gap.

But, and this is the question of interest, will the flux density be the same in the presence of the gap as in its absence? Without the gap $B = B_r$ (Fig. 11.11). Denoting the value of flux density by $B = B_{r1}$ in the presence of the gap, the magnetic field in the gap will be

$$H_g = \frac{B_{r1}}{\mu_0}. \tag{11.29}$$

If you can remember Ampere's law, which states that the line integral of the magnetic field in absence of a current must vanish for a closed path, it follows that

$$H_g\delta + H_m l = 0, \tag{11.30}$$

where H_m is the magnetic field in the material, and δ and l are the lengths of the paths in the gap and in the material respectively. From eqns. (11.29) and (11.30) we get

$$B_{r1} = -\frac{\mu_0 l}{\delta} H_m. \tag{11.31}$$

But remember that the relationship between B_{r1} and H_m is given also by the hysteresis curve. Hence the value of B_{r1} may be obtained by intersecting the hysteresis curve by the straight line of eqn. (11.31) as shown in Fig. 11.11. The aforegoing construction depends on the particular geometry of the permanent magnet we assume, but similar 'demagnetization' will occur for other geometries as well. Hence we may conclude in general that in order to have a large, useful flux density, the B–H curve must be flat. We may therefore adopt as a figure of merit the product B_rH_c or, as it is more usual, the product $(BH)_{max}$ in the second quadrant.

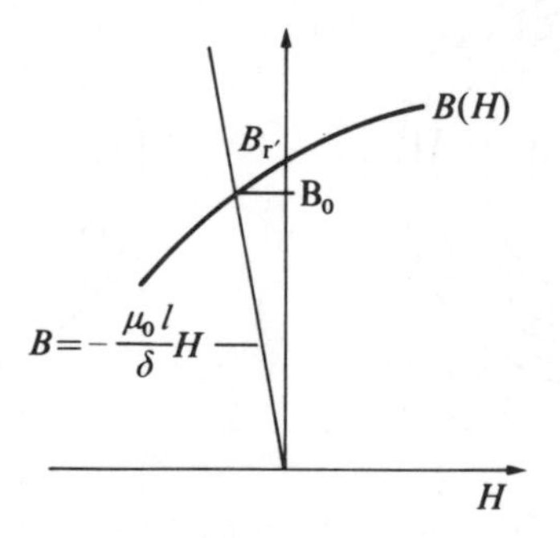

FIG. 11.11. Construction for finding B_0.

How can one achieve a large value of H_c? It is relatively easy to give an answer in principle. All the things which caused the quality of soft materials to deteriorate are good for permanent magnets which, to emphasize the contrast, are referred to as 'hard' magnetic materials.

The simplest permanent magnet one could conceive in principle would be a single crystal of a material which has a large anisotropy and has only one axis of easy magnetization. The anisotropy may be characterized by an effective field H_a which attempts to keep the magnetization along the axis. If a single crystal material is magnetized along this axis and a magnetic field is applied in the opposite direction, nothing should happen in principle until the field H_a is reached, and then, suddenly, the magnetization of the whole crystal should reverse. In practice, H_c is always smaller than H_a because there are always some irregularities which can initiate the reversal process.

What other factors will favour the production of good permanent magnets? We need to include impurities which restrict domain changes. Carbon is an obvious additive to try, and various high-carbon steels have been used as permanent magnets. More recently there has been a new idea, due to Néel, which has led to considerable improvements in permanent magnets. Néel found that a very small particle cannot contain more than one domain. The critical size is about 20 nm for iron. Once magnetized, these are very stable magnetically. The best known of this type of permanent magnet materials is the Alnico series (Fig. 11.9b). The basic formula is approximately Fe_2NiAl. Cobalt and copper are included in some of these alloys. The small needle-like particles of almost pure iron in a nickel and aluminium matrix are formed by controlled temperature gradients as the molten alloy freezes. Another application of the same principle is to bond together magnetic particles smaller than one domain wall thickness. A representative of these powder magnets is 'Ferroxdur' based on barium ferrites. Fine-particles magnets have also been made from Fe or Fe–Co alloy bonded with Pb. They are often referred to as ESD magnets since they are comprised of *e*longated *s*ingle *d*omain particles.

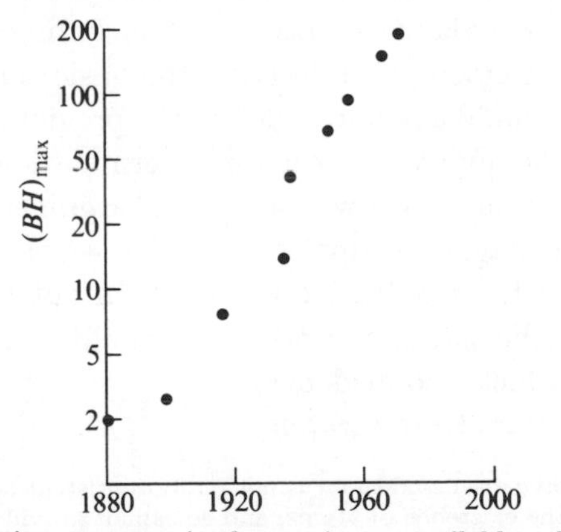

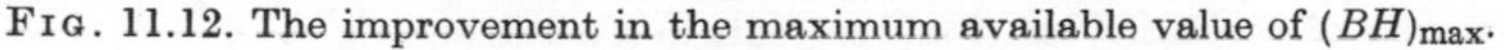
FIG. 11.12. The improvement in the maximum available value of $(BH)_{max}$.

Figures 11.9a and b show that the magnetic field needed to give a roughly similar saturated flux level in the hard and soft materials differs by a factor of approaching one million. This is some measure of how successful the intensive, if largely empirical, materials research on ferromagnetic materials has been. Another measure of the progress is shown in Fig. 11.12 where the maximum available value of $(BH)_{\max}$ is plotted as a function of years. The increase is roughly exponential: an improvement by a factor of a hundred is achieved in about as many years. Not as spectacular as the improvement in optical fibres, but one certainly gains the impression of steady advance.

11.5. Microscopic theory (quantum-mechanical)

Classical theory gives a reasonable physical picture of what is happening in a magnetic material and does give some guidance to people searching for new materials.† The question arises whether we should discuss quantum theory as well. I would like to advise against excessive optimism. Don't expect too much; the situation is not as cheerful as for semiconductors, where the injection of a tiny dose of quantum theory sufficed to explain all the major phenomena. The same is not true for magnetism. The quantum theory of magnetism is much more complicated and much less useful to an engineer. The most important activity, the search for better magnetic materials, is empirical anyway, and there are not many magnetic devices clamouring for quantum theory to solve the riddle of their operation.

Up to 1956 I would have certainly done no more than give a passing reference to quantum theory in a course such as this. I think (and this is just a personal opinion) that the invention of the paramagnetic maser in 1956 tipped the balance in favour of including quantum theory. So I shall make an attempt to describe the basic concepts.

First of all, we should ask how much of the previously outlined theory remains valid in the quantum-mechanical formulation. Not a word of it! There is no reason whatsoever why a classical argument (as, for example, the precession of magnetic dipoles around the magnetic field) should hold water. When the resulting formulae turn out to be identical (as, for example, for the paramagnetic susceptibility at normal temperatures), it is just a lucky coincidence.

So we have to start from scratch.

† The theory we have discussed so far is not really consistent because classical theory cannot even justify the existence of atoms, and so cannot provide any good reasons for the presence of circulating electronic currents in a material.

Let us first talk about the single electron of the hydrogen atom. As we mentioned before, the electron's properties are determined by the four quantum numbers n, l, m_l, and s, which have to obey certain relationships between themselves; as for example, that l must be an integer and may take values between 0 and $n-1$. Any set of these four quantum numbers will uniquely determine the properties of the electron. As far as the specific magnetic properties of the electron are concerned, the following rules are relevant:

(i) The total angular momentum is given by

$$\Pi = \hbar\{j(j+1)\}^{\frac{1}{2}}, \tag{11.32}$$

where $j = l+\frac{1}{2}$, that is a combination of the quantum numbers l and s.

(ii) The possible components of the angular momentum along *any* specified direction† are determined by the combination of m_l (which may take on any integral value between $-l$ and $+l$) and s, yielding

$$j, j-1, \ldots -j+1, -j.$$

Taking as an example a d electron, for which $l = 2$, the total angular momentum is

$$\Pi = \hbar\left(\frac{5}{2}\cdot\frac{7}{2}\right)^{\frac{1}{2}} = \frac{\hbar}{2}\sqrt{35}, \tag{11.33}$$

and its possible components along (say) the z-axis are

$$\frac{5}{2}\hbar, \frac{3}{2}\hbar, \frac{\hbar}{2}, -\frac{\hbar}{2}, -\frac{3}{2}\hbar, -\frac{5}{2}\hbar$$

as shown in Fig. 11.13.

(iii) The quantum-mechanical relationship between magnetic moment and angular momentum is nearly the same as the classical one, represented by eqn. (11.7)

$$\mu_{\text{m}} = g\frac{e}{2m}\Pi. \tag{11.34}$$

† This is sheer nonsense classically because according to classical mechanics once the angular momentum is known about three axes perpendicular to each other, it is known about any other axes (and it will not therefore take necessarily integral multiples of a certain unit). In quantum mechanics we may know the angular momentum about several axes but *not simultaneously*. Once the angular momentum is measured about one axis, the measurement will alter the angular momentum about some other axis in an unpredictable way. If it were otherwise we would get into trouble with the uncertainty relationship. Were we to know the angular momentum in all directions, it would give us the plane of the electron's orbit. Hence we would know the electron's velocity in the direction of the angular momentum vector (it would be zero), and also the position (it would be in the plane perpendicular to the angular momentum in line with the proton). But this is forbidden by the uncertainty principle, which says that it is impossible to know both the velocity and the position coordinate in the same direction as the velocity.

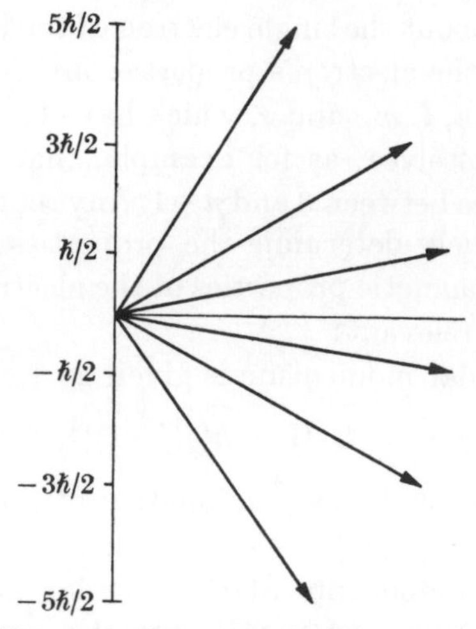

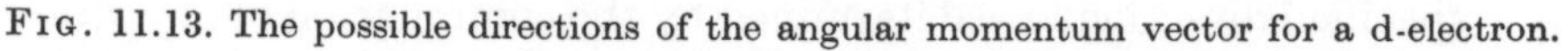
FIG. 11.13. The possible directions of the angular momentum vector for a d-electron.

The only difference is the factor g (admirably called the g-factor). For pure orbital motion its value is 1; for pure spin motion its value is 2; otherwise it is between 1 and 2.

(iv) The energy of a magnetic dipole in a magnetic field H (taken in the z-direction) is

$$\begin{aligned} E_{\text{mag}} &= -(\mu_{\text{m}})_z \mu_0 H \\ &= -ge\Pi_z \mu_0 H/2m. \end{aligned} \tag{11.35}$$

The term—$e\hbar/2m$ is called a *Bohr magneton* and denoted by μ_{mB}. We may rewrite eqn. (11.32) in the form

$$E_{\text{mag}} = g\mu_{\text{mB}}\Pi_z \mu_0 H/\hbar, \tag{11.36}$$

where $\Pi_z/\hbar$, as we have seen before, may take the values $j, j-1$, etc. down to $-j$.

We know now everything about the magnetic properties of an electron in the various states of the hydrogen atom. In general, of course, the hydrogen atom is in its ground state, for which $l = 0$ and $m_l = 0$, so that only the spin of the electron counts. The new quantum number j comes to $\frac{1}{2}$, and the possible values of the angular momentum in any given direction are $\hbar/2$ and $-\hbar/2$. Furthermore $g = 2$ and the magnetic moment is

$$\mu_{\text{m}} = \mu_{\text{mB}}, \tag{11.37}$$

that is, the magnetic moment of hydrogen happens to be one Bohr magneton.

We can get the magnetic properties of more complicated atoms by combining the quantum numbers of the individual electrons. There exist a set of rules (known as Hund's rules) that tell us how to combine the spin and orbital quantum numbers in order to get the resultant quantum number J. The role of J for an atom is exactly the same as that of j for an electron. Thus, for example, the total angular momentum is given by

$$\Pi = \hbar\{J(J+1)\}^{\frac{1}{2}}, \tag{11.38}$$

and the possible components of the angular momentum vector along any axis by

$$\hbar J,\ \hbar(J-1),\ \ldots\ \hbar(-J+1),\ -\hbar J.$$

The general rules are fairly complicated and can be found in textbooks on magnetism. I should just like to note two specific features of the magnetic properties of atoms:

(i) Atoms with filled shells have no magnetic moments (this is because the various electronic contributions cancel each other);

(ii) The spins arrange themselves so as to give the maximum possible value consistent with the Pauli principle.

It follows from (i) that helium and neon have no magnetic moments; and stretching the imagination a little one may also conclude that hydrogen, lithium, and silver, for example, possess identical magnetic properties (because all of them have one outer electron).

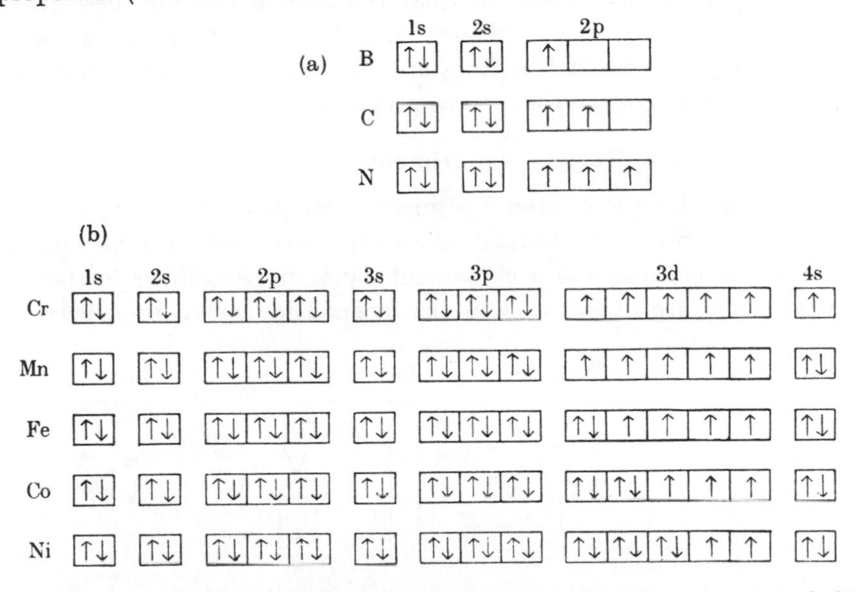

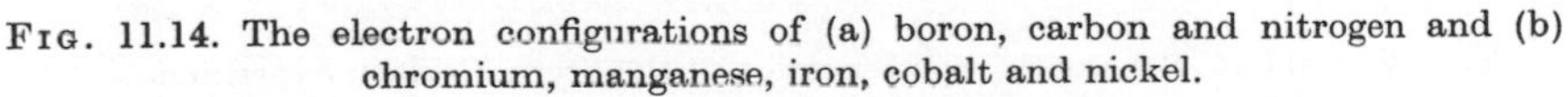

FIG. 11.14. The electron configurations of (a) boron, carbon and nitrogen and (b) chromium, manganese, iron, cobalt and nickel.

The consequences of (ii) are even more important. It follows from there that states with identical spins are occupied first. Thus, boron with a configuration $1s^22s^22p^1$ has one electron with spin 'up' in the outer shell (see Fig. 11.14a); carbon has two electrons with spin up and nitrogen has three. Similarly all five electrons of chromium and manganese in the 3d shell have spins up, and the states with opposite spins start to fill up only later when there is no alternative. This is shown in Fig. 11.14b, where the electronic configurations are given for chromium, manganese, iron, cobalt, and nickel.

We shall return to the spins of the 3d electrons a little later; first let me summarize the main points of the argument. The most important thing to realize is that electrons in an atom do not act individually. We have no right to assume (as we did in the classical treatment) that all the tiny electronic currents are randomly oriented. They are not. They must obey Pauli's principle, and so within an atom they all occupy different states that do bear some strict relationships to each other. The resultant angular momentum of the atom may be obtained by combining the properties of the individual electrons, leading to the quantum number J, which may also be zero. Thus an atom that contains many 'magnetic' electrons may end up without any magnetic moment at all.

You may ask at this stage what is the evidence for these rather strange tenets of quantum theory? Are the magnetic moments of the atoms really quantized? Yes, they are. The experimental proof actually existed well before the theory was properly formulated.

11.6. The Stern–Gerlach experiment

The proof for the existence of discrete magnetic moments was first obtained by Stern and Gerlach in an experiment shown schematically in Fig. 11.15. Atoms of a chosen substance (it was silver in the first experiment) are evaporated in the oven. They move then with the

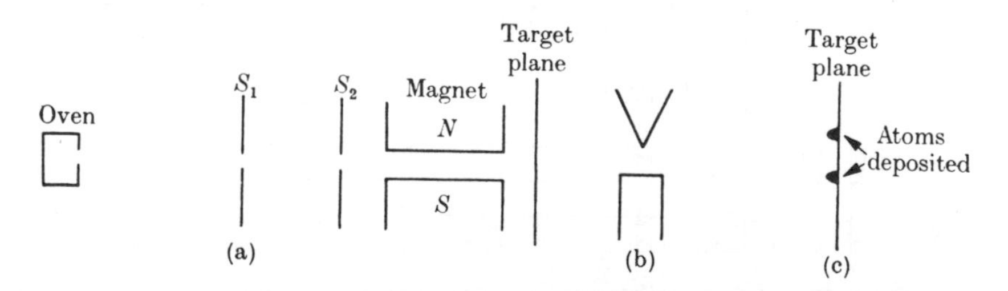

FIG. 11.15. Schematic representation of the Stern–Gerlach experiment.

average thermal velocity, and those crossing the diaphragms S_1 and S_2 may be expected to reach the target plane in a straight line—provided they are non-magnetic. If, however, they do possess a magnetic moment, they will experience a force (obtained by differentiating eqn. (11.35)) expressed by

$$F = (\mu_{\mathrm{m}})_z \mu_0 \frac{\partial H}{\partial z}. \tag{11.39}$$

Thus the deflection of the atoms in the vertical plane depends on the magnitude of this force. $\partial H/\partial z$ is determined by the design of the magnet (a strong variation in the z-component of the magnetic field may be achieved by making the upper pole piece wedge-shaped as shown in Fig. 11.15b) and is a constant in the experiment. Hence the actual amount of deflection is a measure of $(\mu_{\mathrm{m}})_z$.

Were the magnetic moments entirely randomly oriented, the trace of the atoms on the target plane would be a uniform smear along a vertical line. But that is not what happens in practice. The atoms in the target plane appear in distinct spots as shown in Fig. 11.15c.

For silver $J = \frac{1}{2}$, and the beam is duly split into two, corresponding to the angular momenta $\Pi_z = \hbar/2$ and $-\hbar/2$. Repeating the experiment with other substances the result is always the same. One gets a discrete number of beams corresponding to the discrete number of angular momenta the atom may have.

Paramagnetism. We are now in a position to work out with the aid of quantum theory the paramagnetic susceptibility of a substance containing atoms with quantum numbers $J \neq 0$. When we apply a magnetic field all the atoms will have some magnetic moments in the direction of the magnetic field. The relative number of atoms possessing the same angular momentum is determined again by Boltzmann statistics. The mathematical procedure for obtaining the average magnetic moment is analogous to the one we used for electric dipoles but must now be applied to a discrete distribution.

The possible magnetic moments are

$$M_J g \mu_{\mathrm{mB}} \quad \text{where} \quad M_J = J, J-1, \ldots -J+1, -J.$$

Hence their energies are

$$E_{\mathrm{mag}} = -M_J g \mu_{\mathrm{mB}} \mu_0 H, \tag{11.40}$$

and the average magnetic moment may be obtained in the form

$$\langle \mu_{\mathrm{m}} \rangle = \frac{\sum_{-J}^{J} M_J g \mu_{\mathrm{mB}} \exp(M_J g \mu_{\mathrm{mB}} \mu_0 H/kT)}{\sum_{-J}^{J} \exp(M_J g \mu_{\mathrm{mB}} \mu_0 H/kT)}. \tag{11.41}$$

The macroscopic magnetic moment may now be calculated by multiplying $\langle \mu_{\mathrm{m}} \rangle$ by the number of atoms per unit volume.

Equation (11.41) turns out to be a very accurate formula† for describing the average magnetic moment as shown in Fig. 11.16, where

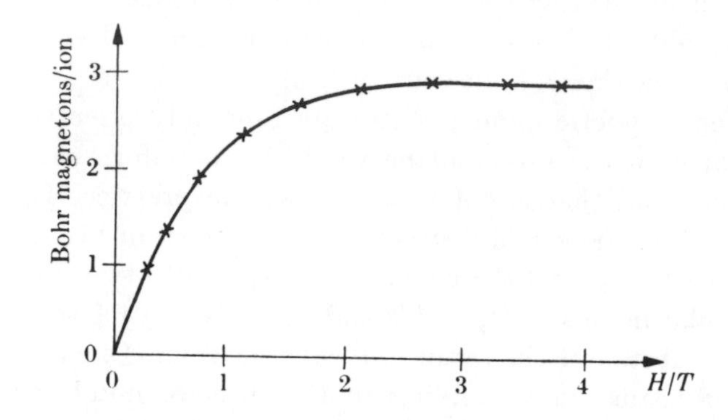

Fig. 11.16. The magnetic moment as a function of H/T for potassium chromium alum (after Henry).

it is compared with the experimental results of Henry on potassium chromium alum. The vertical scale is in Bohr magnetons per *ion*. Note that experimental results for paramagnetic properties are often given for ions embedded in some salt. The reason is that in these compounds the ions responsible for magnetism (Cr^{3+} in the case of potassium chromium alum) are sufficiently far from each other for their interaction to be disregarded.

† We need not be too much impressed by these close agreements between theory and experiments. The theoretical curve was *not* calculated from first principles, in the sense that the value of J was arrived at by semi-empirical considerations. The problem is far too difficult to solve exactly. The usual approach is to set up a simple model and modify it (e.g. by taking account of the effect of neighbouring atoms) until theory and experiment agree. It is advisable to stop rather abruptly at that point because further refinement of the model might increase the discrepancy.

If the exponent is small enough, the exponential function may be expanded to give

$$\langle \mu_{\mathrm{m}} \rangle = -g\mu_{\mathrm{mB}} \frac{\sum_{-J}^{J} M_J(1 - M_J g \mu_{\mathrm{mB}} \mu_0 H/kT)}{\sum_{-J}^{J} (1 - M_J g \mu_{\mathrm{mB}} \mu_0 H/kT)}$$

$$= \frac{g^2 \mu_{\mathrm{mB}}^2 \mu_0 H}{(2J+1)kT} \sum_{-J}^{J} M_J^2 \tag{11.42}$$

because

$$\sum_{-J}^{J} M_J = 0. \tag{11.43}$$

The summation in eqn. (11.42) is one of the simpler ones to perform, yielding

$$\tfrac{1}{3}J(J+1)(2J+1),$$

which gives finally

$$\langle \mu_{\mathrm{m}} \rangle = g^2 \mu_{\mathrm{mB}}^2 J(J+1) \mu_0 H/3kT. \tag{11.44}$$

We may now express the above equation in terms of the *total* angular momentum

$$\Pi = \hbar\{J(J+1)\}^{\frac{1}{2}} \tag{11.45}$$

and total magnetic moment

$$\mu_{\mathrm{m}} = ge\Pi/2m \tag{11.46}$$

to get

$$\langle \mu_{\mathrm{m}} \rangle = \mu_{\mathrm{m}}^2 \mu_0 H/3kT, \tag{11.47}$$

in agreement with the classical result.†

Paramagnetic solids. As we have seen, the magnetic properties of electrons combine to produce the magnetic properties of atoms. These properties can be measured in a Stern–Gerlach apparatus where each atom may be regarded as a separate entity. This is because the atoms in the vapour are far enough from each other not to interact. However, when the atoms aggregate in a solid, the individual magnetic properties of atoms combine to produce a resultant magnetic moment. The electrons that are responsible for chemical bonding are usually responsible for the magnetic properties as well. When for example, sodium atoms and chlorine atoms combine to make up the ionic solid, NaCl, then the valence electron of the sodium atom moves over to the chlorine atom and fills up the shell. Hence both the sodium and the chlorine ions

† This perhaps shows the power of human imagination. If one has a fair idea how the final conclusion should look, one can get a reasonable answer in spite of following a false track.

have filled shells and consequently solid NaCl is non-magnetic. A similar phenomenon occurs in the covalent bond, where electrons of opposite spin strike up a durable companionship, and as a result the magnetic moments cancel again.

How then can solids have magnetic properties at all? Well, there is first the metallic bond, which does not destroy the magnetic properties of its constituents. It is true that the immobile lattice ions have closed shells and hence no magnetic properties, but the pool of electrons do contribute to magnetism owing to their spin. Some spins will be 'up' (in the direction of the magnetic field); others will be 'down'. Since there will be more up than down, the susceptibility of all metals has a paramagnetic component, of the order of 10^{-5}. This is about the same magnitude as that of the diamagnetic component; hence some metals are diamagnetic.

Another possibility is offered by salts of which potassium chromium alum is a typical example. There again, as mentioned above, the atoms, responsible for the magnetic properties, being far away from each other, do not interact. In these compounds, however, the atoms lose their valence electrons; they are needed for the chemical bond. Hence the compound will have magnetic properties only if some of the ions remain magnetic. This may happen in the so called 'transition elements', which have unfilled inner shells. The most notable of them is the 3d shell but Table 4.1 shows that the 4d, 4f, 5d, and 5f shells have similar properties.

Taking chromium again as an example, it has a valency of two or three; hence in a chemical bond it must lose its 4s electron (see Fig. 11.14b) and one or two of its 3d electrons. The important thing is that there are a number of 3d electrons left that have identical spins, being thus responsible for the paramagnetic properties of the salt.

Antiferromagnetism. Let us now study the magnetic properties of solid chromium. From what we have said so far it would follow that chromium is a paramagnetic solid with a susceptibility somewhat larger than that of other metals because as well as the contribution of the free electrons, the lattice ions are magnetic as well. These expectations are not entirely false and this is what happens above a certain temperature, the *Néel temperature* (475 K for chromium). Below this temperature, however, a rather odd phenomenon occurs. The spins of the neighbouring atoms suddenly acquire an ordered structure; they become antiparallel as shown in Fig. 11.17. This is an effect of the 'exchange interaction', which is essentially just another name for Pauli's principle.

According to Pauli's principle two electrons cannot be in the same state unless their spins are opposite. Hence two electrons close to each other have a tendency to acquire opposite spins. Thus, the electron-pairs

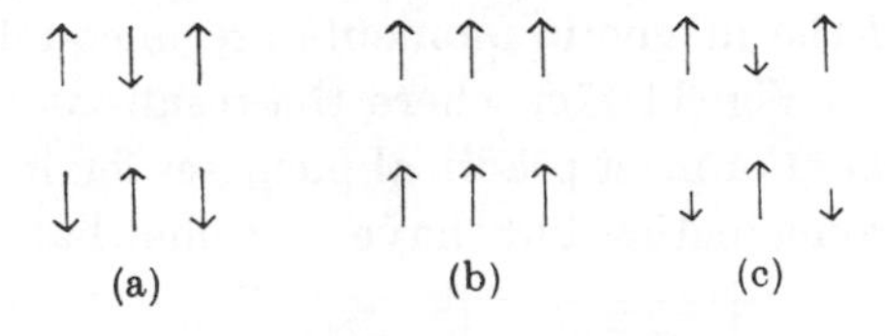

FIG. 11.17. The angular momentum vector for (a) antiferromagnetic, (b) ferromagnetic and (c) ferrimagnetic materials.

participating in covalent bonds have opposite spins, and so have the electrons in neighbouring chromium atoms. Besides chromium there are a number of compounds like MnO, MnS, FeO, etc. and another element, manganese, (Néel temperature 100 K) that have the same antiferromagnetic properties.

Antiferromagnetics display an ordered structure of spins; so in a sense they are highly magnetic. Alas, all the magnetic moments cancel each other (in practice *nearly* cancel each other) and there are therefore no external magnetic effects.

Ferromagnetism. Leaving chromium and manganese we come to iron, cobalt, and nickel, which are ferromagnetic. In a ferromagnetic material the spins of neighbouring atoms are parallel to each other (Fig. 11.17b). Nobody quite knows why. There seems to be general agreement that the exchange interaction is responsible for the lining-up of the spins (as suggested first by Heisenberg in 1928) but there is no convincing solution yet. The simplest explanation (probably as good as any other) is as follows.

Electrons tend to line up with their spins antiparallel. Hence a conduction electron passing near a 3d electron of a certain iron atom will acquire a tendency to line up antiparallel. When this conduction electron arrives at the next iron ion, it will try to make the 3d electron of that atom antiparallel to itself; that is, parallel to the 3d electron of the previous iron atom. Hence all the spins tend to line up.

In Weiss's classical picture the magnetic moments are lined up by a long-range internal field. In the quantum picture they are lined up owing to nearest-neighbour interaction. 'One is reminded,' writes Keffer† 'of the situation when, as the quiet of evening descends, suddenly all the dogs in a town get to barking together, although each dog responds only to the neighbouring dogs.'

† F. Keffer, Magnetic properties of materials, *Scientific American*, September 1967.

Ferrimagnetism and ferrites. This type of magnetism occurs in compounds only, where the exchange interaction causes the electrons of each set of atoms to line up parallel but the two sets are antiparallel to each other. If the magnetic moments are unequal then we get the situation shown in Fig. 11.17c, where the resultant magnetic moment may be quite large. For most practical purposes ferrimagnetic materials behave like ferromagnetics but have a somewhat lower saturation magnetization.

The ferrites (a group of ferrimagnetics) are of considerable commercial interest since they have the commendable property of being insulators. Thus they will reduce eddy-current losses and can be used at high frequencies. Their main application is at high frequencies for transformers and television scanning coils. Special types have been developed as non-reciprocal elements in microwave circuits and for computer memory cores. The various chemical arrangements allow a range of properties to be covered with various sizes and shapes of hysteresis loop.

The general chemical formula of a ferrite is $MO.Fe_2O_3$ where M is a metal, typically Ni, Al, Zn, or Mg. If the metal M is iron, the material is iron ferrite, Fe_3O_4. So the earliest known magnetic material is, in fact, ferrimagnetic.

The main properties of two ferrites are listed in Table 11.1 among those of other magnetic materials. One is a manganese-zinc ferrite that is used for high-frequency, low-loss transformers; the other is a barium ferrite (trade name 'Ferroxdur') that is used for permanent magnets.

Garnets. This is the name for a class of compounds crystallizing in a certain crystal structure. As far as magnetic properties are concerned their most interesting representative is yttrium-iron garnet ($Y_3Fe_5O_{12}$), which happens to be ferromagnetic for a rather curious reason. The spin of the yttrium atoms is opposite to the spin of the iron atoms so the magnetic moments would line up alternately—if the orbital magnetic moments were small. But for yttrium the orbital magnetic moment is large, larger actually than the spin, and is in the opposite direction. Hence the total magnetic moment of the yttrium atom is in the same direction as that of iron, making the compound ferromagnetic.

Helimagnetism. You may wonder why the magnetic moments of neighbouring atoms in an ordered structure are either parallel or anti-parallel. One would expect quantum mechanics to produce a larger variety. In actual fact, there *are* some materials in which the spins in a given atomic layer are all in the same direction but the spins of

TABLE 11.1. *Some typical magnetic materials*

(a) Soft magnetic materials

Material	Saturated flux density, B_{sat} Wb/m^2	Maximum rel. permeability $(\mu_r)_{max}$	Resistivity, ρ Ω m	Hysteresis loss per cycle J/kg Hz
Iron, Fe	2·15	5000	$1{\cdot}0\ 10^{-7}$	0·03
Fe–4 %Si	1·97	7000	$6{\cdot}0\ 10^{-7}$	0·02
Grain oriented	2·00	30,000	$5{\cdot}5\ 10^{-7}$	0·005
78 Permalloy Fe 22 %, Ni 78 %	1·08	100,000	$1{\cdot}6\ 10^{-7}$	0·0005
Supermalloy	0·79	1,000,000	$6{\cdot}0\ 10^{-7}$	0·0001
Manganese-zinc ferrite $MnZn(Fe_2O_3)_2$	—	2500	0·2	0·001

(b) Hard magnetic materials

Material	H_c A/m	B_r Wb/m^2	$(BH)_{max}$ AWb/m^3
Carbon steel 0·9 %C, 1 %Mn	$4{\cdot}0.10^3$	0·9	8.10^2
Alnico 5 8 %Al, 24 %Co, 3 %Cu, 14 %Ni	$4{\cdot}6.10^4$	1·25	2.10^4
'Ferroxdur' $(BaO)(Fe_2O_3)_6$	$1{\cdot}6.10^5$	0·35	$1{\cdot}2.10^4$
ESD Fe–Co	$8{\cdot}2.10^4$	0·9	4.10^4
Alnico 9	$1{\cdot}3.10^5$	1·05	10^5

adjacent layers lie at an angle (e.g. 129° in MnO_2 below a certain temperature), producing a kind of helix. For the moment this is a scientific curiosity with no practical application.

11.7. Magnetic resonance

Paramagnetic resonance. The possible energies of an atom in a magnetic field are given by eqn. (11.40). There are $2J+1$ energy levels with separations of $\Delta E = g\mu_{mB}\mu_0 H$ as shown in Fig. 11.18 for $J = 1$.

We now put a sample containing magnetic atoms (e.g. a paramagnetic salt) into a waveguide and measure the transmission of the electromagnetic waves as a function of frequency. When $f = \Delta E/h$, the incident photon has just the right energy to excite the atom from a lower energy level into a higher energy level. Thus some of the photons

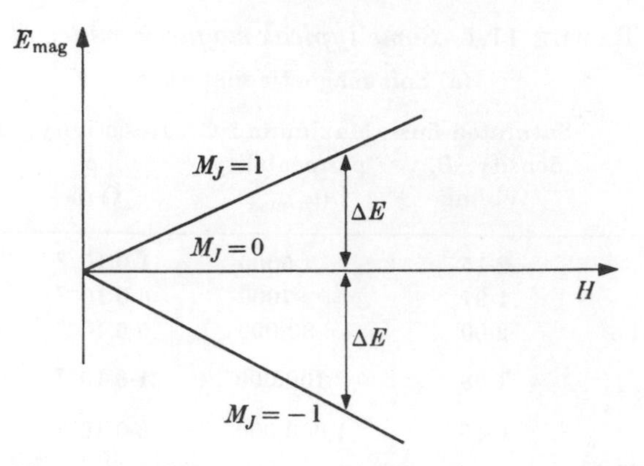

FIG. 11.18. The energy of an atom as a function of magnetic field for $J = 1$.

transfer their energies to the atomic system; this means loss of photons, or in other words absorption of electromagnetic energy. Hence there is a dip in the transmission spectrum as shown in Fig. 11.19. Since the absorption occurs rather sharply in the vicinity of the frequency $\Delta E/h$, it is referred to as *resonant absorption*, and the whole phenomenon is known as *paramagnetic resonance*.

In practice the energy diagram is not quite like the one shown in Fig. 11.18 (because of the presence of local electric fields) and a practical measuring apparatus is much more complicated than our simple waveguide (in which the absorption would hardly be noticeable); but the principle is the same.

Electron spin resonance. This is really a special case of paramagnetic resonance when only the spin of the electron matters. It is mainly used

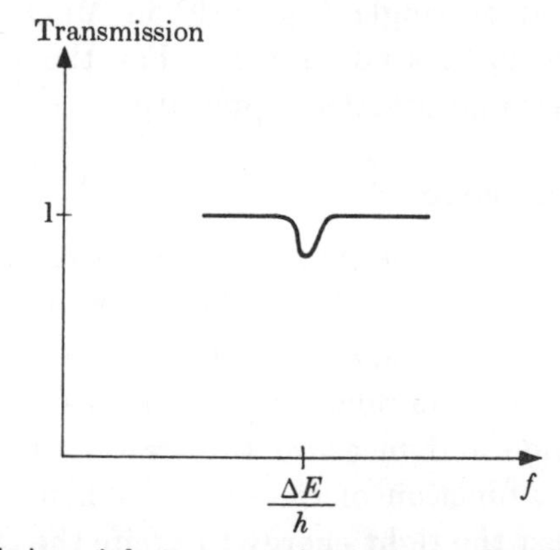

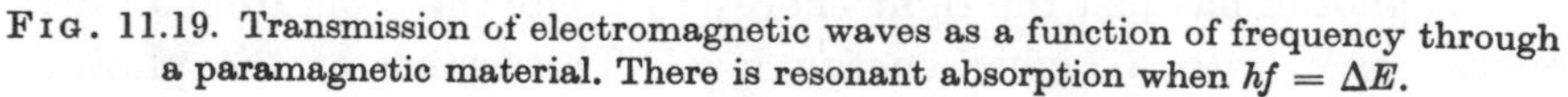

FIG. 11.19. Transmission of electromagnetic waves as a function of frequency through a paramagnetic material. There is resonant absorption when $hf = \Delta E$.

by organic chemists as a tool to analyse chemical reactions. When chemical bonds break up, electrons may be left unpaired, that is the 'fragments' may possess a net spin (in which case they are called free radicals). The resonant absorption of electromagnetic waves indicates the presence of free radicals, and the magnitude of the response can serve as a measure of their concentration.

Ferromagnetic, antiferromagnetic and ferrimagnetic resonance. When a crystal with ordered magnetic moments is illuminated by an electromagnetic wave the mechanism of resonant absorption is quite complicated owing to the interaction of the magnetic moments. The resonant frequencies cannot be predicted from first principles (though semiclassical theories exist) but they have been measured under various conditions for all three types of materials.

Nuclear magnetic resonance. If electrons, by virtue of their spins, can cause resonant absorption of electromagnetic waves one would expect protons to behave in a similar manner. The main difference between the two particles is in mass and in the sign of the electric charge; so the analogous formula

$$f = \frac{1}{2\pi} g \frac{e}{2m_p} \mu_0 H \qquad (11.48)$$

(where m_p is the mass of the proton) should apply.

The linear dependence on magnetic field is indeed found experimentally but the value of g is not 2 but 5·58, indicating that the proton is a more complex particle than the electron.

Neutrons also possess a spin; so they can also be excited from spin 'down' into spin 'up' states. Although they are electrically neutral, the resonant frequency can be expressed in the same way and the measured g-factor is 3·86.

The resonance is sharp in liquids but broader, by a few orders of magnitude, in solids. The reason for this is that the nuclear moments are affected by the local fields, which may vary in a solid from place to place but average to zero in a liquid.

Since both the shape of the resonant curve and the exact value of the resonant frequency depend on the environment in which a nucleus finds itself, nuclear magnetic resonance can be used as a tool to investigate the properties of crystals. An important application in a different direction is the precision measurement of magnetic fields. The proton resonance of water is generally used for this purpose. The accuracy that can be achieved is about 1 part in 10^6.

Cyclotron resonance. We have already discussed the phenomenon of cyclotron resonance from a classical point of view and we shall now consider it quantum mechanically. For resonant absorption one needs at least two energy levels or, even better, many energy levels equally spaced from each other. What are the energy levels of an electron in a solid? Remember that in our earlier model we neglected the interaction between electrons and simply assumed that the solid may be regarded as an infinite potential well. The possible energy levels were then given by eqn. (6.2)

$$E = \frac{\hbar^2}{2m}(k^2+k_y^2+k_z^2)$$

$$= \frac{h^2}{8m(2a)^2}(n_x^2+n_y^2+n_z^2),$$

where n_x, n_y, n_z are integers.

When a magnetic field is applied in the z-direction, then the above equation modifies to†

$$E = (\lambda+\tfrac{1}{2})\hbar\omega_c+\frac{\hbar^2}{2m}k_z^2, \tag{11.49}$$

where λ is an integer and ω_c is the cyclotron frequency. For constant k_z the difference between the energy levels (called *Landau levels*) is $\hbar\omega_c$. Hence we may look upon cyclotron resonance as a process in which electrons are excited by the incident electromagnetic wave from one energy level to the next.

11.8 Some applications

Until about 30 years ago the only significant application of magnetic materials was for electrical machines and transformers. Modern technology brought some new applications, of which I shall describe here three: ferrites and garnets for two different kinds of storage elements and ferrites for microwave isolators.

Ferrites can be used in computer storage systems because they are insulators (exhibiting no eddy-current losses) and because they have two stable states capable of storing a 'bit' of information. In the most widely used arrangement the tiny rings of ferrite are knitted

† The effect of the magnetic field may be taken into account by replacing p^2 by $(p - eA)^2$ in the Hamiltonian of Schrödinger's equation (where A is the vector potential).

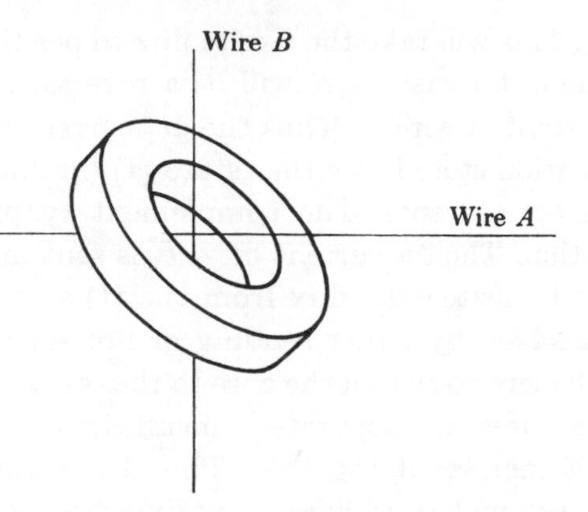

FIG. 11.20. A ferrite core used in computers for storing one bit of information.

on to delicately wired three-dimensional patterns often combining hundreds of thousands of cores. (It is interesting to note that this device, the heart of the computer-controlled revolution, is itself hand-made by nimble fingered young women who in other eras might have been weaving tapestry or embroidering silks.) We shall just consider one of these cores (Fig. 11.20) and how its square hysteresis loop (Fig. 11.21) can be used to store digital information. The core is threaded by two wires. Assume for the moment that no current flows through either wire; then the ferrite will be in either state (1) or (2) in Fig. 11.21. These states indicate 'yes' and 'no' (or 0 and 1) in the binary logic system. There are two basic processes: *reading* the stored information and *writing* it into the store. In the first process a current of $2I_1$ is passed

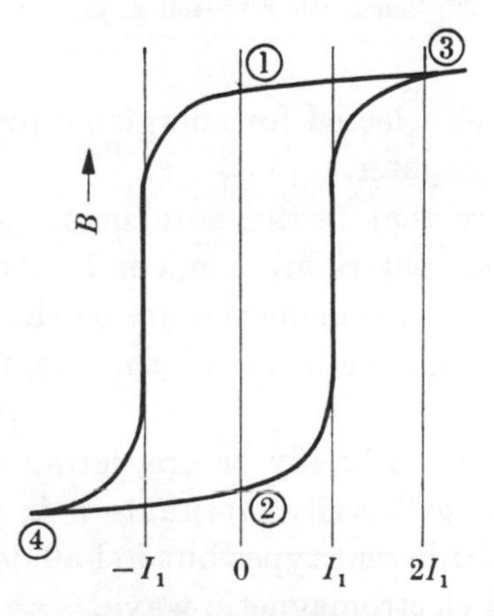

FIG. 11.21. Diagram showing the magnetism states of the ferrite core for the 'read' and 'write' operations.

through wire A. This will take the ferrite flux to position (3) either from (1) or (2). In the latter case there will be a reversal of the flux, which will induce a current in wire B. Thus the $2I_1$ current has (destructively) read the information stored as either state (1) (nothing happens) or (2) (flux change gives current). The complementary process is to *write* in this information. Then a current of $-I_1$ is sent into wire A. This is just insufficient to switch the flux from the (1) state to (2). The state of the core is decided by either sending or not sending into wire B a pulse that is sufficient to switch the core to the (4)-(2) flux line. Both the currents in the wires are separately insufficient to switch the flux; they must both energize it together. Thus by intricate external connections to a large matrix of wires containing cores at each junction,

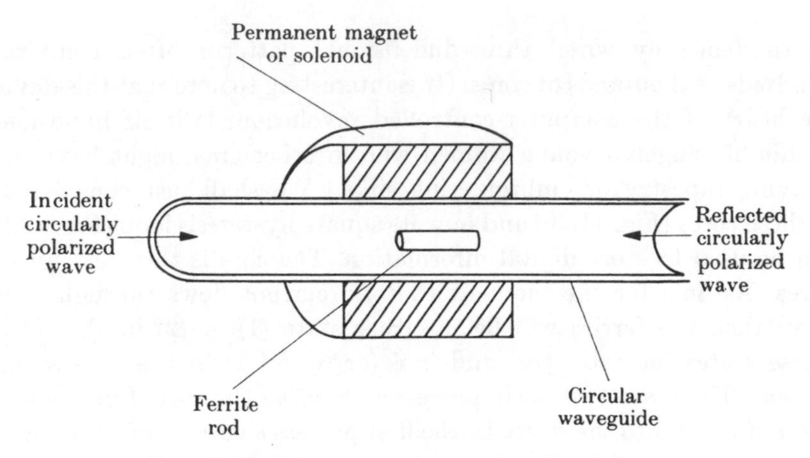

FIG. 11.22. Schematic representation of an isolator.

individual cores can be selected for energizing (or not) by the suitable coded input to the computer.

I have to note here that ferrite core memories have passed their heyday. The latest computers are semiconductor memories, but that does not mean that magnetic memories are on their way out. Their new champion, just entering the ring, is the magnetic bubble domain memory.

Isolators, as I mentioned briefly before, let an electromagnetic wave pass in one direction but heavily attenuate it in the reverse direction. There are a number of different types but I shall describe only one using a circularly polarized electromagnetic wave.

The ferrite is placed into a waveguide and biased by the magnetic field of a permanent magnet (Fig. 11.22). The input circularly polarized wave may propagate unattenuated but the reflected circularly polarized wave (which is now rotating in the opposite direction) is absorbed. Thus the operation of the device is based on the different attentuations of circularly polarized waves that rotate in opposite directions.

The usual explanation is given in classical terms. We have seen that a magnetic dipole will precess in a constant magnetic field. Now if in addition to the constant magnetic field in the z-direction there appears a magnetic vector in the x-direction (Fig. 11.23), then there is a further

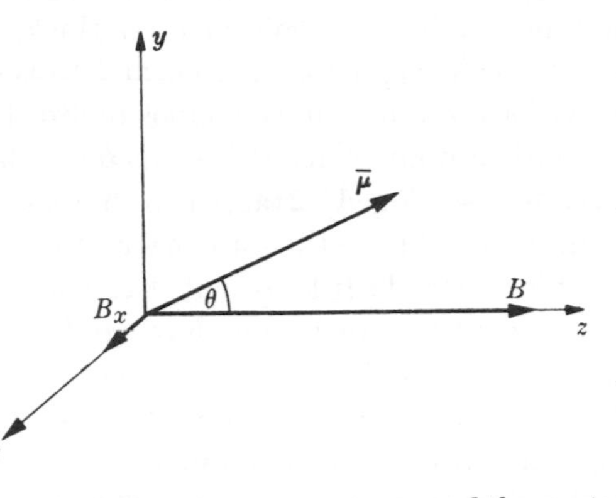

FIG. 11.23. The magnetic dipole moment precesses around the constant magnetic field, B. An additional magnetic field, B_x, gives an extra torque trying to increase the angle of precession.

torque acting upon the magnetic dipole. The effect of this torque is insignificant except when the extra magnetic field rotates with the speed of precession—and, of course, in the right direction. But this is exactly what happens for one of the circularly polarized waves when its frequency is equal to the frequency of precession. The interaction is then strong and energy is taken out of the electromagnetic wave in order to increase the angle of precession. Hence for one given frequency (the resonance frequency) and one sense of rotation (that of the reflected wave) the electromagnetic wave is absorbed.

The quantum explanation is based on the resonance phenomena discussed in the last section. The electromagnetic wave is absorbed because its energy is used to sponsor transitions between the respective energy levels. Unfortunately quantum mechanics provides no intuitive description of the effect of circularly polarized waves. You either believe that the result comes out of the mathematical description of the problem or, alternatively, you stick to the classical picture. This is what many quantum physicists do but to ease their conscience they put the offending noun between inverted commas. They do not claim that anything is really precessing, nonetheless, they talk of 'precession'.

Magnetic bubbles. So far I have mentioned domains as part of the explanation of ferromagnetism (section 11.4) and as a cause of jerks in the *B–H* loop (Fig. 11.5). The fact that small regions of magnetic materials can have differing magnetic alignments within a uniform physical shape did not have a use until a computer application was demonstrated by Brobeck in 1967. The technique is to grow very thin films epitaxially of either orthoferrites or garnets on a suitable substrate.† The film is only a few micrometres thick. All the domains can be aligned in a weak magnetic field normal to the film. Then by applying a stronger localized field in the opposite direction it is possible to produce a cylindrical domain (called a 'magnetic bubble') with its magnetic axis inverted (see Fig. 11.24a, which shows several).

An important question is, Does the bubble stay there when the strong field that created it is removed? It turns out that with suitable materials the domain-wall coercivity is great enough to produce a stable bubble, and that the most stable bubble size is when the radius is about equal to the garnet film thickness, i.e. a few micrometres. The next thing is to move the bubble about in a controlled manner.

One way of achieving controlled motion is by printing a pattern of small permalloy bars on the surface. The usual manufacturing technique for this is photoengraving using a photo-resist material similar to the process described for integrated circuits in section 9.20.

The actual devices look fairly complicated. Our intention is just to show the basic principles of how the bubbles can be persuaded to move from one place to another. So let us consider just two typical permalloy bars on the surface and assume the presence of a bubble with its north pole upwards, as shown in Fig. 11.24b.

In the absence of a magnetic field the permalloy bars are unmagnetized and have no effect upon the bubble. However, if a magnetic field (as shown in Fig. 11.24c) is applied then bar 1 becomes magnetized and the north pole of the bubble will move to the south pole of bar 1. How can we move the bubble to bar 2? We only need to change the direction of the magnetic field, as shown in Fig. 11.24d. Then bar 2

† The usual material is garnet with the general formula $R_3Fe_5O_{12}$, where R represents yttrium or a combination of rare-earth ions. Sometimes gallium or aluminium is substituted for some of the iron, to lower the saturation magnetization. In these ways the magnetic properties are bespoke by the chemists, and a typical successful composition is $Eu_1\ Er_2\ Ga_{0\cdot 7}\ Fe_{4\cdot 3}O_{12}$. Chemistry was never like this when I was at school.

becomes magnetized, and the bubble moves to the south pole of bar 2.

Bubbles can be detected by making them pass under a strip of indium antimonide which has a high *magneto-resistance*, i.e. its resistance is changed by a magnetic field. This property is closely connected with the large Hall effect in InSb, already mentioned in section 9.19.

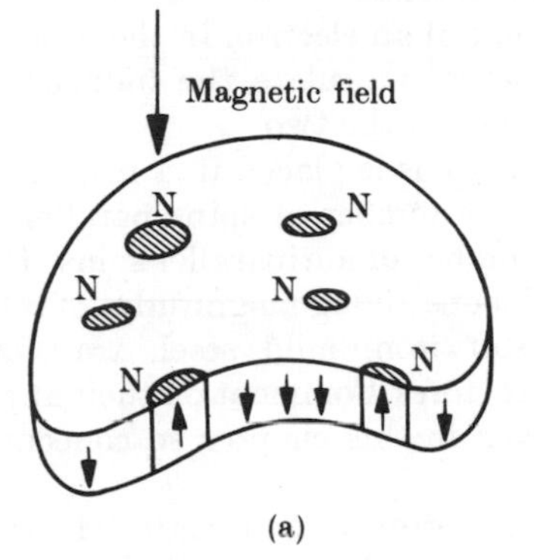

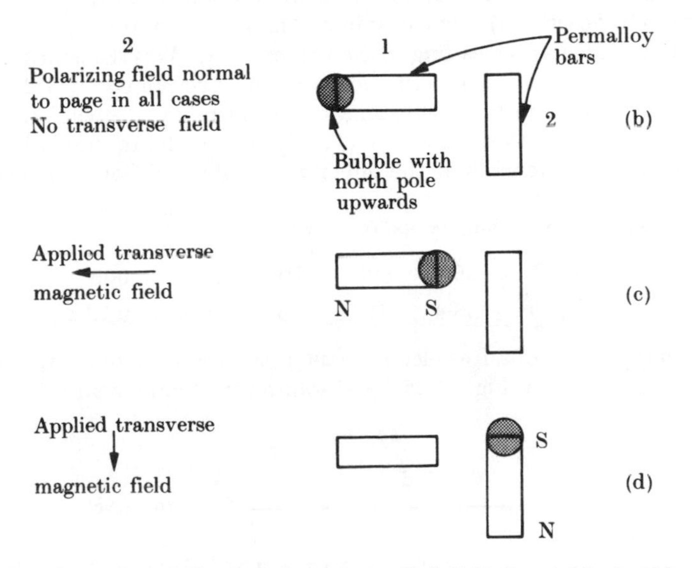

FIG. 11.24. Magnetic bubble domains (a) Applied field for stable bubbles. (b)–(d) illustrates how a bubble can be moved.

So you see that bubbles, by their presence or absence, may be used for storing binary information, and that the information can be read out. What is the advantage of using magnetic bubbles? Mainly size and speed. As the bubble dimensions are about 2 μm diameter, it is quite conceivable to have a mega-bit store on a 1 cm^2 slice, with a bit-speed (controlled by the domain velocity) of a few megahertz.

Examples.

1. Check whether eqn. (11.11) is dimensionally correct. Take reasonable values for N_a, Z and r and calculate the order of magnitude of the diamagnetic susceptibility in solids.
2. The magnetic moment of an electron in the ground state of the hydrogen atom is 1 Bohr magneton. Calculate the induced magnetic moment in a field of 1 weber/m^2. Compare the two.
3. A system of electron spins is placed in a magnetic field $B = 2$ weber/m^2 at a temperature T. The number of spins parallel to the magnetic field is twice as large as the number of antiparallel spins. Determine T.
4. Sketch typical B-H loops giving magnitudes of relevant quantities for the following materials: soft iron, mild steel, transformer laminations, and ferrite used in computer stores. Comment on their applications. (N.B. Much of this information is given in this chapter so concentrate on finding supplementary data.)
5. How have domains in ferromagnetic materials been observed?
6. In a magnetic flux density of 0·1 weber/m^2 at about what frequencies would you expect to observe (i) electron spin resonance, (ii) proton spin resonance?
7. Check the calculation leading to the values of the Weiss constant, magnetic moment and saturation magnetization for iron given in eqn. (11.27).
8. Show that the data for the magnetic susceptibility of nickel given below is consistent with the Curie law (eqn. (11.25)) and evaluate the Curie constant and temperature. Hence find the effective number of Bohr magnetons per atom.

   Atomic weight 58·7, density 8850 kg/m^3.

T(°C)	500	600	700	800	900
$\chi_m 10^5$	38·4	19·5	15	10·6	9·73

9. The energy levels of a free electron gas in the presence of an applied magnetic field are shown in Fig. 11.25 for absolute zero temperature. The relative

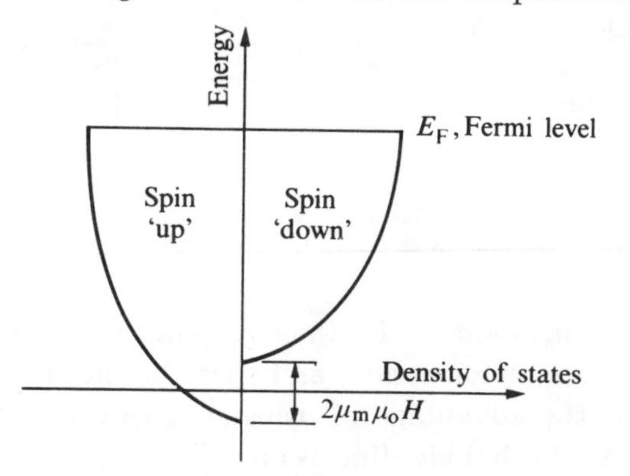

FIG. 11.25.

numbers of electrons with spins 'up' and 'down' will adjust so that the energies are equal at the Fermi level. Show that the paramagnetic susceptibility is given by the approximate expression

$$\chi_m = \mu_m^2 \mu_0 Z(E_F)$$

where μ_m is the magnetic moment of a free electron, μ_0 the free space permeability, and $Z(E_F)$ the density of states at the Fermi level. Assume that $\mu_m \mu_0 H \ll E_F$.

12. Introduction to masers and lasers

What goes up must come down.

It's all done with mirrors.

19th century aphorisms

12.1. Equilibrium

We have several times arrived at useful results by using the concept of equilibrium. It is a pretty basic tenet of science and like a similar idea, conservation of energy, it is always coming in handy. When we say that the electrons in a solid have a Fermi–Dirac distribution of energies we are really saying two things: first, that the system is in equilibrium; second, that it has a particular temperature. Temperature is a statistical concept and is bound up with the idea of equilibrium. On the one hand we cannot meaningfully speak of the temperature of a single particle; on the other, if we have a system of particles that is perturbed from equilibrium, say by accelerating *some* of them, then for a transient period the temperature cannot be specified, since there is no value of T that will make the Fermi function describe the actual distribution. Of course for electrons in a solid, or atoms in the gaseous state the effect of collisions rapidly flattens out the perturbation and the whole system returns to its equilibrium state and the *idea* of temperature becomes valid again, although its actual value may have changed.

We have on one or two occasions considered perturbed equilibrium. We saw, for example, in Chapter 1 that large currents may flow in a conductor with a very slight change in the energy distribution. Thus we could describe low field conduction in metals and semiconductors without departing from the equilibrium picture.

Lasers and masers are different. They have massively perturbed population distributions that are nevertheless in some kind of equilibrium. But when we come to consider what temperature corresponds to that equilibrium, it turns out to be negative. Now you know that 0 K is a temperature that can never quite be obtained by the most elaborate refrigerator; so how can we get a negative temperature? It's not inconsistent really because, as we shall show, a negative temperature is hotter than the greatest positive temperature. But before going further into Erewhon† let us return to earth and start lasers and masers from the beginning.

12.2. Two-state systems

Let us consider a material in which the electrons have only two narrow allowed energy levels as illustrated in Fig. 12.1. Provided

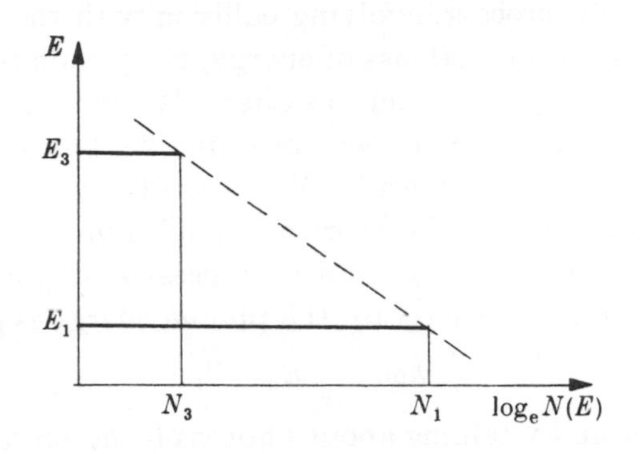

FIG. 12.1. Number of electrons in a natural two-state system as a function of energy. The dotted line shows the Boltzmann function, decaying exponentially with increasing energy.

that the whole system containing the material is in thermal equilibrium, the two allowed levels will be populated corresponding to a dynamic energy equilibrium between the bulk of the material and the electrons. The population of the energy levels is therefore accurately described by the temperature T of the system, and its appropriate statistics. For simplicity, and because it is usually correct, we shall consider a Boltzmann statistical system. This is true for gaseous atoms, and is a high energy approximation for solids.‡

The two levels we are considering are labelled E_1 and E_3 in Fig. 12.1 Later on we shall see what happens in a three-level system, with the third level called E_2; but for the moment do not be put off by this notation; we are still talking of only two levels. The numbers of electrons N_1, N_3 in the levels E_1, E_3 are related by the Boltzmann function so that they will be of the general form

$$N = N_0 \exp(-E/kT), \tag{12.1}$$

† Erewhon (approximately 'nowhere' backwards) was a country in the book of the same name by Samuel Butler where all habits and beliefs were the opposite of ours and were justified with impeccable logic and reasonableness.

‡ This is the same approximation that we made when the Fermi energy level was several times kT from the band edge in semiconductors.

where N_0 is a constant. Therefore

$$N_3 = N_1 \exp\left(-\frac{E_3 - E_1}{kT}\right). \tag{12,2}$$

As I said above, the electrons are in dynamic equilibrium, which means that the number of electrons descending from E_3 to E_1 is the same as the number leaping from E_1 to E_3. An electron at E_3 can lose the energy E_3–E_1 either by a process involving collision with the lattice or other particles, that is, a thermal loss of energy, or by radiation of a photon. I shall consider only the latter case here. We must make the mental reservation that not all processes are radiative, but all those that are good for lasers and masers must be. When a radiative transition between E_3 and E_1 occurs during the thermal equilibrium process, it is called *spontaneous* emission for the 'down' process and photon absorption for the 'up' process. In each case the photon energy is given by

$$\hbar\omega_{31} = E_3 - E_1. \tag{12.3}$$

What do we mean by talking about photons being present? It is a very basic law of physics that every body having a finite temperature will radiate thermal or 'black body' radiation. This radiation comes from the sort of internal transitions that I have just mentioned. As we saw in Chapter 2, the whole business of quantum theory historically started at this point. Planck found it necessary to say that atomic radiation was quantized in order to derive a radiation law that agreed with experiments. This famous radiation equation is

$$U(\omega)\,\mathrm{d}\omega = \frac{\hbar\omega^3\,\mathrm{d}\omega}{\pi^2 c^2\{\exp(\hbar\omega/kT) - 1\}}, \tag{12.4}$$

where $U(\omega)$ is the energy radiated from the body at temperature T in a band of the frequency spectrum of width $\mathrm{d}\omega$ and at an angular frequency ω. The derivation can be found in many textbooks.

So far we have talked about photons generated within the material. Now if photons of energy $\hbar\omega_{31}$ are shone on to the system from outside, a process called *stimulated* emission occurs. Either the photon gets together with an electron in a lower (E_1) state and pushes it up to E_3; or, less obviously, it *stimulates* the emission by an E_3-state electron of a photon ($\hbar\omega_{31}$). In the latter case one photon enters the system and two photons leave it. It was one of Einstein's many remarkable contributions to physics to recognize, as early as 1917, that both these events must be occurring in a thermodynamical equilibrium; he then went on to prove that the probabilities of a photon stimulating an 'up' or a 'down' transition were exactly equal. The proof is simple and elegant so we will give it here.

Consider our system, remembering that we have two states in equilibrium. The rate of stimulated transitions ($R_{1\to3}$) from the lower to the upper state will be proportional to both the number of electrons in the lower state and the number of photons that can cause the transition. So we can write

$$R_{1\to3} = N_1 B_{13} U(\omega_{31})\,\mathrm{d}\omega, \tag{12.5}$$

where the constant of proportionality B_{13} is the stimulated emission probability, often referred to as the Einstein B-coefficient. For the reverse transition, from E_3 to E_1 we have a similar expression for stimulated emission, except that we will write the Einstein B-coefficient as B_{31}. There is also spontaneous emission. The rate for this to occur will be proportional only to the number of electrons in the upper state, since the spontaneous effect is not dependent on external stimuli. The constant of proportionality or the probability of each electron in the upper state spontaneously emitting is called the Einstein A-coefficient, denoted by A_{31}. Hence

$$R_{3\to1} = N_3\{A_{31}+B_{31}U(\omega_{31})\}\,\mathrm{d}\omega. \tag{12.6}$$

In equilibrium the rates are equal

$$R_{1\to3} = R_{3\to1}, \tag{12.7}$$

that is,

$$N_1 B_{13} U(\omega_{31})\,\mathrm{d}\omega = N_3\{A_{31}+B_{31}U(\omega_{31})\}\,\mathrm{d}\omega. \tag{12.8}$$

After a little algebra, using eqn. (12.1) to relate N_3 to N_1, we get

$$U(\omega_{31})\,\mathrm{d}\omega = \frac{A_{31}\,\mathrm{d}\omega}{B_{13}\exp(\hbar\omega_{31}/kT)-B_{31}} \tag{12.9}$$

Comparing this with eqn. (12.4), which is a universal truth as far as we can tell, we find that our (or rather Einstein's) B-coefficients must be equal

$$B_{13} = B_{31}, \tag{12.10}$$

that is, stimulated emission and absorption are equally likely.

Now we can say what happens when light of frequency ω_{31} shines on the two-state system. The probabilities of 'up' and down' transitions are exactly equal but as there are more electrons in the lower state (in accordance with eqn. (12.1)) there will be more transitions from E_1 to E_3. In other words, there will be net absorption of photons. This we often see in nature. For example, many crystalline copper salts have two energy bands separated by photon energy corresponding to yellow light. Thus, when viewed in white light, the yellow part is absorbed and the crystal transmits and reflects the complementary colour, blue. Ruby (chromium ions in crystalline alumina) has an absorption band in the green by this mechanism, and hence looks red in white light.

We usually describe absorption of light in a crystal by an absorption coefficient, α, which defines the exponential decay of intensity, I, as the light goes through the crystal, viz.:

$$I = I_0 \exp(-\alpha x). \tag{12.11}$$

From the physical argument used above, it follows that the net rate of absorption will be proportional both to the B-coefficient and to the excess population in the lower level compared with the upper one. Hence the loss of intensity in a time $\mathrm{d}t$ may be written as

$$\mathrm{d}I \sim I(N_1 - N_3)B\,\mathrm{d}t, \tag{12.12}$$

or, for a travelling wave, as

$$\mathrm{d}I \sim I(N_1 - N_3)B\,\mathrm{d}x, \tag{12.13}$$

leading to the relationship

$$\alpha \sim B(N_1 - N_3). \tag{12.14}$$

When light is absorbed in this way the population of the upper level is increased, as more transitions go up than come down. Normally this perturbation from the equilibrium condition is small. But if we have an increasingly intense 'pump' light source, the number in level 3 will go on increasing, by the same amount as those in level 1 decrease. Fairly obviously, there is a limit when the levels are equally populated and the pump is infinitely strong. This is illustrated in Fig. 12.2. For intense pumping the non-equilibrium level populations (denoted by an asterisk) become almost equal:

$$N_1^* \simeq N_3^* \simeq \frac{N_1 + N_3}{2}. \tag{12.15}$$

Now let us consider a 3-level system, with the third level E_2 between E_1 and E_3, also shown in Fig. 12.2. The pumping will have no effect on its population, which is the equilibrium value N_2. So with the 3-level system strongly pumped, the number of electrons in the three states are N_1^*, N_2, and N_3^*. Suppose that some photons come along with energy

$$\hbar\omega_{32} = E_3 - E_2. \tag{12.16}$$

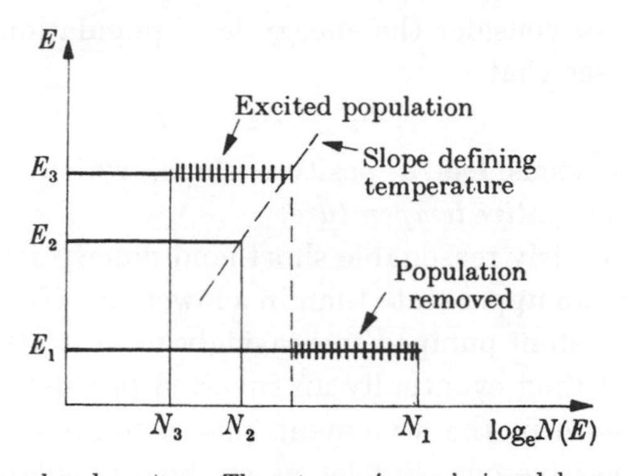

FIG. 12.2. The three level system. The strong 'pump' signal has equalized levels E_1 and E_3 so that E_3 now has a greater population than E_2. The dotted line shows how population is changing with energy, as in Fig. 12.1, but now it has a positive slope.

They will clearly interact with the system, causing stimulated emission by transitions from E_3 to E_2, and absorption by transitions from E_2 to E_3. But now we have an unnatural occurrence: there are more electrons in the upper state (E_3) than in the lower (E_2). So instead of there being a net *absorption* of photons of energy $\hbar\omega_{32}$, there will be a net *emission.* The 3-level system will amplify a photon of frequency ω_{32}, which is called the *signal frequency.* If the signal frequency is in the microwave region, this is the basis of a *maser*, an acronym for *m*icrowave *a*mplification by *s*timulated *e*mission of *r*adiation. If the signal is in the optical range, substitute 'light' for 'microwave' and the acronym is *laser.*

When there are more electrons in an upper than a lower level as in the case of E_3, and E_2 in Fig. 12.2, it is justifiable jargon to speak of an 'inverted population'. The other point we should clear up before describing some real system with inverted populations is the one concerned with temperature. From eqn. (12.1) the locus of the line representing the populations of the various energy levels

$$\frac{\mathrm{d}E}{\mathrm{d}N} = -\frac{kT}{N} \tag{12.17}$$

(shown as a dotted curve in Fig. 12.1) has a negative slope proportional to T/N. Now look at Fig. 12.2. First consider the populations N_3^* and N_1^*. They are in a steady state in the sense that as long as the pump continues steadily, they do not change with time. But for these two levels with a finite energy difference there is virtually no difference in population. Therefore if we regard eqn. (12.17) as a way of defining temperature, for a well-pumped two-level system the temperature is infinite. If we now consider the energy-level populations at E_3 and E_2 in Fig. 12.2, we see that

$$N_3^* > N_2 \tag{12.18}$$

and the dE/dN locus has a positive slope, which by eqn. (12.17) corresponds to a *negative temperature.*

Again this is a fairly reasonable shorthand description of there being more electrons in an upper state than in a lower one. Now if you imagine a natural-state system pumped increasingly until it attains an infinite temperature and then eventually an inverted population, you will see there is *some* sense in the statement that a negative temperature is hotter than a positive one. But let us see how population inversion is achieved in practice, before going on to discuss a few more properties of stimulated radiation.

12.3. Resonators

If the energy levels were infinitely sharp then the frequency of the photon would be perfectly defined. In practice this is not so. The population inversion is usually achieved in a narrow but finite band resulting in the emission of photons of somewhat different frequency. For single-frequency emission all the excited states should decay in unison. But how would an electron in one corner of the material know when its mate in the other corner decides to take the plunge? They need some kind of coordinating agent or—in the parlance of the electronic engineer—a feedback mechanism. What could give the required feedback? The photons themselves. They stimulate the emission of further photons as discussed in the previous section, and also ensure that the emissions should occur at the right time. If we want to form a somewhat better physical picture of this feedback mechanism it is advisable to return to the language of classical physics and talk of waves and relative phases. Thus instead of a photon being emitted we may say that an electromagnetic wave propagates in which any two points bear strict phase relationships relative to each other. This phase information will be retained if we put perfect reflectors in the path of the waves on both sides, constructing thereby a resonator. The electromagnetic wave will then bounce to and fro between the two reflectors establishing standing waves, which also implies that the region between the two reflectors must be an integer multiple of half wavelengths. Thus in a practical case we have a relatively wide frequency band in which population inversion is achieved, and the actual frequencies of oscillation within this band are determined by the possible resonant frequencies of the resonator. This is, indeed, the usual situation in a laser. Unless we do something clever the emission is at a large number of closely spaced frequencies. The phenomenon can, incidentally, be useful as well, offering the possibility of obtaining very short pulses, as will be discussed in Section 12.5.

12.4. Some practical laser systems

How can we build a practical laser? We need a material with suitable energy levels, a pump, and a resonator. Is it easy to find a combination of these three factors which will result in laser oscillation? It is like many other things; it seems prohibitively difficult before you've done it, and exceedingly easy afterwards. By now thousands of 'lasing' materials have been reported and there must be millions in which laser oscillations are possible.

There are all kinds of lasers in existence; they can be organic or inorganic, crystalline or non-crystalline, insulator or semiconductor, gas or liquid, they can be of fixed frequency or tunable, high power or low power, CW or pulsed. They may be pumped by another laser, by fluorescent lamps, by electric arcs, by electron irradiation, by injected electrons or by entirely non-electrical means as in a chemical laser. You can see that a mere enumeration of the various realizations could easily take up all our time. I shall be able to do no more than describe a few of the better known lasers.

Solid state lasers. The first laser constructed in 1960 was a ruby laser. The energy-level diagram for the transitions in ruby (Cr ions in an Al_2O_3 lattice) is given in Fig. 12.3. I remarked above that ruby owed its characteristic red colour to absorption bands of the complementary colour, green. This absorption is used in the pumping process. A typical arrangement is sketched in Fig. 12.4, which shows how the light from a xenon discharge flash tube 'pumps' the ruby to an excited state. Now the emission process is somewhat different here from that which I sketched previously for three-level systems. The atoms go from level 3 into level 2 by giving up their energy to the lattice in the form of heat. They spend a long enough time† in level 2 to permit the population there to become greater than that of level 1. So laser action may now take place between levels 2 and 1, giving out red light.

† Energy levels in which atoms can pause for a fairly long time (a few milliseconds in the present case) are called 'metastable'.

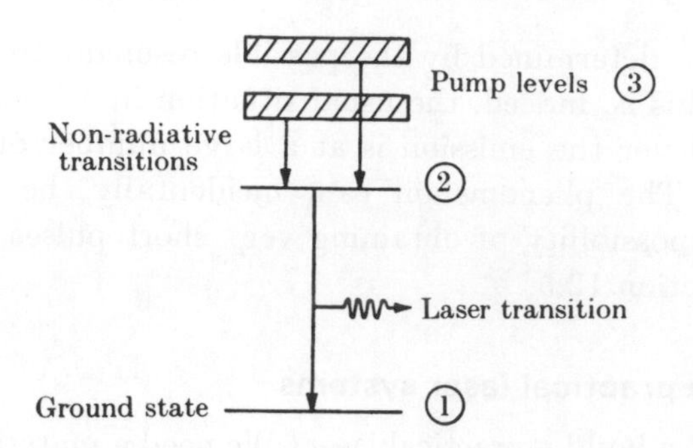

Fig. 12.3. Energy levels of the Cr^{3} ion in ruby. The pump levels are broad bands in the green and blue which efficiently absorb the flash tube light. Level 2 is really a doublet (two lines very close to each other) so that the laser light consists of the two red lines of wavelengths 694·3 nm and 692·9 nm.

The ruby itself is an artificially grown single crystal that is usually a cylinder with its ends polished optically flat. The ends have dielectric (or metal) mirrors evaporated on to them. Thus, as envisaged in the previous section, the resonator comprises two reflectors. At optical frequencies these so-called open resonators are very efficient. Some power is certainly lost by diffraction but these losses are small provided the dimensions of the mirror are much larger than the wavelength. It needs to be noted that one of the mirrors must be imperfect in order to get the power out.

Another notable representative of solid-state crystalline lasers is Nd^{3+}:YAG, that is neodymium ions in an yttrium-aluminium-garnet. It is a four-level laser radiating at a wavelength of 1·06 μm pumped by a tungsten or mercury lamp.

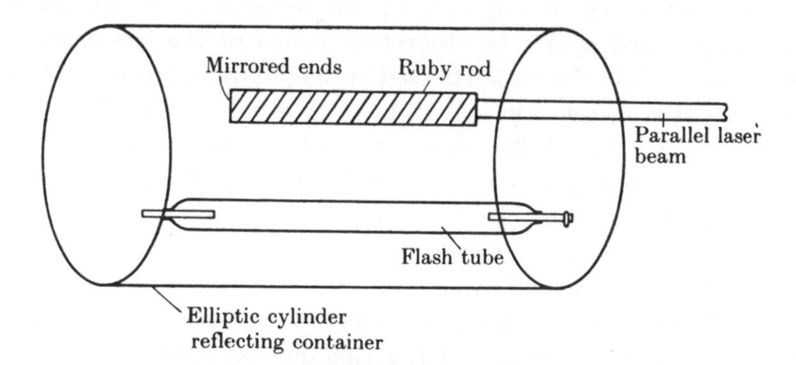

Fig. 12.4. General arrangement of a ruby laser. The ruby and the flash tube are mounted along the foci of the elliptic cylinder reflector for maximum transference of pump light.

Laser operation at the same frequency may be achieved by putting the neodymium ions into a *glass* host material. Glasses have several advantages in comparison with crystals: they are isotropic, they can be doped at high concentrations with excellent uniformity, they can be fabricated by a variety of processes (drilling, drawing, fusion, cladding), they can have indices of refraction in a fairly wide range, and last but not least, they are considerably cheaper than crystalline materials. Their disadvantage is low thermal conductivity, which makes glass lasers unsuitable for high average power applications.

The gaseous discharge laser. When a current is passed through a gas, as happens in a fluorescent lamp or a neon sign, most of the charged particles making up the current come from gas atoms that have been ionized by collision. But as well as atoms being completely dispossessed of their electrons, the collisional process causes some bound electrons to gain extra energy and go into a higher state, that is, a state described by higher quantum numbers. You will remember that we had a formula for the simplest gas, hydrogen, in Chapter 4:

$$E_n = -\frac{13{\cdot}6}{n^2}. \tag{12.19}$$

This shows that there is an infinite number of excited states above the ground state at $-13{\cdot}6$ eV, getting closer together as the ionization level (0 eV) is approached.

In the helium–neon laser the active 'lasing' gas is neon but there is about 7–10 times as much helium as neon present. Consequently there are quite a lot of helium atoms excited to states about 20 eV above the ground state (Fig. 12.5). Now helium atoms in these particular states can get rid of their energy in one favourable way—by collision with other atoms that also have levels at the same energies. Since neon happens to have suitably placed energy levels, it can take over the extra energy making the population of the upper levels ($3a'$, $3b'$) more numerous than that of the lower level ($2'$), and thus laser action may occur. It is of course necessary to adjust gas pressures, discharge tube dimensions, and current quite critically to get the inverted population; in particular it is obtained only in a fairly narrow range of gas pressures around 1 Torr.

The reflectors are external to the tube as shown in Fig. 12.6. Note that the windows are optical flats oriented at the Brewster angle θ_B in order to minimize reflections for the desired polarization. The advantage of spherical mirrors is that their adjustment is not critical and they also

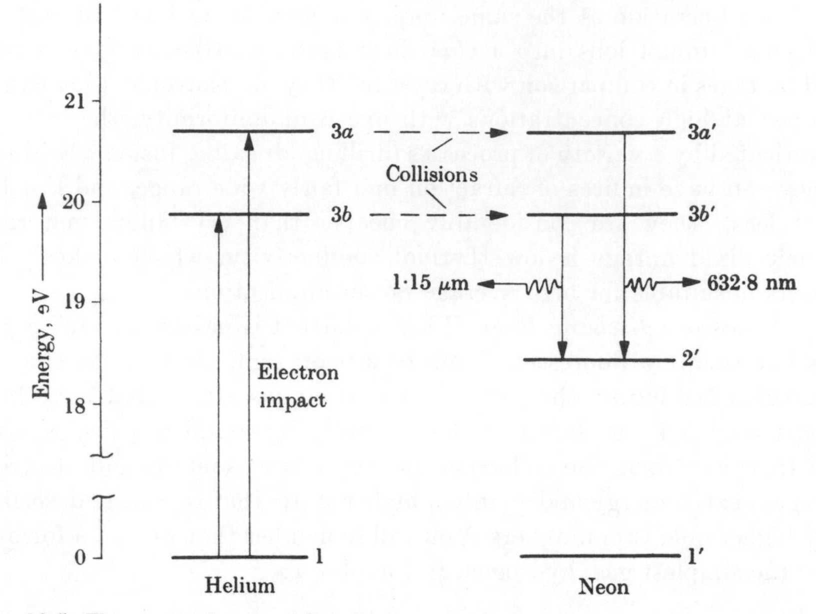

FIG. 12.5. The energy levels of interest for a helium–neon laser. Helium atoms get excited to levels 3*a* and 3*b* due to the impact of accelerated electrons. Neon atoms, which happen to have the same energy levels (3*a*′, 3*b*′) collide with helium atoms and take over the extra energy. Laser action may now occur at two distinct wavelengths corresponding to radiative transitions from levels 3*a*′ and 3*b*′ to a lower level 2′.

improve efficiency. Dielectric mirrors are also used not only because they give better reflections than metal mirrors but because also they can select the wavelength required from the two possible transitions shown in Fig. 12.5.

A close, though more powerful, relative of the He–Ne laser is the *argon ion* laser operating in a pure Ar discharge. The pumping into the upper level is achieved by multiple collisions between electrons and argon ions. It can deliver CW power up to about 40 watts at 488 nm

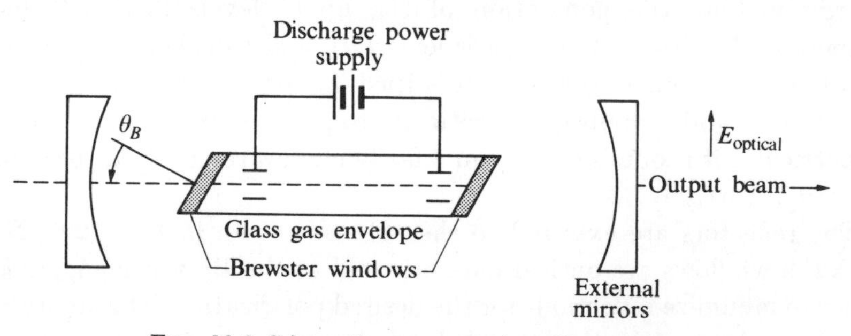

FIG. 12.6. Schematic representation of a gas laser.

and 514 nm wavelengths. It is, in the company of the He–Ne laser, the one most often seen on laboratory benches.

The CO_2 laser is capable of delivering even higher power (tens of kW) at the wavelength of 10·6 μm. It is still a discharge laser but the energy levels of interest are different from those discussed up to now. They are due to the internal vibrations of the CO_2 molecule. All such molecular lasers oscillate in the infrared, some of them (e.g. the HCN laser working at 537 μm) approaching the microwave range.

Dye lasers. This is an interesting class of lasers employing fluorescent organic dyes as the active material. Their distinguishing feature is the broad emission spectrum which permits the tuning of the laser oscillations.

The energy levels of interest are shown in Fig. 12.7a. The heavy lines represent vibrational states and the lighter lines represent the rotational fine structure which provides a near continuum of states. The pump (flashlamp or another laser) will excite states in the S_1 band ($A \to b$ transition) which will decay non-radiatively to B and will then make a radiative transition ($B \to a$) to an energy level in the S_0 band. Depending on the endpoint a a wide range of frequencies may be emitted. Finally, the cycle is closed by the non-radiative $a \to A$ transition. Unfortunately, at any given frequency of operation, there are some other competing non-radiative processes indicated by the dotted lines. A photon may be absorbed by exciting some state in the higher S_2 band or there might be a non-radiative decay to the ground state via some other energy levels. There is net gain (meaning the gain of the wave during a single transit between the reflectors) if the absorptive processes are weaker than the fluorescent processes.

The tuning range of a specific dye laser (rhodamine 6G) is shown in Fig. 12.7b by the shaded area where the fluorescent and absorption curves are also plotted as a function of wavelength. Laser action becomes possible when the absorption curve intersects the fluorescence curve. At the long wavelength extreme the gain of the laser (meaning the gain of the wave during a single transit between the reflectors) becomes too small for oscillation as a result of the decrease in fluorescence efficiency.

How can we tune the laser? An ingenious solution is shown in Fig. 12.7c where one of the mirrors is a rotatable diffraction grating. The oscillation frequency of the laser will be determined by the angular position of the grating which will reflect a different frequency at each position. The tuneable range is a respectable 7 %. Note that this is not

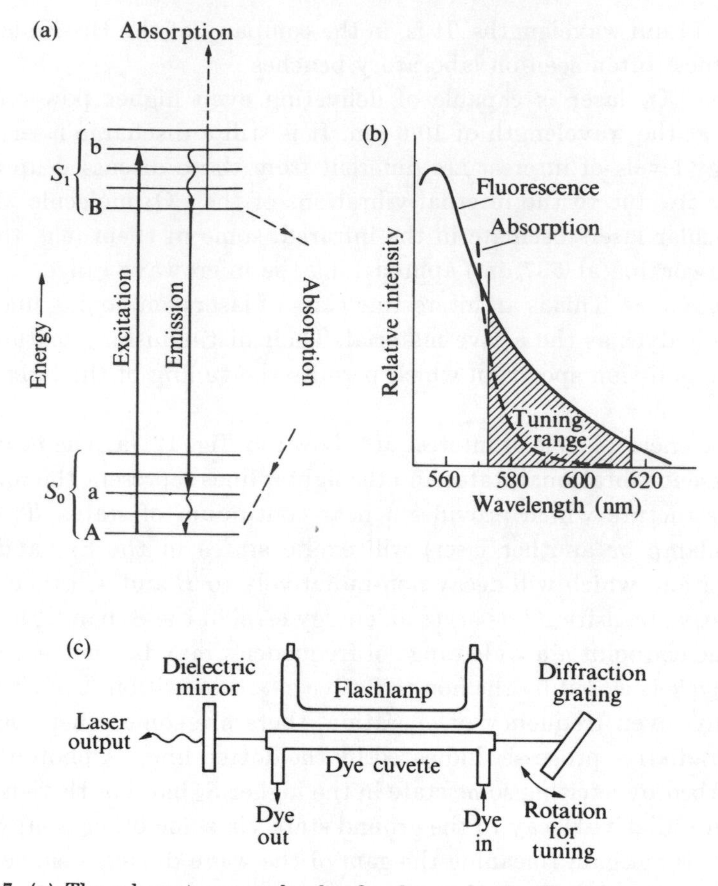

FIG. 12.7. (a) The relevant energy levels of a dye molecule. The wavy arrows from b to B and from a to A represent non-radiative transitions. The broken lines leading to the right also represent non-radiative transitions in which some other states are involved. (b) The tuning range of rhodamine 6G as a function of wavelength. (c) Schematic representation of a tuneable dye laser.

the end of the dye laser's tuneability. By choosing the appropriate dyes any frequency within the visible range may be obtained.

Parametric oscillators. In principle this is the same thing as already explained in connection with varactor diodes in Section 9.13. The main differences are that in the present case (i) the nonlinear capacitance is replaced by a nonlinear optical medium, (ii) the dimensions are now large in comparison with the wavelength; hence wave propagation effects need to be taken into account, and (iii) instead of amplifiers we are concerned here with oscillators. What is the advantage of parametric oscillators? Why should we worry about three separate frequencies when we can easily build oscillators at single frequencies? The reason is that we can have tuneable outputs.

A schematic diagram of the optical parametric oscillator is shown in Fig. 12.8. The parametric pump (not to be confused with the pump needed to make the laser work) is a laser oscillating at ω_3. There is also a resonator which has resonant frequencies of ω_1 and ω_2. If the waves at these frequencies satisfy both the $\omega_3 = \omega_1 + \omega_2$ and the $k_3 = k_1 + k_2$ (k = propagation coefficient as in our classical studies of Chapter 1) conditions then there is a parametric interaction between the waves. The power at ω_1 and ω_2 builds up from the general noise background at the expense of pump power. Thus we have output at all three frequencies. But why is this set-up tuneable? Because of the particular properties of the chosen nonlinear medium. It is a crystal in which the dielectric constant is dependent on the direction of propagation. By rotating the crystal the matching condition for the propagation constants is satisfied at another set of frequencies ω_1' and ω_2' still obeying $\omega_3 = \omega_1' + \omega_2'$.

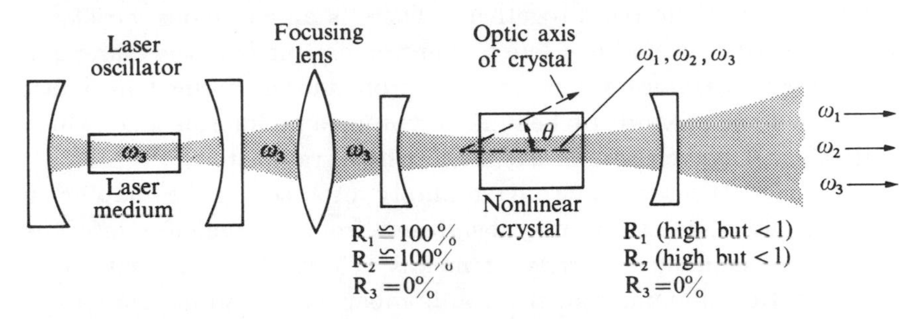

FIG. 12.8. Schematic representation of a tuneable parametric oscillator. R_1, R_2, and R_3 are the reflectivities of the mirrors at frequencies ω_1, ω_2, ω_3 respectively.

The tuning range is quite considerable as may be seen in Fig. 12.9 which shows results obtained experimentally in the set-up of Fig. 12.8 employing as the nonlinear medium an ADP (ammonium dihydrogen phosphate) crystal.

Gasdynamic lasers. The essential difference between these lasers and all the others discussed so far is that no electrical input is needed. One starts with a hot gas (e.g. CO_2) in the so called stagnation region. Then most of the energy is associated with the random translation and rotation of the gas molecules and only about one tenth of the energy is associated with vibration. Next the gas is expanded through a supersonic nozzle causing the translational and rotational energies to change into the directed kinetic energy of the flow. The vibrational energy would disappear entirely if it remained in equilibrium with the decreasing gas temperature. But the vibrational relaxation times are long in comparison with the expansion time, hence the population of the vibrational levels remains practically unchanged. At the same time the lower level population diminishes rapidly with the expansion, leading to significant population inversion after a few centimetres downstream. For CO_2 gas the emission wavelength is again 10·6 μm, using other gases the typical range is from 8 μm to 14 μm although oscillations may be achieved at much shorter wavelengths as well.

The advantage of gas-dynamic lasers is the potential for high average powers because waste energy can be removed quickly by high-speed flow.

Chemical lasers. As the name suggests the population inversion comes about as a result of chemical reactions. In some of the types the reaction is initiated by an electrical discharge but it is also possible to take two commercially available bottled gases and obtain laser oscillations just by letting them react together.† There is an enormous variety of chemical lasers. I would just like to mention one made in the Clarendon Laboratory at Oxford which uses krypton fluoride formed in a fast discharge in a mixture of helium, krypton, and fluorine and which oscillates at a wavelength of 249 nm in the ultraviolet.

Semiconductor lasers. As we have amply discussed in the chapter on semiconductors, electrons may be raised from the valence into the conduction band by a variety of means. Three kinds of excitations, namely optical illumination, electron bombardment and impact ionization by electron–hole pairs under avalanche breakdown may indeed lead to laser action in a bulk semiconductor but the most important semiconductor laser relies on the injection of carriers in a p–n junction. It is similar to the semiconductor lamp mentioned in Chapter 9. Electrons injected into the p-side of the junction cause an excessive local population, which may return to equilibrium by a radiation process. A

† The advent of chemical lasers raises an intriguing problem I have often asked myself. What path would have technology followed if electricity had never been discovered? The question may be posed because electricity and technology developed separately, the former being a purely scientific pastime until the fourth decade of the last century. Had scientists been less interested in electricity or had they been just a bit lazier, it is quite conceivable that the social need for fast communications (following the invention of the locomotive) would have been satisfied by systems based on modulated light. In the search for better light sources the chemical laser could then have been invented by the joint efforts of chemists and communication engineers a century ago.

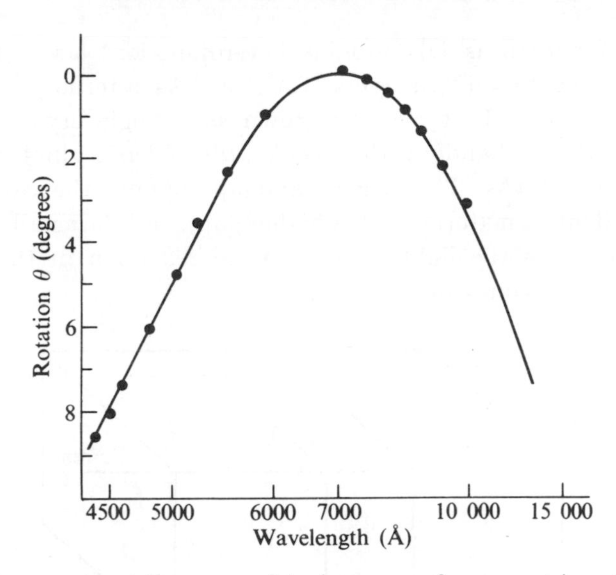

FIG. 12.9. An experimentally measured tuning curve of a parametric oscillator using ADP as the nonlinear medium.

semiconductor lamp becomes a laser if the following conditions hold good.

(i) The diode is pumped so hard that an inverted, rather than just a non-equilibrium, population is achieved. 'Pumping' in this context means simply the injection of electrons, and 'hard pumping' means the injection of large number of electrons. Thus for laser action we need a high starting current.

(ii) The ends of the active area are made into mirrors. This is usually done by cleaving the crystal and simply using the flat boundary between the crystal and the air as the reflector.

(iii) The semiconductor is of the direct-gap type in which transitions from the bottom of the conduction band to the top of the valence band are more likely. Thus the commoner semiconductors, Si and Ge, cannot be used; the logical choice is the next widely investigated semiconductor, GaAs. The resulting radiation corresponds to the energy gap (1·4 eV) just beyond the edge of the visible region.

There are quite a number of semiconductor materials from which p–n junction injection lasers have been made but for CW output they need to be cooled to low temperatures (to 77 K or below).

The practical need for a room temperature device was fulfilled by the ingenious construction of the sandwich structure shown in Fig. 12.10, usually referred to as DH (double heterojunction) laser. The p-type GaAs layer is between two layers of $Al_xGa_{1-x}As$, a ternary alloy chosen for three reasons: (i) it may be grown as a single crystal on GaAs, (ii) it has a larger bandgap than GaAs, (iii) it has a smaller dielectric constant than GaAs. The larger bandgap confines the carriers (they cannot spill into a material with a higher gap), and the smaller dielectric constant confines the light (causing total reflection in the region of higher dielectric constant).

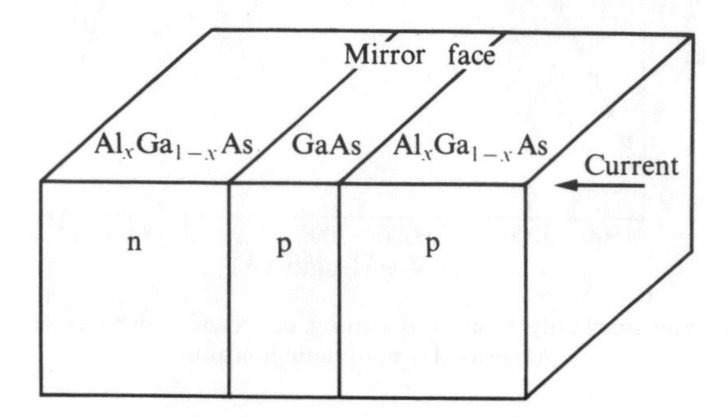

FIG. 12.10. Schematic representation of a double hetero structure (DH) semiconductor injection laser.

The biggest single advantage of semiconductor injection lasers is the potentially low price. They are small, compact, rugged, can be made by the same techniques as their low frequency counterparts, and could have (potentially) the same long life.

The obvious applications of these lasers are in communication systems. But before we finally decide to submerge fibres, repeaters and all into the Ocean, we must be fairly sure that *all* the lasers will have long (of the order of 10^5 hours) lifetimes. Thus the main problem to overcome is the mass production of reliable semiconductor lasers. A further problem of somewhat less importance is to tailor the frequencies (by finding the right bandgap materials) to the requirements of the given fibres.

12.5. Laser modes and mode control techniques

Transverse modes. What will be the amplitude distribution of the electromagnetic wave in the laser resonator? Will it be more or less uniform or will it vary violently over the cross-section? These questions were answered in a classical paper by Kogelnik and Li in 1966 showing both theoretically and experimentally the possible modes in a laser resonator. The experiments were performed in a He–Ne laser producing the mode patterns of Fig. 12.11. For most applications we would like a nice, clean beam as shown in the upper left-hand corner. How can we eliminate the others? By introducing losses for the higher order modes. This may be done, for example, by reducing the size of the reflector. Since the higher order modes have higher diffraction losses (they radiate out more) this will distinguish in favour of the fundamental mode. However this will influence laser operation in the fundamental mode as well; thus a more effective method is to place an iris diaphragm into the resonator which lets through the fundamental mode but 'intercepts' the higher order modes.

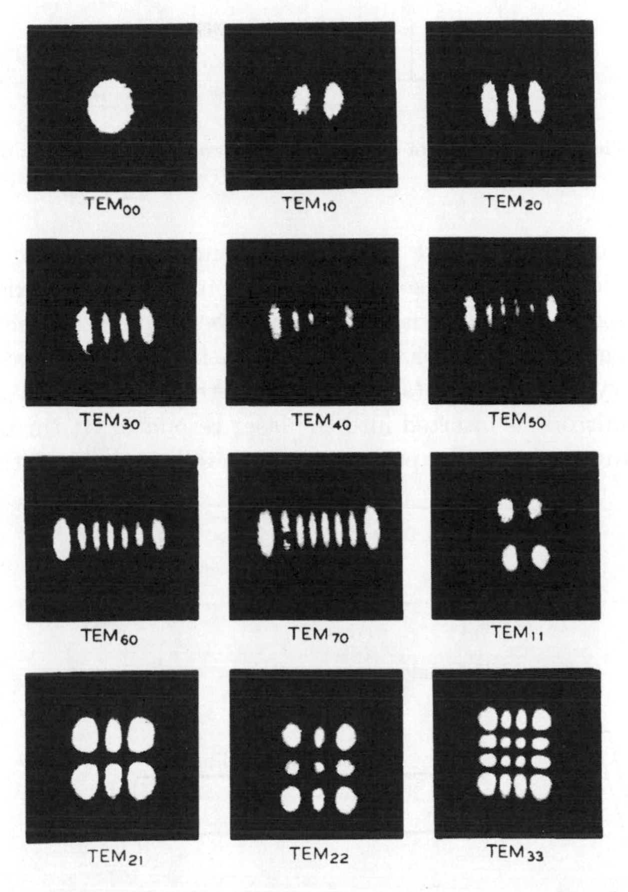

FIG. 12.11. Experimentally measured transverse mode patterns in a He–Ne laser having a resonator of rectangular symmetry. (H. Kogelnik and T. Li, *Proc. IEEE 54*, pp. 1312–1329, Oct. 1966).

Axial modes. As mentioned in Section 12.3 and shown in Fig. 12.12, laser oscillations are possible at a number of axial modes each having an integral number of half wavelengths in the resonator. The frequency difference between the nearest modes is $c_m/2L$ (see Example 2), where L is the length of the resonator and c_m is the velocity of light in the

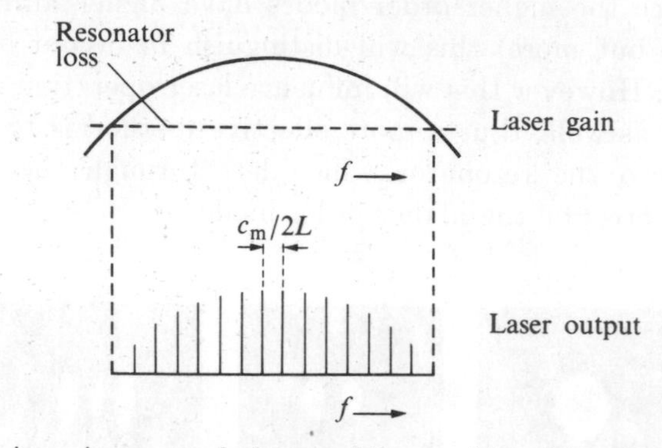

FIG. 12.12. The inversion curve of a laser and the possible axial modes as a function of frequency.

medium. How can we have a single frequency output? One way is to reduce the length of the resonator so that only one mode exists within the inversion range of the laser. Another technique is to use the good offices of another resonator. This is shown in Fig. 12.13, where a so-called Fabry–Perot etalon (a piece of dielectric slab with two partially reflecting mirrors) is inserted into the laser resonator. It turns out that the resonances of this composite structure follow those of the etalon,

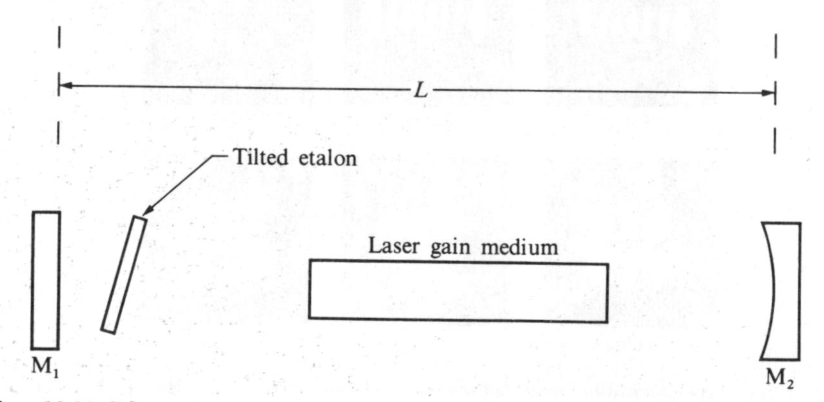

FIG. 12.13. Schematic representation of a laser oscillator in which single mode operation is achieved with the aid of an etalon.

PLATE 2. Lead coated glass sphere supported by the magnetic field produced by currents in two superconducting lead rings (Simon, 1953).

that is the frequency spacing is $c/2d$, where d is the etalon thickness. Since $d \ll L$, single frequency operation becomes possible.

Aren't we losing too much power by eliminating all that many axial modes? No, we lose very little power because the modes are not independent of each other. The best explanation is a kind of optical Darwinism or the survival of the fittest. Imagine a pack of young animals (modes) competing for a certain amount of food (inverted population). If the growth of some of the animals is prevented, the others grow fatter. This is called *mode competition*.

Q switching. This is a method for concentrating a large amount of power into a short time period. It is based on the fact that for the build-up of oscillations a feedback mechanism is needed, usually provided by mirrors. If pumping goes on but we spoil the reflectivity of one of the mirrors (i.e. spoil the Q of the resonator) by some means, then there will be a lot of population inversion without any output. Restoring now the reflectivity (switching the Q) for a short period to its normal value, the laser oscillations can suddenly build up, resulting in a giant pulse output. The pulse duration might be as short as a few nanoseconds, the power as much as 10^{10} watts and the repetition frequency may be up to 100 kHz. The easiest, though not the most practical, way of spoiling the Q is by rotating the mirror. The Q is then high only for the short period the mirrors are nearly parallel.

Cavity dumping. This is another, very similar method for obtaining short pulses also based on manipulating the Q of the resonator (called also 'cavity'; that's where the name comes from). We let the pump work and make the reflectivity 100% for a certain period, so the oscillations can build up but cannot get out. If we now lower the reflectivity to zero all the accumulated energy will be dumped in a time equal to twice the transit time across the resonator. The method may be used up to about a repetition rate of 30 MHz.

Mode locking. We have implied earlier that it is undesirable to have a number of axial modes in a laser. This is not always so. The large number of modes may come useful if we wish to produce very short pulses of the order of picoseconds. The trick is to bring the various axial modes into definite relationships with each other. How will that help in producing short pulses? It's possible to get a rough idea by doing a little mathematics. Let us assume that there are $N+1$ modes oscillating at frequences $\omega_0+l\omega$, where $l = (-N/2, \ldots 0, \ldots N/2)$, that they all have the same phase and amplitude, and they all travel in the positive z direction (the set travelling in the opposite direction will make a similar contribution). The electric field may then be written in the form

$$\mathscr{E}(z,t) = \mathscr{E}_0 \sum_{l=-N/2}^{N/2} \exp[-\mathrm{i}(\omega_0 + l\omega)(t - z/c_\mathrm{m})]$$

$$= \mathscr{E}_0 \exp[-\mathrm{i}\omega_0(t - z/c_\mathrm{m})]F(t - z/c_\mathrm{m}), \qquad (12.20)$$

where $\mathscr{E}_0$ is a constant and

$$F(x) = \frac{\sin(\frac{1}{2}N\omega x)}{\sin(\frac{1}{2}\omega x)}. \qquad (12.21)$$

Eqn. (12.20) represents a travelling wave whose frequency is ω_0 and its shape (envelope) is given by the function F. If $N \gg 1$, F is of the form of a sharp pulse of width $4\pi/N\omega$ and it is repeated with a frequency of ω. Taking $N = 100$, a resonator length of 10 cm and a refractive index of 2, we get a pulsewidth of 27 ps and a repetition frequency of 750 MHz. The situation is of course a lot more complicated in a practical laser but the above figures give good guidance. The shortest pulses to date have been obtained in dye lasers with pulsewidths well below 1 ps.

How can we lock the modes? The most popular method is to put a saturable absorber in the resonator which attenuates at low fields but not at high fields. Why would a saturable absorber lock the modes? A rough answer may be produced by the following argument: When the modes are randomly phased relative to each other, the sum of the amplitudes at any given moment is small, hence they will be adversely affected by the saturable absorber. However, if they all add up in phase, their amplitude becomes large and they will *not* be affected by the saturable absorber. Thus the only mode of operation that has a chance of building up is the one where the modes are locked, consequently that will be the only one to survive in the long run (where long means a few nanoseconds).

12.6. Masers

So far in discussing the practical lasers we have been concerned with what are in fact self-oscillators rather than amplifiers.† Optical amplifiers have been made in the laboratory but they have at present

† If we replace 'amplification' by 'oscillation' in a more accurate acronym for the optical wavelength device, we get 'loser'. This idea has not caught on.

limited interest and application, except for getting very high power from ruby and neodymium glass devices. Conversely, a maser could oscillate but that would not be of great value. At microwave frequencies coherent oscillators are commonplace and what is required is a low-noise amplifier. Most microwave amplifiers rely on the ordered motion of charge carriers in vacuum or in a solid. These are noisy because of the inherent random fluctuations of such currents. What other requirements does an amplifier have? It should be tuneable, so that the receiver can be adjusted to the same frequency as a distant transmitter. This means that it is useless to rely on a naturally occurring energy separation. The most convenient tuneable transition (we came across it when discussing magnetism) is between two paramagnetic levels split by a magnetic field. The possible energies are given by eqn. (11.37):

$$E = -M_J g \mu_{\mathrm{mB}} B, \qquad M_J = J, J-1, \ldots -J. \tag{12.22}$$

Chromium ions in ruby again provide the best illustration of maser operation. For trebly ionized chromium the outer 3d-shell has three electrons of identical spin. Hence its total spin contribution is $\frac{3}{2}$. The contribution from orbital angular momentum is taken as zero,† thus $J = \frac{3}{2}$, leading to the values

$$M_J = \tfrac{3}{2}, \tfrac{1}{2}, -\tfrac{1}{2}, -\tfrac{3}{2}. \tag{12.23}$$

Taking further $g = 2$ (corresponding to pure spin), the energy levels are shown in Fig. 12.14a. The energy levels found experimentally are illustrated in Fig. 12.14b and c for two particular orientations of the magnetic field. The deviations from the simple theory are caused by the local magnetic fields.

As far as the tuning of a maser is concerned it is of no significance that the energy versus magnetic field curve is not necessarily linear. What matters is that the difference between the energy levels (say between 2 and 1 in Fig. 12.14c) is a function of the magnetic field and thus the frequency of operation can be varied. The magnetic fields required are reasonable and can be realized in practice.

† There is some theoretical justification for doing so, but the real reason is that unless orbital momentum is disregarded there is no resemblance at all between theory and experiment.

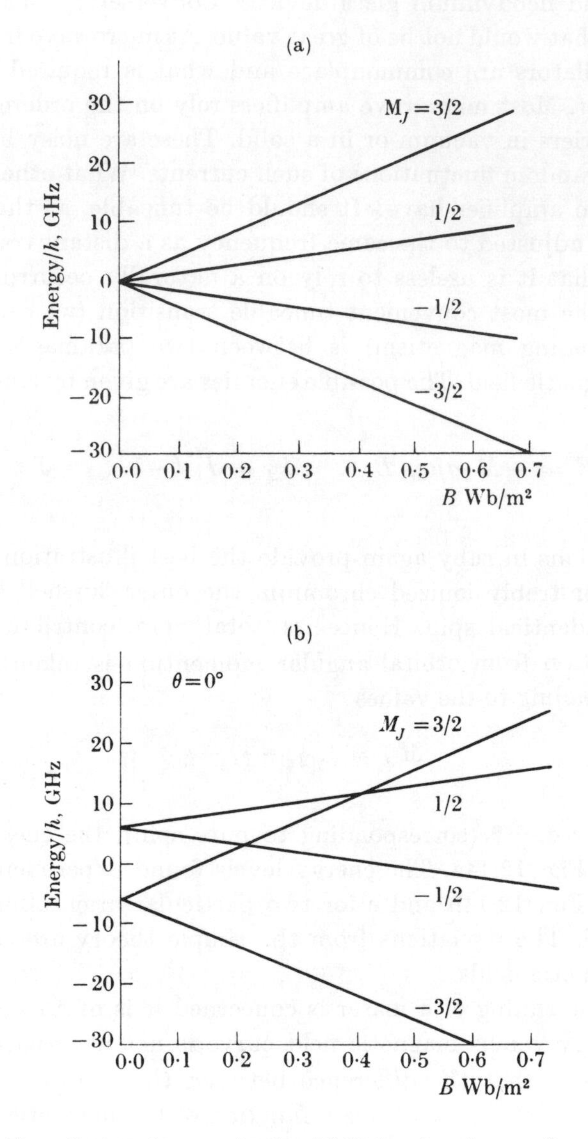

FIG. 12.14. The splitting of energy levels of Cr^{3+} ions in ruby as a function of magnetic field. (a) Plot of eqn. (12.20) for the case when the orbital momentum is quenched and the angular momentum is due to spin only. (b) Experimental curves for B in the direction of the symmetry axis of the ruby crystal, $\theta = 0$. (c) Experimental curves for B perpendicular to the symmetry axis of the ruby crystal, $\theta = 90°$.

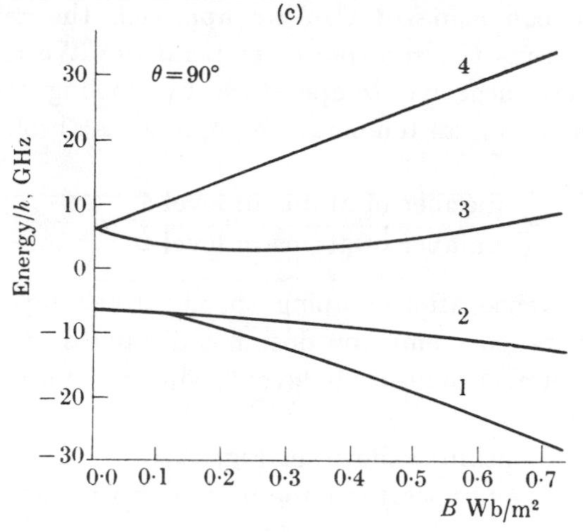

FIG. 12.14. (*cont.*)

12.7. Noise

I have mentioned before that masers have low noise output but the arguments I used were not very convincing. True, in a maser there are no moving charge carriers so we managed to get rid of one source of noise but we have instead another source, namely spontaneous emission.

The amount of noise generated in an amplifier may be characterized by a parameter called 'noise temperature'; a low 'noise temperature' means a small amount of noise. For a maser it can be shown that under ideal conditions† this noise temperature is numerically equal to the negative temperature of the emission mechanism. Hence the aim is to have a low negative temperature, that is, large population inversion. How can we achieve large population inversion? With reference to our three-level maser scheme we have to do two things: (i) pump hard so that the population of levels 3 and 1 become roughly equal; (ii) keep the device at a low temperature‡ so that there are many atoms available for pumping in level 1.

Be careful, we are talking now of three different 'temperatures'. The maser has to work at a low (ordinary) temperature to get a low negative (inversion) temperature, which happens to be equal to the noise temperature of the amplifier. Now what is the minimum noise

† Ideal conditions mean high gain, no ohmic losses, and no reflections from a noisy load.

‡ Low temperatures help incidentally in reducing the ohmic losses as well.

temperature one can achieve? Can we approach the zero negative temperature and thus the zero noise temperature? We can certainly approach the zero negative temperature by cooling the amplifier towards 0 K. As the actual temperature approaches absolute zero, the ratio

$$\frac{\text{number of atoms in level 1}}{\text{number of atoms in level 2}}$$

tends to infinity. Hence, after pumping, the negative temperature tends to zero. But spontaneous emission does not disappear, since it is proportional to the number of atoms in level 3. Thus the noise temperature cannot reach zero.

The correct relationship is shown in Fig. 12.15. It may be seen that for very low negative temperatures the noise temperature tends to the finite value

$$T_{\text{noise}} = \frac{hf}{k}, \tag{12.24}$$

where f is the frequency of operation. Taking $f = 5\ 10^9$ Hz, the limiting noise temperature comes to about 0·25 K. Experimental results on masers cooled to liquid helium temperatures are not far from this value. Noise temperatures around 2 K have actually been measured. This is in contrast to about 150 K obtained in this frequency range by the best 'traditional' microwave amplifier (a specially designed low-noise travelling-wave tube).

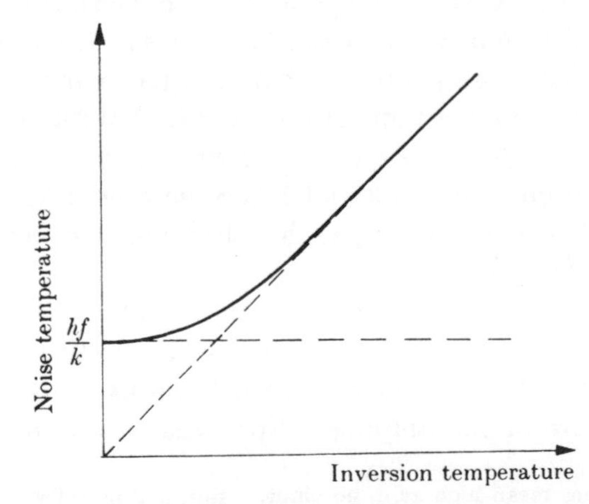

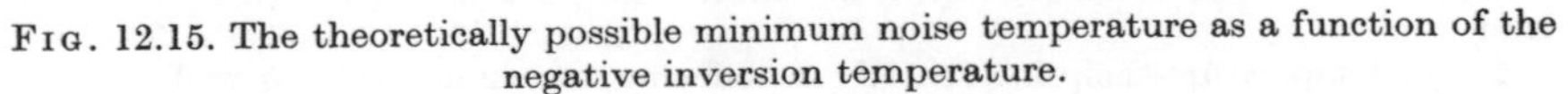
FIG. 12.15. The theoretically possible minimum noise temperature as a function of the negative inversion temperature.

All I have said so far about noise applies to lasers as well, though the numerical values will be radically different. For the argon laser mentioned before, $f = 6{\cdot}6 \times 10^{14}$ Hz, giving $T_{\text{noise}} \sim 30{,}000$ K as the theoretically available minimum.

12.8. Applications

Finally I should like to say a few words about applications. What are masers good for? For low-noise amplification. They are extremely useful in satellite communications and in radio astronomy where minute signals must be detected.

What are lasers good for? Surprisingly, an answer to this question was not expected when the first lasers were put on the bench. It is true to say that never has so much effort been expended on a device with so little regard to its ultimate usefulness. Lasers were developed for their own sake.

I suppose that in the trade most people's reaction was that sooner or later something useful was bound to come out of it. Radio waves have provided some service (even allowing for the fact that radio brought upon us the plague of pop music); microwaves have been useful (how else could you see the Olympic finals in Mexico from your armchair in Tunbridge Wells); so coherent light should also be useful for something.

The military who remembered that radar was useful also hoped that laser would be good for something, and they gave their blessing (and their money too!).

Another powerful factor contributing was the human urge to achieve new records. I could never understand why a man should be happier if he managed to run faster by one tenth of a second than anyone else in the world. But that's how it is. If once a number is attached to some performance, there will be no shortage of men trying to reduce or increase that number (whatever the case may be). And so it is with coherent radiation. Man feels his duty to explore the electromagnetic spectrum and produce coherent waves of higher and higher frequencies.

There may have been some other motives too, but there was no unbridled optimism concerning immediate applications. What can we say 18 years later? Well, the military were apparently right. They got a bomb out of it which can be guided with an accuracy of less than 1 m and there are, very likely, lots of other applications in the pipeline. The ray-gun, that favourite dream of boys, science fiction writers, and generals may not be very far from realization. What about civilian applications? There is enormous potential but very few applications have so far been adopted in practice. If semiconductors would disappear tomorrow our civilization would be in dire trouble. If lasers disappeared tomorrow they would be mourned mostly by scientists. And indeed the scientific applications are those of immediate importance. The laser is a wonderful research tool. A brief list below will indicate some of the scientific applications.

Nonlinear optics. The whole subject, the study of nonlinear phenomena at optical frequencies, was practically born with the laser.

Spectroscopy. An old subject has been given a new lease of life by the invention of tuneable lasers. Spectroscopists have now both power and spectral purity previously unattainable.

Photochemistry. Carefully selected high-energy states may be excited in certain substances and their chemical properties may be studied.

Study of rapid events. With the aid of picosecond light pulses a large number of rapidly occurring phenomena may be studied in physics, chemistry, and biology. The usual technique is to generate a phenomenon by a strong pulse and probe it by another time-delayed pulse. A field in which these techniques have been successfully utilized is the creation and decay of excitons in a semiconductor crystal.

Plasma diagnostics. Many interesting properties of plasmas may be deduced by their scatter of laser light.

Plasma heating. A plasma may be heated to high temperatures by absorbing energy from powerful lasers.

Acoustics. Properties of high-frequency (in the GHz range) acoustic waves in solids may be studied by interacting them with laser light.

Genetics. Chromosomes may be destroyed selectively by illuminating single cells by focused laser beams.

Metrology. The velocity of light may be determined from the relationship $c = f\lambda$ by measuring the frequency and wavelength of certain laser oscillations. The laser is stabilized by locking it to a molecular absorption line and its frequency is measured by comparing it with an accurately known frequency, which is multiplied up from the microwave into the optical range. The wavelength is measured independently by interferometric methods. The accuracy with which we know the velocity of light was improved this way by a factor of hundred.

I am sure there are hundreds of other scientific applications. A laser is a wonderful device for any research worker who has (temporarily, of course) run out of ideas. Whatever he is investigating he may always ask the question: 'And how will this phenomenon be influenced by laser light?'

Let us turn now to some potential applications. I shall start with the one which has the highest potential, namely *laser fusion*. The chances of realizing a fusion reactor with the aid of lasers must be deemed meagre. Nevertheless, a lot of work is going on; several types have already been suggested, discussed, and relevant experiments are being done. The stakes are high. If we manage to crack this particular nut mankind will be grateful to us for the next hundred thousand years.

The principles are simple. As I have already mentioned, a plasma may be heated by absorbing energy supplied by a number of high-power pulsed lasers. The fusion fuel (deuterium and tritium) is injected into the reactor in the form of a solid pellet, evaporated, ionized, and heated instantly by a laser pulse, and the energy of the liberated neutrons is converted into heat by (in one of the preferred solutions) a lithium blanket which also provides the much needed tritium.

The next in importance is another nuclear application, namely *isotope separation*. With the change from fossil to fissile energy sources we shall need more and more enriched uranium. The cost of uranium enrichment in the U.S.A. for the next 20 years has been estimated at over 100,000 million dollars. Thus the motivation for cheaper methods of separation is strong.

The laser-driven process, estimated to be cheaper by a factor of 20, is based on the fact that there is an optical isotope shift in atomic and molecular spectra. Hence the atoms or molecules containing the desired isotope can be selectively excited by laser radiation. The separation of excited atoms may, for example, be achieved by a second excitation in which they become ionized and can be collected by an electric field.

A disadvantage of the process is that once perfected it will enable do-it-yourself enthusiasts (with possibly a sprinkling of terrorists among them) to make their own atomic bombs.

Another high-potential application is in *communications*. The fantastic advance in the purity of glass fibres combined with the advent of the double heterojunction semiconductor injection laser has made optical transmission by far the cheapest for inter-city communications. It is bound to come, and bound to be followed by cheap intercontinental communications. In 10 years' time you will probably be able to call Uncle Billy in New York for ten pence.

There is a wide range of applications concerned with *machining* and *welding*, owing to the possibility that a powerful laser beam may be focused to a small spot of the order of a micrometer. How widespread these applications will be depends purely on the state of competing technologies.

I should mention radar as well since the methods of microwave radar may be easily extended into the optical region. The advantage of *optical radar* is that it can give much higher accuracy (for the moon–earth distance for example) because it employs shorter wavelengths and shorter pulses. It can furthermore determine the position of objects (e.g. clouds, layers of air turbulence, agents of pollution) which do not give sufficient reflection at microwave frequencies.

An interesting application is for navigation, which necessitates the measurement of rotation. Lasers can detect rotational movement as low as a thousandth degree per hour. The basic principles may be understood from Fig. 12.16. This is a so-called ring laser in which resonance is achieved by a ray biting its own tail. The condition of resonance is now that the total length around the ring should be an integral multiple

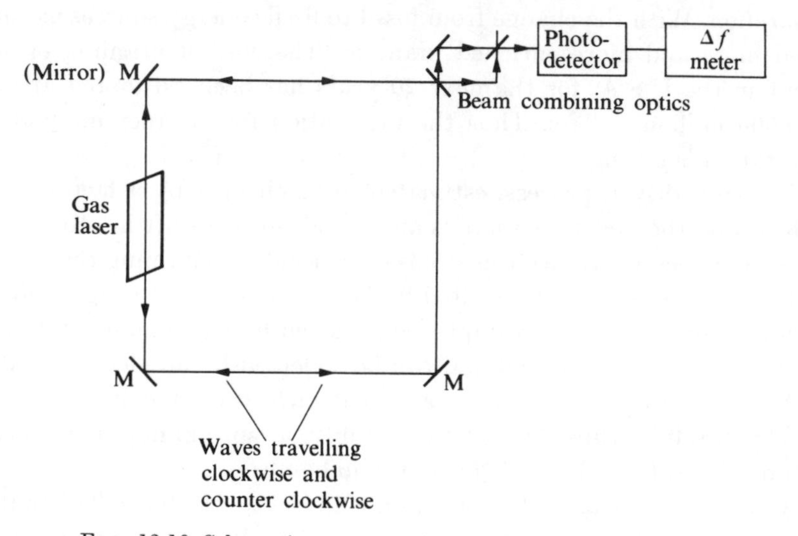

FIG. 12.16. Schematic representation of a laser rotation sensor.

of the wavelength. When the system is at rest (or moving with uniform velocity) the clockwise and anti-clockwise paths are equal and consequently the resonant wavelengths are equal too. However, angular rotation of the whole system makes one path shorter than the other one leading to different frequencies of oscillation. The two beams are then incident upon a photodetector which produces a current at the difference frequency. By measuring this difference frequency the rate of rotation may be deduced.

A further application likely to come is the recording and replay of optical information by so called *video discs*. Its advantage over magnetic recording is that it is potentially much cheaper. The principle is simple enough. Recording is by a focused laser beam which burns holes in a metal disc. Replay is by another laser beam which gets modulated by the holes.

Laser beams may also be used as surgical knifes or dental drills, making bloodless cuts and causing no pain when drilling. The future will show whether surgeons and dentists will warm to the idea. There is however one medical application in which lasers are uniquely useful and that is the reattachment of the human retina. The laser can provide a short concentrated pulse which gives sufficient heat.

As the last application I would like to mention *holography*, a method of image reconstruction invented by Dennis Gabor in 1948. It is by no means limited to the optical region; it could in principle be used at any frequency in the electromagnetic spectrum, and indeed holography can

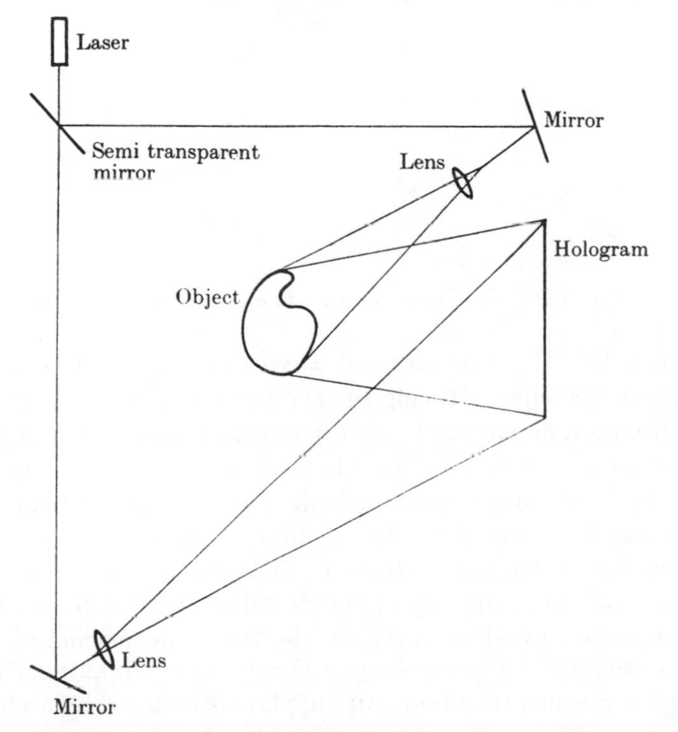

FIG. 12.17. Schematic representation of taking a hologram.

be produced by all kinds of waves including acoustic and electron waves. Nevertheless holography and laser became strongly related to each other, mainly because holographic image reconstruction can most easily be done with lasers at optical frequencies.

The basic set-up is shown schematically in Fig. 12.17. The laser beam is split into two and the object is illuminated by one of the beams. The so-called 'hologram' is obtained by letting the light scattered from the object interfere with the other beam. The pattern that appears depends both on the phase and on the amplitude of the scattered light. Storing this information on a photographic plate we have our 'picture', which bears no resemblance to the object at all. However, when the hologram is illuminated by a laser (Fig. 12.18) the original object will dutifully

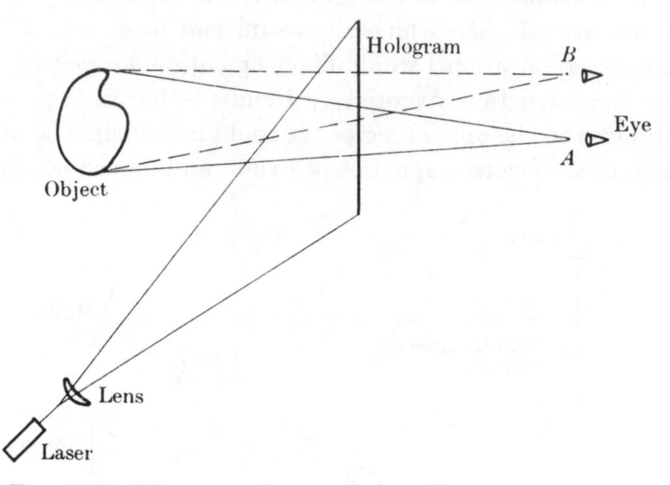

FIG. 12.18. Schematic representation of viewing a hologram.

spring into life. The reconstructed wave forms appear to diverge from an image of the object. Moving the eye from A to B means viewing the object from a different angle, and it looked different indeed just as in reality. So the picture we obtain is as good as the object itself, if not better. For examining small biological specimens, for example, the picture may be better than the original because the original will not sit motionless under the microscope. A hologram can be investigated at leisure without losing any of the details and one can actually focus the microscope to various depths in the three-dimensional picture.

A variation on the same theme offering some advantages is *volume holography* in which the photographic plate is replaced by a photosensitive medium. The medium (see Section 10.12 on one of the representatives of this class) can store an interference pattern in the form of dielectric constant variation. One of the advantages of volume holograms is that the holographic reproduction is strongly wavelength dependent (this is because radiations from various parts of the medium must be added in the correct phase at the reconstruction stage). Hence the image may be viewed in white light, the hologram selecting the wavelength it can respond to from the broad spectrum available.

The entertainment provided by holography has so far been confined to the laboratory with the exceptions of a few exhibitions. But with the advent of volume holography and with some further improvement in materials we may soon see the day when every teenager's bedroom will be decorated by the three-dimensional lifesize image of a pop singer.

I have not had time to talk about the less popular aspects of holography. Let me just mention two of them. The ability to convert wavefronts into each other makes them suitable for all kinds of optical instruments (lenses, mirrors, diffraction gratings, etc.) including couplers between optical guiding structures. The ability to capture the interference between a reference light and light reflected by a vibrating surface makes them powerful analysers of the vibration modes of various structures (e.g. turbine blades).

Examples.

1. Determine the wavelength of a GaAs junction laser assuming that the radiative transition occurs between the bottom of the conduction band and the top of the valence band.

Define carefully what is meant by an inverted population in a junction laser.

2. Assuming that the inversion curve of a laser is wide enough to permit several axial modes, determine the frequency difference between nearest modes. Assume further that L, the length of the resonator, is much larger than the wavelength.

3. A microwave cavity of resonant frequency ω is filled with a material having a two-level system of electron spins, of energy difference $\hbar\omega$. The microwave magnetic field strength H can be considered uniform throughout the cavity volume V. Show that the rate of energy loss to the cavity walls is

$$\frac{\omega\mu_0 H^2 V}{Q},$$

where Q is the quality factor of the cavity.

If the probability for induced transitions between levels per unit time is αH^2 and spontaneous emission is negligible, show that the condition for maser oscillation is

$$\Delta N > \frac{\mu_0}{\alpha \hbar Q}$$

where ΔN is the excess of upper level population density over lower.

(Hint: $Q = \omega \times$energy stored/energy lost per second)

4. Discuss briefly why gas lasers working on ionic, atomic, and molecular transitions occur in different frequency regions. What are the difficulties in making lasers in the far-ultraviolet and in the X-ray region?

5. What causes the laser beam on a screen to appear as if it consisted of a large number of bright points, and why do these points appear to change their brightness as the eye is moved?

13. Superconductivity

I go on forever.

TENNYSON *The Brook*

13.1. Introduction

SUPERCONDUCTIVITY has so far been a scientific curiosity but, as the price of liquid helium is decreasing and the critical temperature is climbing up, it might not stay that way for very long. In the laboratory it has already proved useful in providing magnetic fields as high as 15 Wb m^{-2} and helping to measure voltages as small as 10^{-14} V. Its industrial application might also be around the corner in the form of memory elements for computers, and there is some, even if not very serious, talk about using superconductors for power transmission. So there are good reasons why an engineer should know something about superconductors; but besides this I feel that some acquaintance with superconductivity should be part of modern engineering education. Superconductivity is after all such an extraordinary phenomenon, so much in contrast with everything we are used to. It is literarily out of this world. Our world is classical but superconductivity is a quantum phenomenon—a quantum phenomenon on macroscopic scale. The wave functions, for example, that lead an artificial existence in quantum mechanics proper appear in superconductivity as measurable quantities.

The discovery of superconductivity was not very dramatic. When Kamerlingh Onnes succeeded in liquefying helium in 1908 he looked round for something worth measuring at that temperature range. His choice fell upon the resistivity of metals. He tried platinum first and found that its resistivity continued to decline at lower temperatures, tending to some small but finite value as the temperature approached the absolute zero. He could have tried a large number of other metals with similar prosaic results. But he was in luck. His second metal, mercury, showed quite unorthodox behaviour. Its resistivity (as shown in Fig. 13.1) suddenly decreased to such a small value that he was unable to measure it—and no one has succeeded in measuring it ever since. The usual technique is to induce a current in a ring made of superconducting material and measure the magnetic field due to this current. In a normal metal the current would decay in about 10^{-12} second.

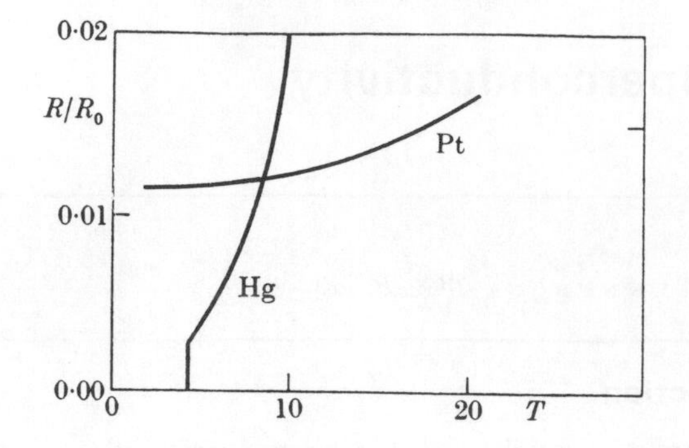

FIG. 13.1. The resistance of samples of platinum and mercury as a function of temperature (R_0 is the resistance at 0°C).

In a superconductor the current can go round for a considerably longer time—measured not in picoseconds but in years. One of the longest experiments was made somewhere in the United States; the current was going round and round for three whole years without any detectable decay. Unfortunately, the experiment came to an abrupt end when a research student forgot to fill up the Dewar flask with liquid nitrogen. So the story goes anyway.

Thus for all practical purposes we are faced with a real lossless phenomenon. It is so much out of the ordinary that no one quite knew how to approach the problem. Several phenomenological theories were born but its real cause remained unknown for half a century. Up to 1957 it defied all attempts; so much so, that it gave birth to a new theorem, namely that 'all theories of superconductivity are refutable'. These heroic times are over now. In 1957 Bardeen, Cooper, and Schrieffer produced a theory (nowadays called the BCS theory) that managed to explain all the major properties of superconductivity. The essence of the theory is that superconductivity is caused by electron–lattice interaction and that the superconducting electrons consist of ordinary electrons paired up.

There is not much point in going into the details of this theory; it is far too complicated, but a rough idea can be provided by the following qualitative explanation due to Little.

Figure 13.2a shows the energy–momentum curve of an ordinary conductor with seven electrons sitting discreetly in their discrete energy levels. In the absence of an electric field the current from electrons moving to the right is exactly balanced by that from electrons moving to the left. Thus the net current is zero.

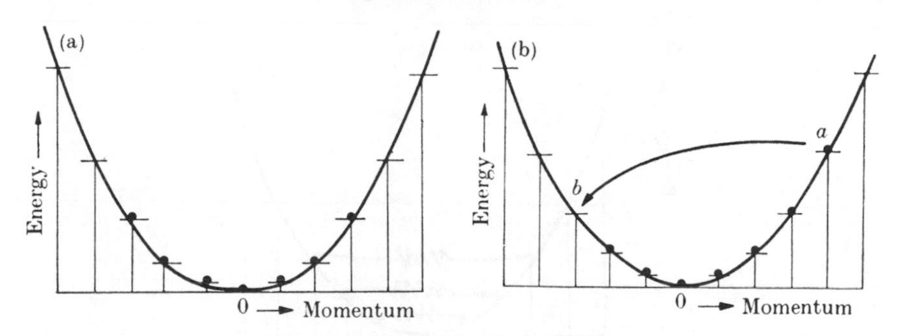

FIG. 13.2. A one-dimensional representation of the energy–momentum curve for seven electrons in a conductor. (a) All electrons in their lowest energy states, the net momentum is zero. (b) There is a net momentum to the right as a consequence of an applied electric field.

When an electric field is applied all the electrons acquire some extra momentum, and this is equivalent to shifting the whole distribution in the direction of the electric field, as shown in Fig. 13.2b. Now what happens when the electric field is removed? Owing to collisions with the vibrating lattice, with impurity atoms, or with any other irregularity, the faster electrons will be scattered into lower energy states until the original distribution is re-established. For our simple model it means that the electron is scattered from the energy level a into energy level b.

In the case of a superconductor it becomes energetically more favourable for the electrons to seek some companionship. Those of opposite momenta (the spins incidentally must also be opposite, but since spins play no further role they are generally disregarded when discussing superconductivity) pair up to form a new particle called a *superconducting electron* or, after its discoverer, a *Cooper pair*.† This link between two electrons is shown in Fig. 13.3a by an imaginary mechanical spring.

† In actual fact, the first man to suggest the pairing of electrons was R. A. Ogg. According to Gamow's limerick:

> In Ogg's theory it was his intent
> That the current keep flowing, once sent;
> So to save himself trouble,
> He put them in double,
> And instead of stopping, it went.

Ogg preceded Cooper by about a decade but his ideas were put forward in the language of an experimental chemist, which is unforgivable. No one believed him, and his suggestion faded into oblivion. This may seem rather unfair to you but that's how contemporary science works. In every discipline there is a select band of men whose ideas are taken up and propagated, so if you want to invent something great, try to associate yourself with the right kind of people.

Don't try to make any contribution to theoretical physics unless you are a trained theoretical physicist, and don't meddle in theology unless you are a bishop.

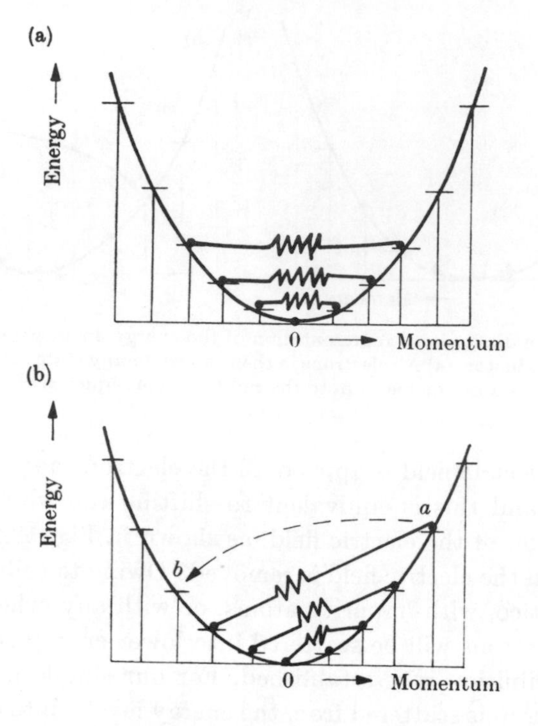

FIG. 13.3. The energy–momentum curve for seven electrons in a superconductor. Those of opposite momenta pair up that is represented here by a mechanical spring. (a) All pairs in their lowest energy states, the net momentum is zero. (b) There is a net momentum to the right as a consequence of an applied electric field.

We may ask now a few questions about our newly born composite particle. First of all, what is its velocity? The two constituents of the particle move with v and $-v$ respectively; thus the velocity of the center of mass is zero. Remembering the de Broglie relationship ($\lambda = h/p$), this means that the wavelength associated with the new particle is infinitely long. And this is valid for *all* superconducting electrons.

It does not quite follow from the above simple argument (but it comes out from the theory) that all superconducting electrons behave in the same way. This is for our electrons a complete break with the past. Up to now, owing to the rigour of the Pauli principle, all electrons had to be different. In superconductivity they acquire the right to be the same. So we have a large number of identical particles all with infinite wavelength; that is, we have a quantum phenomenon on a macroscopic scale.

An applied electric field will displace all the particles again as shown in Fig. 13.3b but when the electric field disappears there is no change.

Scattering from energy level a to energy level b is no longer possible because then the electrons both at b and c would become pairless, which is energetically unfavourable. One may imagine a large number of simultaneous scatterings that would just re-establish the symmetrical distribution of Fig. 13.3a, but that is extremely unlikely. So the asymmetrical distribution will remain; there will be more electrons going to the right than to the left, and this current will persist for ever—or, at least, for three years.

13.2. The effect of a magnetic field

The critical magnetic field. One of the applications of superconductivity coming immediately to mind is the production of a powerful electromagnet. How nice it would be to have high magnetic fields without any power dissipation. The hopes of the first experimenters were soon dashed. They found out that above a certain magnetic field the superconductor became normal. Thus in order to have zero resistance not only the temperature but also the magnetic field must be kept below a certain threshold value. Experiments with various superconductors have shown that the dependence of the critical magnetic field on temperature is well described by the formula

$$H_c = H_0\{1-(T/T_c)^2\}. \tag{13.1}$$

This relationship is plotted in Fig. 13.4. It can be seen that the material is normal above the curve and superconducting below the curve. H_0 is defined as the magnetic field that destroys superconductivity at absolute zero temperature. The values of H_0 and T_c for a number of

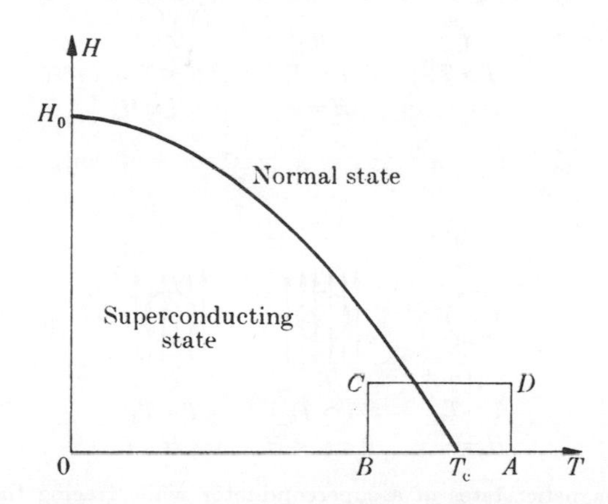

FIG. 13.4. The critical magnetic field as a function of temperature.

TABLE 13.1. *The critical temperature and critical magnetic field of a number of superconducting elements.*

Element	T_c(K)	$H_0 \times 10^{-4}$(A/m)	Element	T_c(K)	$H_0 \times 10^{-4}$(A/m)
Al	1·19	0·8	Pb	7·18	6·5
Ga	1·09	0·4	Sn	3·72	2·5
Hg α	4·15	3·3	Ta	4·48	6·7
Hg β	3·95	2·7	Th	1·37	1·3
In	3·41	2·3	V	5·30	10·5
Nb	9·46	15·6	Zn	0·92	0·4

superconductors are given in Table 13.1. For certain alloys T_c may be as high as 19 K, and much higher magnetic fields can also be achieved; but this is a more complicated problem that we shall tackle later.

The Meissner effect. We have seen that below a certain temperature and magnetic field a number of materials lose their electrical resistivity completely. How would we expect these materials to behave if taken from point A to C in Fig. 13.4 by the paths ABC and ADC respectively? At point A there is no applied magnetic field and the temperature is higher than the critical one (Fig. 13.5a). Going from A to B the temperature is reduced below the critical temperature; so the material loses its resistivity but nothing else happens. Going from B to C means switching on the magnetic field. The changing flux creates an electric field that sets up a current opposing the applied magnetic field. This is just

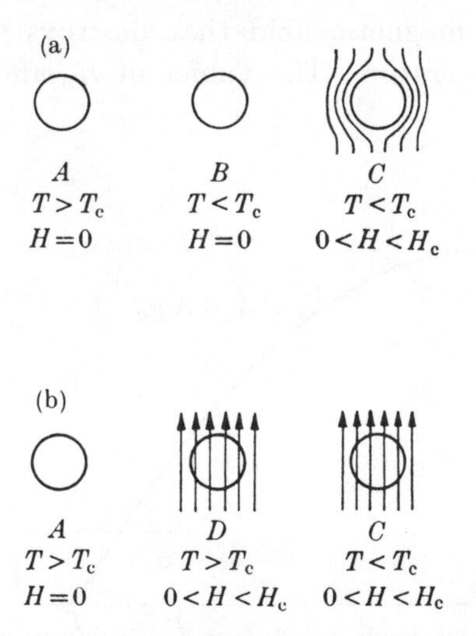

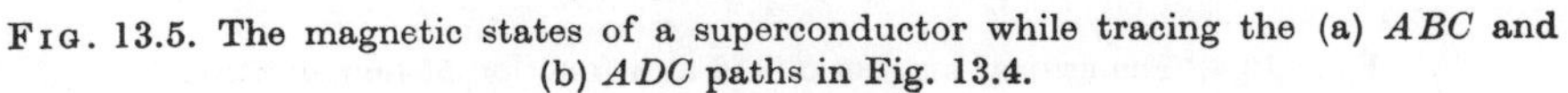

FIG. 13.5. The magnetic states of a superconductor while tracing the (a) ABC and (b) ADC paths in Fig. 13.4.

Lenz's law and in the past we have referred to such currents as eddy currents. The essential difference now is the absence of resistivity. The eddy currents do not decay; they produce a magnetic field that completely cancels the applied magnetic field inside the material. Thus we may regard our superconductor as a perfect diamagnet.

Starting again at A with no magnetic field (Fig. 13.5b) and proceeding to D puts the material into a magnetic field at a constant temperature. Assuming that our material is nonmagnetic (superconductors are in fact slightly paramagnetic above their critical temperature, as follows from their metallic nature), the magnetic field will penetrate. Going from D to C means reducing the temperature at constant magnetic field. The material becomes superconducting at some point but there is no reason why this should imply any change in the magnetic field distribution. At C the magnetic field should penetrate just as well as at D. Thus the distribution of the magnetic field at C depends on the path we have chosen. If we go via B the magnetic field is expelled; if we go via D the magnetic field is the same inside as outside. The conclusion is that for a perfect conductor (meaning a material with no resistance) the final state depends on the path chosen. This is quite an acceptable conclusion; there are many physical phenomena exhibiting this property. What is interesting is that superconductors do *not* behave in this expected manner. A superconductor cooled in a constant magnetic field will set up its own current and expel the magnetic field when the critical temperature is reached.

The discovery of this effect by Meissner in 1933 showed superconductivity in a new light. It became clear that superconductivity is a new kind of phenomenon that does not obey the rules of classical electrodynamics.

13.3. Microscopic theory

The microscopic theory is well beyond the scope of an engineering undergraduate course and, indeed, beyond the grasp of practically anyone. It is part of quantum field theory and has something to do with Green's functions and has more than its fair share of various operators. I shall not say much about this theory, but I should just like to indicate what is involved.

The fundamental tenet of the theory is that superconductivity is caused by a second-order interaction between electrons and the vibrating lattice. This is rather strange. After all, we do know that thermal vibrations are responsible for the presence of resistance and not for its absence. This is true in general; the higher the temperature the larger the electrical resistance. Below a certain temperature, however, and for a select group of materials, the lattice interaction plays a different role. It is a sort of intermediary between two appropriately placed electrons. It results in an apparent attractive force between the two electrons, an attractive force larger than the repulsive force owing to the Coulomb interaction. Hence the electron changes its character. It stops obeying Fermi–Dirac statistics, and any number of electrons (or more correctly any number of electron pairs) can be in the same state.

Do we have any direct experimental evidence that superconductivity is caused by electron–lattice interaction? Yes, the so-called isotope effect. The critical temperature of a superconductor depends on the total mass of the nucleus. If we add a neutron (that is, use an isotope of the material) the critical temperature decreases.

Is electron–lattice interaction absolutely essential? Could some other interactions play a similar role? Perhaps they could. There have been a few recent suggestions about such mechanisms but no experimental proofs are available. I feel it would be unfair of Nature to tuck away such a tremendously important phenomenon into a dark corner of physics. I think the next decade will bring new superconducting mechanisms and with them a very strong upsurge of interest in the applications of superconductivity.

13.4. Thermodynamical treatment

Let us look again at Fig. 13.4. Above the curve our substance behaves in the normally accepted way. It has the same sort of properties it had at room temperature. Its magnetic properties are the same and its electric properties are the same; true, the electrical resistivity is smaller than at room temperature but there is nothing unexpected in that. However, as soon as we cross the curve the properties of the substance become qualitatively different. Above the curve the substance is non-magnetic, below the curve it becomes diamagnetic; above the curve it has a finite electrical resistance, below the curve the electric resistance is zero.

If you think about it a little you will see that the situation is very similar to that you have studied under the name of 'phase change' or 'phase transition' in thermodynamics. Recall, for example, the diagram showing the vaporization of water (Fig. 13.6). The properties of the substance differ appreciably above and below the curve and we do not

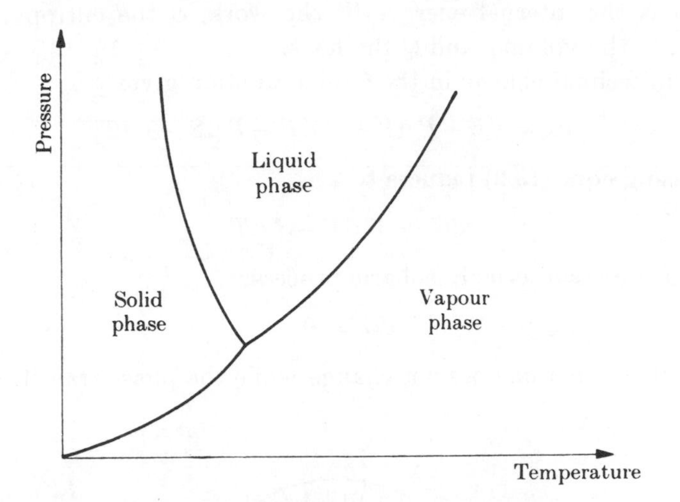

FIG. 13.6. The pressure against temperature diagram for water.

need elaborate laboratory equipment to tell the difference. Our senses are quite capable of distinguishing steam from water. It is quite natural to call them by different names and refer to the state above the curve as the liquid phase, and to the state below the curve as the vapour phase. Analogously, we may talk about normal and superconducting phases when interpreting Fig. 13.4.

Thus the road is open to investigate the properties of superconductors by the well-established techniques of thermodynamics. Well, is the road open? We must be careful; thermodynamics can be applied only if the change is reversible. Is the normal to superconducting phase change reversible? Fortunately it is. Had we a perfect conductor instead of a superconductor the phase change would *not* be reversible and we should not be justified in using thermodynamics. Thanks to the Meissner effect, thermodynamics *is* applicable.

Let us now review the thermodynamical equations describing the phase transitions. First there is the first law of thermodynamics:

$$\begin{aligned} \mathrm{d}E &= \mathrm{d}Q - \mathrm{d}W \\ &= T\,\mathrm{d}S - P\,\mathrm{d}V. \end{aligned} \tag{13.2}$$

Then there is the Gibbs function (to which we shall also refer as the Gibbs free energy) defined by

$$G = E + PV - TS, \tag{13.3}$$

where E is the internal energy, W the work, S the entropy, P the pressure, V the volume, and Q the heat.

An infinitesimal change in the Gibbs function gives

$$dG = dE + P\,dV + V\,dP - T\,dS - S\,dT, \tag{13.4}$$

which, using eqn. (13.2) reduces to

$$dG = V\,dP - S\,dT. \tag{13.5}$$

Thus, for an isothermal, isobaric process

$$dG = 0,$$

i.e. the Gibbs function does not change while the phase transition takes place.

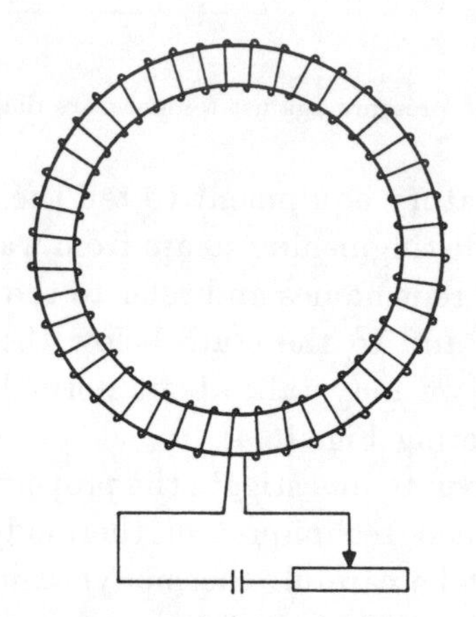

FIG. 13.7. The magnetization of magnetic material in a toroid (for working out the magnetic energy).

In the case of the normal-to-superconducting phase-transition the variations of pressure and volume are small and play negligible roles, and so we can just as well forget them but, of course, we shall have to include the work due to magnetization.

In order to derive a relationship between work and magnetization let us investigate the simple physical arrangement shown in Fig. 13.7. You know from studying electricity that work done on a system in a time dt is

$$dW = -UI\,dt, \tag{13.6}$$

where U is the voltage and I the current, and the negative sign comes from the accepted convention of thermodynamics that the work done *on* a system is negative. Further using Faraday's law, we have

$$U = NA\,\frac{\mathrm{d}B}{\mathrm{d}t}. \tag{13.7}$$

From Ampere's law

$$HL = NI, \tag{13.8}$$

where A is the cross-section of the toroid, N the number of turns, and L the mean circumference of the toroid. We then get

$$dW = -NA\frac{\mathrm{d}B}{\mathrm{d}t}I\,\mathrm{d}t = -NIA\,\mathrm{d}B = -HLA\,\mathrm{d}B = -VH\,\mathrm{d}B \tag{13.9}$$

According to eqn. (11.3)

$$B = \mu_0(H+M).$$

Therefore

$$\begin{aligned}\mathrm{d}W &= -VH\mu_0(\mathrm{d}H+\mathrm{d}M)\\ &= -\mu_0 VH\,\mathrm{d}H-\mu_0 VH\,\mathrm{d}M.\end{aligned} \tag{13.10}$$

The first term on the right-hand side of equation (13.10) gives the increase of energy in the vacuum and the second term is due to the presence of the magnetic material. Thus the work done *on* the material is

$$\mathrm{d}W = -\mu_0 VH\,\mathrm{d}M. \tag{13.11}$$

Hence for a paramagnetic material the work is negative but for a diamagnetic material (where M is opposing H) the work is positive, which means that the system needs to do some work in order to reduce the magnetic field inside the material.

Now to describe the phase transition in a superconductor we have to define a 'magnetic Gibbs function'. Remembering that $P\,\mathrm{d}V$ gives positive work (an expanding gas does work) and $H\,\mathrm{d}M$ gives negative work, we have to replace PV in eqn. (13.3) by $-\mu_0 VHM$. Our new Gibbs function takes the form

$$G = E-\mu_0 VHM-TS \tag{13.12}$$

and

$$\mathrm{d}G = \mathrm{d}E-\mu_0 VH\,\mathrm{d}M-\mu_0 VM\,\mathrm{d}H-T\,\mathrm{d}S-S\,\mathrm{d}T. \tag{13.13}$$

Taking account of the first law for magnetic materials (again replacing pressure and volume by the appropriate magnetic quantities),

$$\mathrm{d}E = T\,\mathrm{d}S+\mu_0 VH\,\mathrm{d}M. \tag{13.14}$$

Equation (13.13) reduces to

$$\mathrm{d}G = -S\,\mathrm{d}T-\mu_0 VM\,\mathrm{d}H. \tag{13.15}$$

This is exactly what we wanted. It follows immediately from the above equation that for a constant temperature and constant magnetic field process

$$dG = 0, \tag{13.16}$$

that is, G remains constant while the superconducting phase transition takes place.

For a perfect diamagnet

$$M = -H, \tag{13.17}$$

which substituted into eqn. (13.15) gives

$$dG = -S\,dT + \mu_0 VH\,dH. \tag{13.18}$$

Integrating at constant temperatures we get

$$G_s(H) = G_s(0) + \tfrac{1}{2}\mu_0 H^2 V, \tag{13.19}$$

where $G_s(0)$ is the Gibbs free energy at zero magnetic field and the subscript s refers to the superconducting phase.

Since superconductors are practically nonmagnetic above their critical temperatures, we can write for the normal phase

$$G_n(H) = G_n(0) = G_n. \tag{13.20}$$

In view of eqn. (13.16) the Gibbs free energy of the two phases must be equal at the critical magnetic field, H_c, that is,

$$G_s(H_c) = G_n(H_c). \tag{13.21}$$

Substituting eqns. (13.19) and (13.20) into eqn. (13.21) we get

$$G_n = G_s(0) + \tfrac{1}{2}\mu_0 H_c^2 V. \tag{13.22}$$

Comparison of eqns. (13.19) and (13.22) clearly shows that at a given temperature (below the critical one) the conditions are more favourable for the superconducting phase than for the normal phase, provided that the magnetic field is below the critical field. There are three cases:

(i) If $H < H_c$ then $G_n > G_s(H)$. (13.23)

(ii) If $H > H_c$ then $G_n < G_s(H)$. (13.24)

(iii) If $H = H_c$ then $G_n = G_s(H_c)$. (13.25)

Now our substance will prefer the phase for which the Gibbs free energy is smaller; that is, in case (i) it will be in the superconducting phase, in case (ii) in the normal phase, and in case (iii) just in the process of transition.

If the transition takes place at temperature $T+\mathrm{d}T$ and magnetic field $H_c+\mathrm{d}H_c$ then it must still be valid that

$$G_s+\mathrm{d}G_s = G_n+\mathrm{d}G_n, \tag{13.26}$$

whence

$$\mathrm{d}G_s = \mathrm{d}G_n. \tag{13.27}$$

This, using eqn. (13.15), leads to

$$-S_s\,\mathrm{d}T-\mu_0 VM_s\,\mathrm{d}H_c = -S_n\,\mathrm{d}T-\mu_0 VM_n\,\mathrm{d}H_c. \tag{13.28}$$

But, as suggested before,

$$M_s = -H_c \quad \text{and} \quad M_n = 0, \tag{13.29}$$

and this reduces eqn. (13.28) to

$$S_n-S_s = -\mu_0 VH_c\frac{\mathrm{d}H_c}{\mathrm{d}T}. \tag{13.30}$$

The latent heat of transition may be written in the form

$$L = T(S_n-S_s), \tag{13.31}$$

and with the aid of eqn. (13.30) this may be expressed as

$$L = -\mu_0 TVH_c\frac{\mathrm{d}H_c}{\mathrm{d}T}. \tag{13.32}$$

After much labour we have, at last, arrived at a useful relationship. We now have a theory that connects the independently measurable quantities L, V, H_c, and T. After measuring them, determining $\mathrm{d}H_c/\mathrm{d}T$ from the H_c–T plot, and substituting their values into eqn. (13.32), the equation should be satisfied; and it is satisfied to a good approximation thus giving us experimental proof that we are on the right track.

It is interesting to note that L vanishes at two extremes of temperature, namely, at $T = 0$ and at $T = T_c$ where the critical magnetic field is zero. A transition which takes place with no latent heat is called a second-order phase transition. In this transition entropy remains constant and the specific heat is discontinuous.

Neglecting the difference between the specific heats at constant volume and constant pressure we can write in general for the specific heat

$$c = T\frac{\mathrm{d}S}{\mathrm{d}T}. \tag{13.33}$$

Substituting from eqn. (13.30)

$$c_n - c_s = T\left(\frac{dS_n}{dT} - \frac{dS_s}{dT}\right)$$

$$= -VT\mu_0\left\{H_c\frac{d^2H_c}{dT^2} + \left(\frac{dH_c}{dT}\right)^2\right\}. \quad (13.34)$$

At $T = T_c$ where $H_c = 0$

$$c_n - c_s = -\left\{VT\mu_0\left(\frac{dH_c}{dT}\right)^2\right\}_{T=T_c}. \quad (13.35)$$

this is negative because the experimentally established H_c–T curves have finite slopes at $T = T_c$. It follows that in the absence of a magnetic

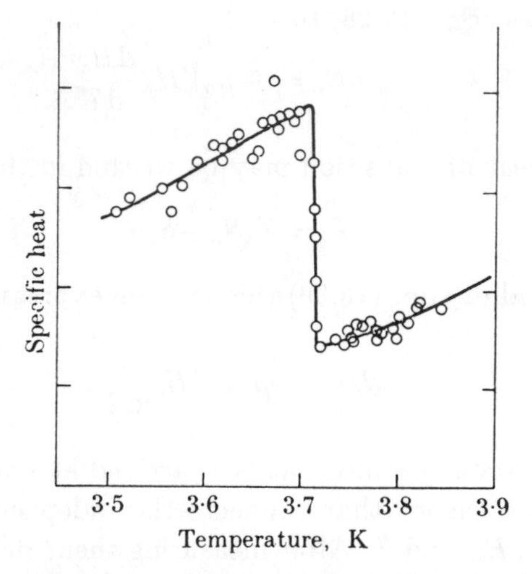

FIG. 13.8. Temperature dependence of the specific heat of tin near the critical temperature (after Keesom and Kok, 1932).

field the specific heat has a discontinuity. This is borne out by experiments as well, as shown in Fig. 13.8, where the specific heat of tin is plotted against temperature. The discontinuity occurs at the critical temperature $T_c = 3{\cdot}72$ K.

13.5. Surface energy

The preceding thermodynamical analysis was based on perfect diamagnetism, that is, we assumed that our superconductor completely expelled the magnetic field. In practice this isn't so, and it can't be so. The currents that are set up to exclude the magnetic field must occupy a finite volume, however small that might be. Thus the magnetic field can also penetrate the superconductor to a small extent. But now we encounter a difficulty. If the magnetic field can penetrate to a finite distance, the Gibbs free energy of that particular layer will decrease, because it no longer has to perform work to exclude the magnetic field. The magnetic field is admitted and we get a lower Gibbs free energy. Carrying this argument to its logical conclusion it follows that the optimum arrangement for minimum Gibbs free energy (of the whole solid at a given temperature) should look like that shown in Fig. 13.9,

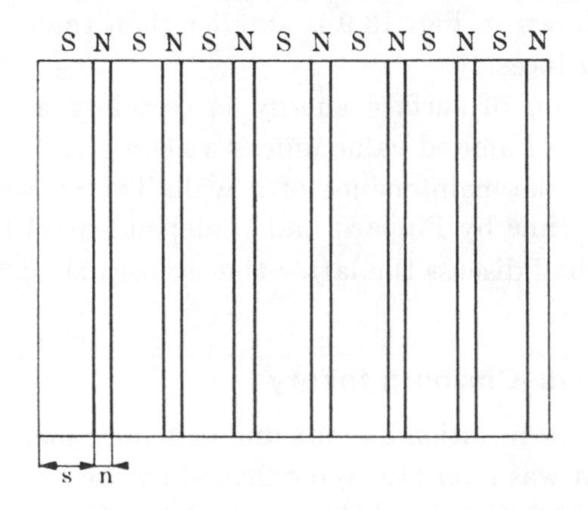

FIG. 13.9. Alternating superconducting and normal layers.

where normal and superconducting layers alternate. The width of the superconducting layers, s, is small enough to permit the penetration of magnetic field, and the width of the normal region is even smaller, $n \ll s$. In this way the Gibbs free energy of the superconducting domains is lower because the magnetic field can penetrate, while the contribution of the normal domains to the total Gibbs free energy remains negligible because the volume of the normal domains is small in comparison with the volume of the superconducting domains.

Thus a consistent application of our theory leads to a superconductor in which normal and superconducting layers alternate. Is this conclusion correct? Do we find these alternating domains experimentally? For some superconducting materials we do; for some other superconducting materials we do not. Incidentally, when the first doubts arose about the validity of the simple thermodynamical treatment, all the experimental evidence available at the time suggested that no break-up could occur. We shall restrict the argument to this historically authentic case for the moment. Theory suggests that superconductors should break up into normal and superconducting domains; experiments show that they do not break up. Consequently the theory is wrong. The theory cannot be completely wrong, however, for it predicted the correct relationship for specific heat. So instead of dismissing the theory altogether, we modify it by introducing the concept of *surface energy*. This would suggest that the material does not break up because maintaining boundaries between the normal and superconducting domains is a costly business. It costs energy†. Hence the simple explanation for the absence of domains is that the reduction in energy resulting from the configuration shown in Fig. 13.9 is smaller than the energy needed to maintain the surfaces.

The introduction of surface energy is certainly a way out of the dilemma but it is of limited value unless we can give some quantitative relationships for the maintenance of a wall. The answer was given at about the same time by Pippard and (independently) by Landau and Ginzburg. We shall discuss the latter theory because it is a little easier to follow.

13.6. The Landau–Ginzburg theory

With remarkable intuition Landau and Ginzburg suggested (in 1950) a formulation that was later (1958) confirmed by the microscopic theory. I shall give here the essence of their arguments, though in a somewhat modified form to fit into the previous discussion.

(i) In the absence of a magnetic field, below the critical temperature, the Gibbs free energy‡ is $G_s(0)$.

(ii) If a magnetic field H_a is applied and is expelled from the interior of the superconductor, the energy is increased by $\frac{1}{2}\mu_0 H_a^2$ per unit volume. This may be rewritten with the aid of flux density as $\frac{1}{2}(1/\mu_0)B_a^2$. If we now abandon the idea of a perfect diamagnet, the magnetic field can penetrate the superconductor, and the flux density at a certain point is B instead of zero. Hence the flux density expelled is not B_a but only $B_a - B$, and the corresponding increase in the Gibbs free energy is

$$\frac{1}{2}\frac{1}{\mu_0}(B_a - B)^2. \tag{13.36}$$

† This is really the same argument that we used for domains in ferromagnetic materials. On the one hand, the more domains we have the smaller is the external magnetic energy. On the other hand, the more domains we have the larger is the energy needed to maintain the domain walls. So the second consideration will limit the number of domains.

‡ From now on, for simplicity, all our quantities will be given per unit volume.

(iii) All superconducting electrons are apparently doing the same thing. We, therefore describe them by the same wave function, ψ where

$$|\psi(x, y, z)|^2 = N_s, \tag{13.37}$$

the density of superconducting electrons. In the absence of an applied magnetic field the density of superconducting electrons is everywhere the same.

(iv) In the presence of a magnetic field the density of superconducting electrons may vary in space, that is, $\nabla\psi \neq 0$. But, you may remember, $-i\hbar\nabla\psi$ gives the momentum of the particle. Hence the kinetic energy of our superconducting electrons,

$$\text{K.E.} = \frac{1}{2m}|-i\hbar\nabla\psi|^2, \tag{13.38}$$

will add to the total energy. It follows then that the appearance of alternating layers of normal and superconducting domains is energetically unfavourable because it leads to a rapid variation of ψ, giving a large kinetic energy contribution to the total energy.

Equation (13.38) is not quite correct. It follows from classical electrodynamics† that in the presence of a magnetic field the momentum is given by $\mathbf{p}-e\mathbf{A}$ where $\mathbf{A}$ is the magnetic vector potential. Hence the correct formula for the kinetic energy is

$$\text{K.E.} = \frac{1}{2m}|-i\hbar\nabla\psi-2e\mathbf{A}\psi|^2, \tag{13.39}$$

where $2e$ is the charge on a superconducting electron.

We may now write the Gibbs free energy in the form

$$G_s(B) = G_s(0)+\frac{1}{2\mu_0}(B_a-B)^2+\frac{1}{2m}|-i\hbar\nabla\psi-2e\mathbf{A}\psi|^2. \tag{13.40}$$

(v) The value of the Gibbs function at zero magnetic field should depend on the density of superconducting electrons, among other things. The simplest choice is a polynomial of the form

$$G_s(0) = G_n(0)+a_1|\psi|^2+a_2|\psi|^4 \tag{13.41}$$

where the coefficients may be determined from empirical considerations. At a given temperature the density of superconducting

† For a discussion, see *The Feynman Lectures on Physics*, vol. 3, pp. 21–5.

electrons will be such as to minimize $G_s(0)$, that is,

$$\frac{\partial G_s(0)}{\partial\, |\psi|^2} = 0, \tag{13.42}$$

leading to

$$|\psi|^2 \equiv |\psi_0|^2 = -\frac{a_1}{2a_2}. \tag{13.43}$$

Substituting this value of $|\psi|^2$ back into eqn. (13.41) we get

$$G_s(0) = G_n - \frac{a_1^2}{4a_2}. \tag{13.44}$$

Let us go back now to eqn. (13.22) (rewritten for unit volume):

$$G_s(0) = G_n - \tfrac{1}{2}\mu_0 H_c^2. \tag{13.45}$$

Comparing the last two equations we get

$$H_c = -a_1/(2a_2\mu_0)^{-\frac{1}{2}}, \tag{13.46}$$

where it is assumed that $a_2 > 0$ and $a_1 < 0$. According to experiment H_c varies linearly with temperature in the neighbourhood of the critical temperature. Thus, for this temperature range we may make eqn. (13.46) agree with the experimental results by choosing

$$a_1 = c_1(T - T_c) \quad \text{and} \quad a_2 = c_2, \tag{13.47}$$

where c_1 and c_2 are independent of temperature.

If you now believe that eqn. (13.41) was a reasonable choice for the $G_s(0)$, we may substitute it into eqn. (13.40) to get our final form for Gibbs free energy

$$G_s(B) = G_n(0) + a_1\,|\psi|^2 + a_2\,|\psi|^4 + \frac{1}{2\mu_0}(\nabla\times\mathbf{A} - B_a)^2 + \\ + \frac{1}{2m}\,|\mathrm{i}\hbar\nabla\psi - 2e\mathbf{A}\psi|^2, \tag{13.48}$$

where the relationship

$$\bar{B} = \nabla\times\mathbf{A} \tag{13.49}$$

has been used.

The arguments used above are rather difficult. They come from various sources (thermodynamics, quantum mechanics, electrodynamics, and actual measured results on superconductors) and must be carefully combined to give an expression for the Gibbs free energy.

(vi) The Gibbs free energy for the entire superconductor may be obtained by integrating eqn. (13.48) over the volume

$$\int_V G_s(B)\,\mathrm{d}V.$$

The integrand contains two undetermined functions $\psi(x, y, z)$ and $A(x, y, z)$ which, according to Ginzburg and Landau, may be obtained from the condition that the integral should be a minimum.

The problem belongs to the realm of variational calculus. Be careful; it is not the minimum of a *function* we wish to find. We want to know how A and ψ vary as functions of the coordinates x, y, and z in order to minimize the definite integral in (vi).

We shall not solve the general problem here but shall restrict the solution to the case of a half-infinite superconductor that fills the space to the right of the $x = 0$ plane. We shall also assume that the applied magnetic field is in the z-direction and is independent of the y- and z-coordinates, reducing the problem to a one-dimensional one where x is the only independent variable.

In view of the above assumptions,

$$B_z = \frac{\mathrm{d}A_y}{\mathrm{d}x}, \tag{13.50}$$

where A_y is the only component of $\mathbf{A}$, and will be simply denoted by A. Since $\nabla\psi$ is a vector in the x-direction, it is perpendicular to $\mathbf{A}$, so that

$$\mathbf{A}.\,\nabla\psi = 0. \tag{13.51}$$

Under these simplifications the integrand takes the form†

$$G_s(B) = G_n + a_1\psi^2 + a_2\psi^4 + \frac{1}{2\mu_0}\left(B_a - \frac{\mathrm{d}A}{\mathrm{d}x}\right)^2 + \frac{1}{2m}\left\{\hbar^2\left(\frac{\partial\psi}{\partial x}\right)^2 + 4e^2A^2\psi^2\right\}. \tag{13.52}$$

The solution of the variational problem is now considerably easier. As shown in Appendix II, $\psi(x)$ and $A(x)$ will minimize the integral if they satisfy the following differential equations:

$$\frac{\partial G_s(B)}{\partial\psi} - \frac{\mathrm{d}}{\mathrm{d}x}\frac{\partial G_s(B)}{\partial(\partial\psi/\partial x)} = 0 \tag{13.53}$$

and

$$\frac{\partial G_s(B)}{\partial A} - \frac{\mathrm{d}}{\mathrm{d}x}\frac{\partial G_s(B)}{\partial(\partial A/\partial x)} = 0. \tag{13.54}$$

† Taking ψ real reduces the mathematical labour and, fortunately, does not restrict the generality of the solution.

Substituting eqn. (13.53) into eqn. (13.54) and performing the differentiations we get

$$2a_1\psi+4a_2\psi^3+\frac{1}{2m}8e^2A^2\psi-\frac{\mathrm{d}}{\mathrm{d}x}\frac{1}{2m}\hbar^2 2\frac{\partial\psi}{\partial x}=0, \qquad (13.55)$$

which after rearrangement yields

$$\frac{\mathrm{d}^2\psi}{\mathrm{d}x^2}=\frac{m}{\hbar^2}2a_1\left(1+\frac{2e^2}{a_1 m}A^2\right)\psi+\frac{4m}{\hbar^2}a_2\psi^3. \qquad (13.56)$$

Similarly, substituting eqn. (13.53) into eqn. (13.55) we get

$$\frac{\mathrm{d}^2A}{\mathrm{d}x^2}=\frac{4e^2\psi^2\mu_0}{m}A, \qquad (13.57)$$

which must be solved subject to the boundary conditions

$$B=B_{\mathrm{a}}=\mu_0H_{\mathrm{a}}, \qquad \mathrm{d}\psi/\mathrm{d}x=0 \quad \text{at} \quad x=0 \qquad (13.58)$$

$$B=0, \qquad \psi^2=\psi_0^2, \qquad \mathrm{d}\psi/\mathrm{d}x=0 \quad \text{at} \quad x=\infty. \qquad (13.59)$$

The boundary conditions for the flux density simply mean that at the boundary with the vacuum the flux density is the same as the applied flux density, and it declines to zero far away inside the superconductor. The condition for $\mathrm{d}\psi/\mathrm{d}x$ comes from the more stringent general requirement that the normal component of the momentum should vanish at the boundary. But since in the one-dimensional case $\mathbf{A}$ is parallel to the surface, $\mathbf{A}.\mathbf{i}_x$ is identically zero, and the boundary condition reduces to the simpler $\mathrm{d}\psi/\mathrm{d}x=0$. Since A is determined except for a constant factor, we can prescribe its value at any point. We shall choose $A(\infty)=0$.

Introducing the new parameters

$$\lambda^2=\frac{m}{4e^2\psi_0^2\mu_0} \qquad (13.60)$$

and

$$\kappa=\lambda^2\frac{2^{\frac{3}{2}}eH_c\mu_0}{\hbar}, \qquad (13.61)$$

and making use of eqns. (13.43) and (13.46), we can rewrite eqns. (13.56) and (13.57) in the forms

$$\frac{\mathrm{d}^2\psi}{\mathrm{d}x^2}=\frac{\kappa^2}{\lambda^2}\left\{-\left(1-\frac{A^2}{2H_c^2\lambda^2\mu_0^2}\right)\psi+\frac{\psi^3}{\psi_0^3}\right\} \qquad (13.62)$$

and

$$\frac{\mathrm{d}^2A}{\mathrm{d}x^2}-\frac{1}{\lambda^2}\frac{\psi^2}{\psi_0^2}A=0. \qquad (13.63)$$

In the absence of a magnetic field, $A \equiv 0$; eqn. (13.62) gives $\psi = \psi_0$ as it should. In the presence of a magnetic field the simplest approximation we can make is to take $\kappa = 0$, which still gives $\psi = \psi_0$. From eqn. (13.63)

$$A = A(0)\mathrm{e}^{-x/\lambda}, \tag{13.64}$$

leading to

$$B = -\frac{1}{\lambda}A(0)\mathrm{e}^{-x/\lambda} = B_{\mathrm{a}}\mathrm{e}^{-x/\lambda}. \tag{13.65}$$

Thus we can see that the magnetic flux density inside the superconductor decays exponentially and λ appears as the penetration depth. Better approximations can be obtained by substituting

$$\psi = \psi_0 + \varphi \tag{13.66}$$

into eqns. (13.62) and (13.63) and solving them under the assumption that φ is small in comparison with ψ_0. Then ψ also varies with distance and B has a somewhat different decay; but these are only minor modifications and need not concern us.

The main merit of the Landau–Ginzburg theory is that by including the kinetic energy of the superconducting electrons in the expression for the Gibbs free energy, it can show that the condition of minimum Gibbs free energy leads to the expulsion of the magnetic field. The expulsion is not complete, as we assumed before in the simple thermodynamic treatment; the magnetic field can penetrate to a distance λ which is typically of the order of 10 nm.

Thus after all there can be no such thing as the break-up of the superconductor into alternating normal and superconducting regions—or can there? We have solved eqn. (13.62) only for the case when κ is very small. There are perhaps some other regions of interest. It turns out that another solution exists for the case when

$$\psi \ll \psi_0 \quad \text{and} \quad B = B_{\mathrm{a}}. \tag{13.67}$$

So we claim now that there is a solution when the magnetic field can penetrate the whole superconducting material, and this happens when the density of superconducting electrons is small. Then (choosing for this case the vector potential zero at $x = 0$)

$$A(x) = B_{\mathrm{a}}x, \tag{13.68}$$

and neglecting the last term in eqn. (13.62) we get

$$\frac{\mathrm{d}^2\psi}{\mathrm{d}x^2} = -\frac{\kappa^2}{\lambda^2}\left(1 - \frac{B_{\mathrm{a}}^2x^2}{2H_{\mathrm{c}}^2\lambda^2\mu_0^2}\right)\psi. \tag{13.69}$$

Now this happens to be a differential equation that has been thoroughly investigated by mathematicians. They maintain that a solution exists only when

$$B_a = \mu_0 H_c \kappa\sqrt{2}/(2n+1) \tag{13.70}$$

where n is an integer; otherwise ψ diverges as $x \to \infty$. The maximum value of B_a occurs at $n = 0$, giving

$$B_a = \mu_0 H_c \kappa\sqrt{2}, \tag{13.71}$$

that is, when $\kappa > 1/\sqrt{2}$ the magnetic field inside the superconductor may exceed the critical field. You may say this is impossible. Haven't we defined the critical field as the field that destroys superconductivity?

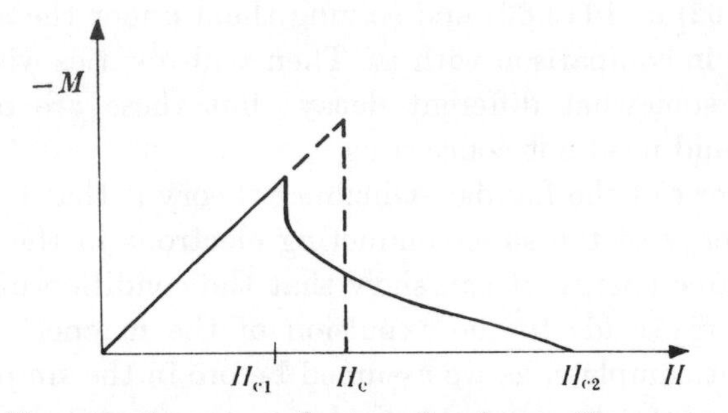

FIG. 13.10. Magnetization curves for type I and type II superconductors. The area under both magnetization curves is the same.

We have, but that was done on the basis of diamagnetic properties. We defined the critical field only for the case when the magnetic field is expelled. Now we say that in certain materials for which $\kappa > 1/\sqrt{2}$ superconductivity may exist in higher magnetic fields. Denoting the former limit by H_{c1} and the new limit by H_{c2}, the relationship is

$$H_{c2} = \kappa\sqrt{2}H_{c1}. \tag{13.72}$$

Up to H_{c1} the superconductor is diamagnetic as shown in Fig. 13.10, where $-M$ is plotted against the applied magnetic field. Above H_{c1} the magnetic field begins to penetrate (beyond the 'diamagnetic' penetration depth), and there is complete penetration at H_{c2} where the material becomes normal. Materials displaying such a magnetization curve are referred to as type II superconductors, while those expelling the magnetic field until they become normal (dotted lines in Fig. 13.10) are called type I superconductors.

FIG. 13.11. The lines of current flow for a two-dimensional type II superconductor. The magnetic field is maximum in the centres of the current vortices. (After Abrikosov, 1957.)

A two-dimensional analysis of a type II superconductor shows that the intensity of the magnetic field varies in a periodic manner with well-defined maxima as shown in Fig. 13.11. Since the current and the magnetic field are uniquely related by Maxwell's equations, the current is also determined. It is quite clear physically that the role of the current is either 'not to let in' or 'not to let out' the magnetic field.

The density of superconducting electrons is zero at the maxima of the magnetic field. Thus, in a somewhat simplified manner, we may say that there is a normal region surrounded by a supercurrent vortex. There are lots of vortices; their distance from each other is about $1\,\mu$m.

The preceding treatment of the theories of superconductivity is not a well-balanced one, neither historically nor as far as their importance is concerned. A comprehensive review would be far too lengthy, so I have just tried to follow one line of thought.

13.7. The energy gap

As you know from electromagnetic theory such optical properties as reflectivity and refractive index are related to the bulk parameters, conductivity and dielectric constant. Thus zero resistivity implies quite radical optical properties, which are not found experimentally. Nothing untoward happens below the critical temperature. Hence we are forced to the conclusion that somewhere between zero and light frequencies the conductivity is restored to its normal value. What is the mechanism? Having learned band theory, we could describe a mechanism that *might* be responsible; this is the existence of an energy gap.

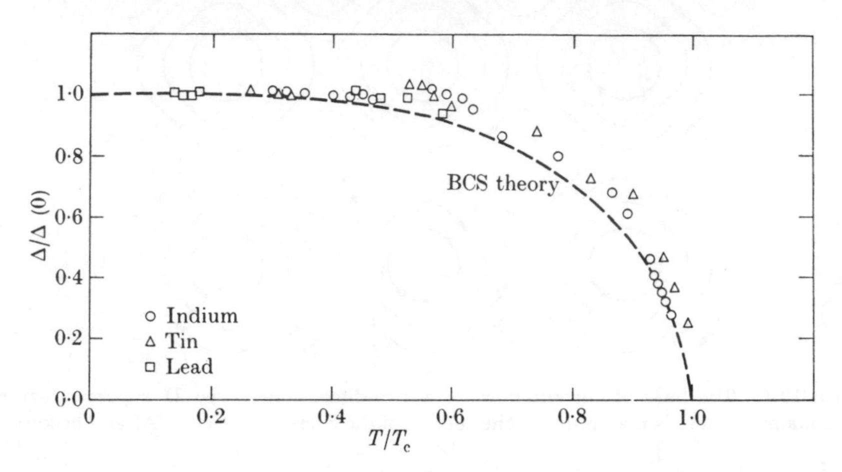

FIG. 13.12. The temperature variation of the energy gap (related to the energy gap at $T = 0$) as a function of T/T_c.

When the frequency is large enough there is an absorption process owing to electrons being excited across the gap. Pairing of electrons is no longer advantageous; all traces of superconductivity disappear. This explanation happens to be correct and is in agreement with the predictions of the BCS theory.

The width of the gap can be deduced from measurements on specific heat, electromagnetic absorption, or tunnelling. Typical values are somewhat below one milli-electronvolt. The gap does not appear abruptly; it is zero at the critical temperature and rises to the value of a few multiples of kT_c at absolute zero temperature. The temperature variation is very well predicted by the BCS theory, as shown in Fig. 13.12 for three superconductors.

When we put a thin insulator between two identical superconductors the energy diagram (Fig. 13.13a) looks very similar to those we encountered when studying semiconductors. The essential difference is that in the present case the density of states is high just above and below the gap, as shown in Fig. 13.13b. An applied voltage produces practically no current until the voltage difference is as large as the gap itself; the situation is shown in Fig. 13.14. If we increase the voltage further, electrons from the left-hand side may tunnel into empty states on the right-hand side, and the current rises abruptly as shown in Fig. 13.15.

An even more interesting case arises when the two superconductors have different gaps†. Since the Fermi level is in the middle of the gap (as for intrinsic semiconductors), the energy diagram at thermal equilibrium is as shown in Fig. 13.16a. There are some electrons above the gap (and holes below the gap) in superconductor A but hardly any (because of the larger gap) in superconductor B. When a voltage is applied a current will flow and will increase with voltage (Fig. 13.17), because more and more of the thermally excited electrons in superconductor A can tunnel across the insulator into the available states of superconductor B. When the applied voltage reaches $\Delta_2 - \Delta_1$ (Fig. 13.16b), it has become energetically possible for all thermally excited electrons to tunnel across. If the voltage is increased further, the current decreases because the number of electrons capable of tunnelling is unchanged but they now face a lower density of states. When the voltage becomes greater than $\Delta_2 + \Delta_1$, the current increases rapidly, because electrons below the gap can begin to flow.

Thus a tunnel junction comprised of two superconductors of different energy gaps may exhibit negative resistance, similarly to the semiconductor tunnel diode. Unfortunately, the superconducting tunnel junction is not as useful because it works only at low temperatures.

The tunnelling we have just described follows the same principles we encountered when discussing semiconductors. There is, however, a tunnelling phenomenon characteristic of superconductors, and of superconductors alone; it is the so-called superconducting or Josephson tunnelling (discovered theoretically by Josephson, a Cambridge graduate student at the time), which takes place when the insulator is very thin (less than 1·5–2 nm). It displays a number of interesting phenomena, of which I shall briefly describe four.

(i) For low enough currents there can be a current across the insulator without any accompanying voltage; the insulator turns into a superconductor. The reason is that Cooper pairs (*not* single electrons) tunnel across.

† I. Giaever, Electron tunnelling between two superconductors, *Phys. Rev. Letters*, vol. 5, p. 464, November 15, 1960.

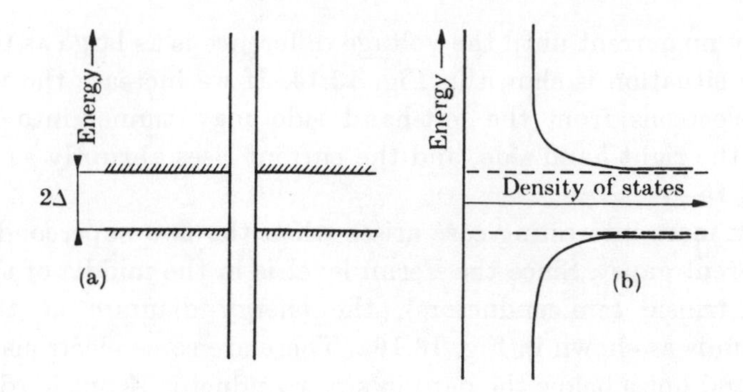

Fig. 13.13. (a) Energy diagram for two identical superconductors separated by a thin insulator. (b) The density of states as a function of energy.

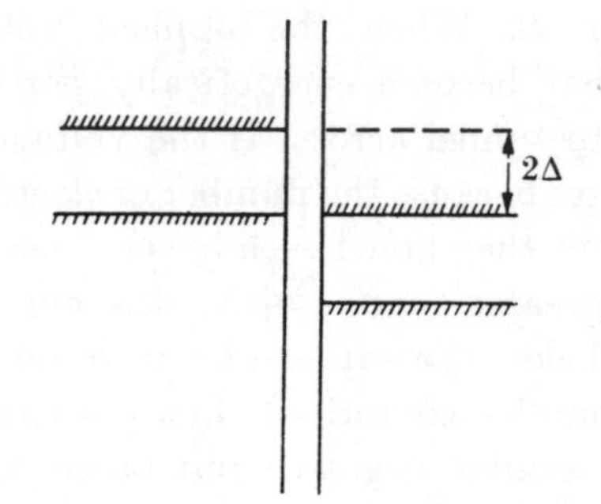

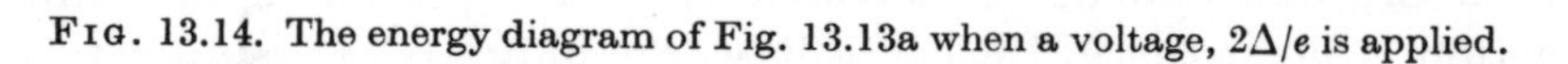

Fig. 13.14. The energy diagram of Fig. 13.13a when a voltage, $2\Delta/e$ is applied.

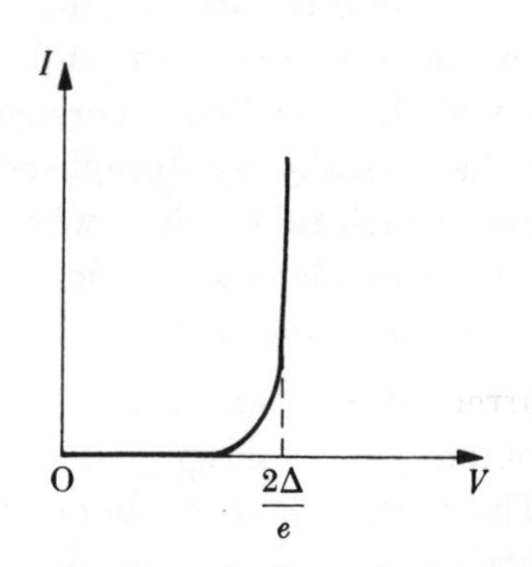

Fig. 13.15. The current as a function of voltage for a junction between two identical superconductors separated by a thin insulator.

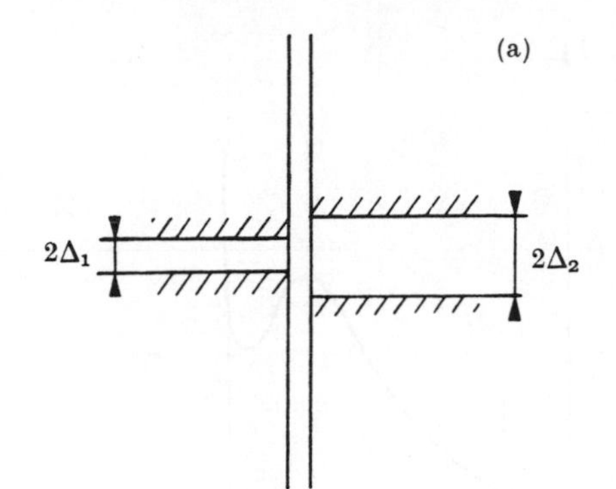

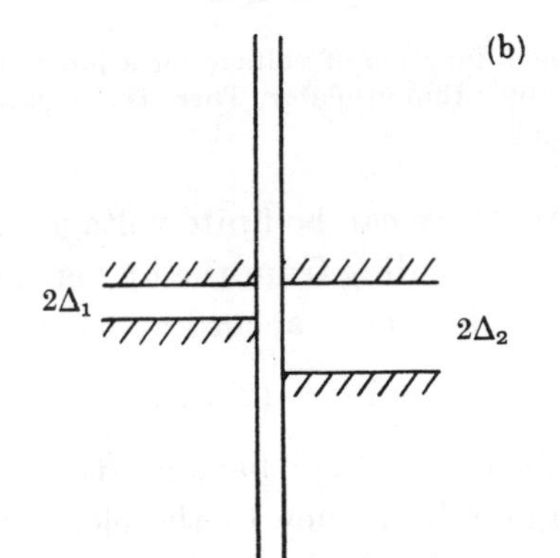

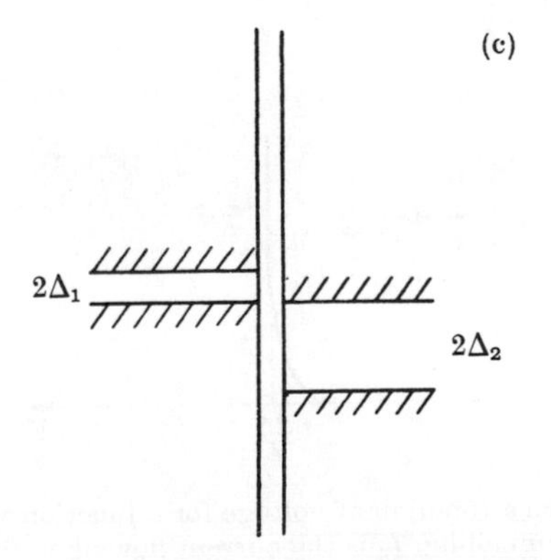

FIG. 13.16. Energy diagrams for two different superconductors separated by a thin insulator. (a) $V = 0$, (b) $V = (\Delta_2 - \Delta_1)/e$, (c) $V = (\Delta_1 + \Delta_2)/e$.

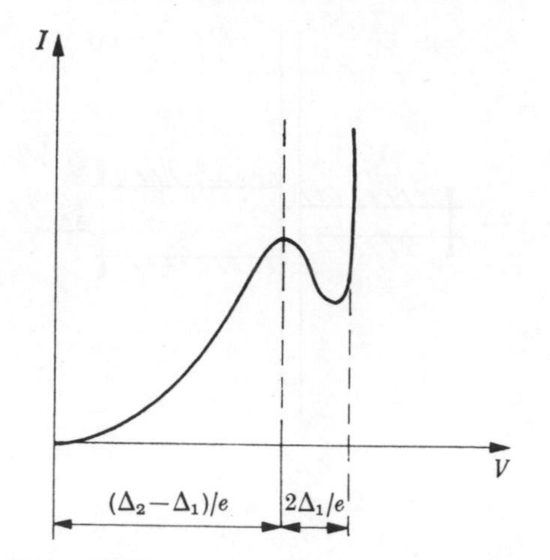

FIG. 13.17. The current as a function of voltage for a junction between two different superconductors separated by a thin insulator. There is a negative resistance region for $(\Delta_2-\Delta_1)/2e<V<(\Delta_1+\Delta_2)/2e$

(ii) For larger currents there can be finite voltages across the insulator. The Cooper pairs descending from the higher potential to the lower one may radiate their energy according to the relationship

$$\hbar\omega = (2e)U_{AB}, \qquad (13.73)$$

where U_{AB} is the d.c. voltage between the two superconductors and ω is the angular frequency of the electromagnetic radiation. Thus we have a very simple form of a d.c. tunable oscillator that could work up to infrared frequencies. It is not, however, very

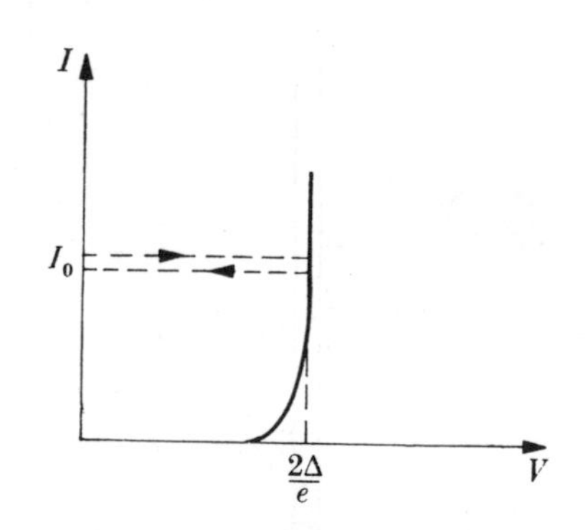

FIG. 13.18. The current as a function of voltage for a junction which may display both 'normal' and Josephson tunnelling. I_0 is the current flowing without any accompanying voltage. The application of a small magnetic field causes a transition between the Josephson and the 'normal' tunnelling characteristics. Removing this extra magnetic field the voltage returns to zero.

practical because, besides the low temperatures needed, it can produce only a minute amount of power.

(iii) A direct transition may be caused between the Josephson characteristics and the 'normal' tunnelling characteristics by the application of a small magnetic field.

(iv) When two Josephson junctions are connected in parallel (Fig. 13.19) the maximum supercurrent that can flow across them is a periodic function of the magnetic flux,

$$I_{\max} = 2I_J\left|\cos\frac{\pi\Phi}{\Phi_0}\right|, \tag{13.74}$$

where I_J is a constant depending on the junction parameters, Φ is the enclosed magnetic flux, and Φ_0 is the so-called flux quantum equal to $h/2e = 2{\cdot}10^{-15}$ Wb.

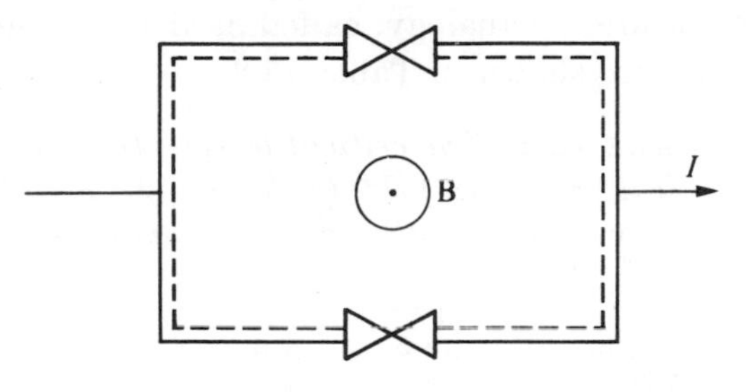

FIG. 13.19. Two Josephson junctions in parallel connected by a superconducting path.

13.8. Some applications

High-field magnets. For the moment the most important practical applications of superconductivity is in producing a high magnetic field.† There is no doubt that for this purpose a superconducting solenoid is superior to conventional magnets. A magnetic field of 7 weber/m² (70,000 gauss if you still prefer cgs units) can be produced by a solenoid not larger than about 12×20 cm. A conventional magnet producing the same magnetic field would look like a monster in comparison, would need a few megawatts of electric power, and at least a few hundred gallons of cooling water per minute.

† Perhaps it is not justified to use the adjective 'practical' because only the various research laboratories are in need of high magnetic fields. Defining, however, 'practical' as referring to something which can be sold for ready money, we may persist in using the word.

What sort of materials do we need for obtaining high magnetic fields? Obviously type II superconductors, as they remain superconducting up to quite high magnetic fields. However, high magnetic fields are allowed only at certain points in the superconductor that are surrounded by current vortices. When a d.c. current flows (so as to produce the high magnetic field in the solenoid) the vortices experience a $\mathbf{J} \times \mathbf{B}$ force that removes the vortices from the material. To exclude the high magnetic field costs energy and the superconductor consequently becomes normal, which is highly undesirable. The problem is to keep the high magnetic fields inside. This is really a problem similar to the one we encountered in producing 'hard' magnetic materials, where the aim was to prevent the motion of domain walls. The remedy is similar; we must have lots of structural defects; that is, we must make our superconductor as 'dirty' and as 'non-ideal' as possible. The resulting materials are, by analogy, called hard superconductors. Some of their properties are shown in Table 13.2.

TABLE 13.2. *The critical temperature and critical magnetic field (at $T = 4{\cdot}2$ K) of the more important hard super-conductors.*

Material	T_c(K)	$H_c \times 10^{-7}$(A/m)
Nb_3Sn	18·5	1·6
V_3Si	17	1·3
V_3Ga	16·8	2·8 (extrapolated)
NbN	16	0·8
$Pb_{0{\cdot}9}Mo_{5{\cdot}1}S_6$	14	4·8
Nb_3Ge	22·3	3·0

There is, however, a further difficulty with vortices. Even if they do not move out of the material, *any* motion represents ohmic loss, causing heating, and making the material become normal at certain places. To avoid this, a good thermal conductor and poor electrical conductor, copper (yes, copper), is used for insulation so that the heat generated can be quickly led away.

Switches and memory elements. The use of superconductors as switches follows from their property of becoming normal in the presence of a magnetic field. We can make a superconducting wire resistive by using the magnetic field produced by a current flowing in another superconducting wire. The switch shown in Fig. 13.20a employs niobium for the control coil so that the magnetic field that makes the tantalum wire become normal should still leave the wire of the control coil superconducting.

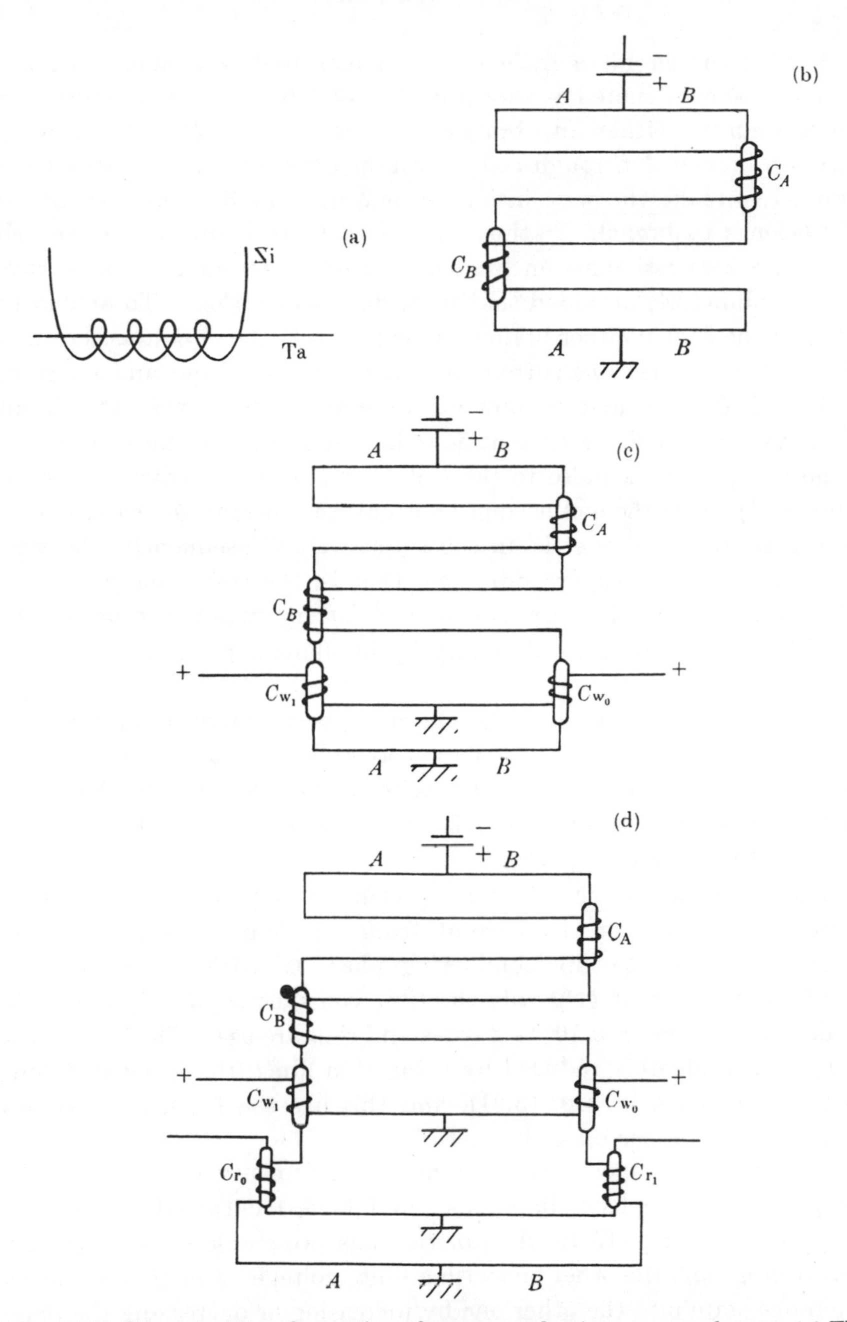

FIG. 13.20. The principles of operation of a superconducting memory element. (a) The tantalum wire inside a coil made of niobium may become resistive when a current is sent through the coil. (b) The information is stored by a current flowing either in branch *A* or in branch *B*. (c) Same as (b) but for two additional coils which serve to switch the current from one branch to the other. (d) Same as (c) but for two additional coils which can read the state of currents.

Making the circuit a little more sophisticated, a memory element can also be constructed as shown in Fig. 13.20b. The current generator feeds a current either into branch A or into branch B. If the current flows in branch A through coil C_A, the magnetic field created is large enough to make the wire inside become normal. But the wire inside C_A belongs to branch B; thus there is a finite resistance in branch B, and a zero resistance in branch A. Hence the current will always flow in branch A, provided that it started to flow there. To enable us to start the flow in either branch, we can modify the circuit as shown in Fig. 13.20c. Let us call a current flowing in branch A 'one' and a current in branch B 'zero', and see how we can write in 'zero' when the circuit displays 'one' (that is, a current flows in branch A). This may be simply done by applying a pulse to the coil C_{w1}. The wire (which is part of branch A) inside the coil becomes normal; the current decreases everywhere in branch A, that is, in coil C_A as well. Consequently, the wire inside C_A becomes superconducting, that is, the resistance in branch B vanishes. Hence the current from the battery switches to branch B; 'one' has been written in. Similarly by applying a pulse to C_{w0}, 'zero' can be written in.

The circuit is still not complete because, besides writing in, we must be able to *read* the state of the circuit as well. This can be achieved by inserting two more coils in the circuit, as shown in Fig. 13.20d. A current in branch A (or in branch B) makes the wire inside normal, which can be sensed in the reading circuit.

The speed with which this memory element can work depends on the time needed to steer the current from one branch into the other. Since the normal-to-superconducting phase transition is rather fast the limiting factor is generally the time constant (L/R) of the circuit, which is of the order of 10^{-5} s if wires and coils are used. The inductance may be considerably reduced by using thin films (the basic switching element is shown in Fig. 13.21), and this has the further advantage of being suitable for mass production.

The latest in superconducting memory elements is based on the properties of Josephson junctions. As I have mentioned before, and may be seen in Fig. 13.18, the junction has two stable states, one with zero voltage and the other one with a finite voltage. It may be switched from one state into the other one by increasing or decreasing the magnetic field threading the junction. The advantage of this Josephson junction memory is that there is no normal to superconducting phase transition necessary, only the *type* of tunnelling is changed, which is a much faster process. Switching times as short as 10 ps have been measured.

Will it ever be worthwhile to go to the trouble and expense of cooling memory stores to liquid helium temperatures? So far computer manufacturers have been rather reluctant (understandably, it's a high risk business) to introduce superconducting memories. Will Josephson junctions fare better? Provided that all the technological problems (mainly reliability and reproducibility) are solved I think they have an excellent chance because they possess three properties ultra-fast computers must have, fast switching speed, high density, and low power dissipation.

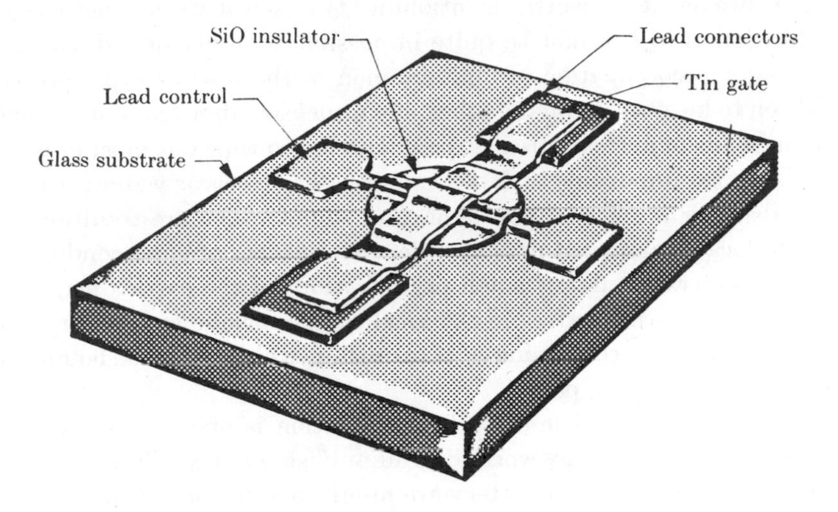

FIG. 13.21. The switch of Fig. 13.19a in thin film form.

It must be remembered that once we are down to switching speeds of 10 ps, the major factor limiting the speed of computer operation is the finite velocity of light. In 10 ps the signal can travel, at best, 3 mm. Hence we need to put our units close to each other, which would normally bring heat dissipation difficulties. With a density of 10^5 gates/cm^2 Josephson junction memories will only dissipate about 10 mW/cm^2, many orders of magnitude lower than that of their competitors.

Magnetometers. A further important application of Josephson junctions is in a magnetometer called SQUID (Superconducting Quantum Interference Device). Its operation is based on the previously mentioned property that the maximum supercurrent through the two junctions in parallel is dependent on the magnetic flux enclosed by the loop. It follows from eqn. (13.74) that there is a complete period in I_{max} while Φ varies from 0 to Φ_0. Thus if we can tell to an accuracy of 10% the magnitude of the supercurrent, and we take a loop area of 1 cm^2, the smallest magnetic field that can be measured is 10^{-11} Wb/m^2. The best result achieved in a laboratory is about 10^{-7} Wb/m^2. Commercially available devices (working on roughly the same principle) can offer a sensitivity of about 10^{-6} Wb/m^2 with a time constant of 1 sec, making it the most sensitive magnetometer available.

Although the Josephson junction does many things superlatively well, like other topics in superconductivity, its applications (so far) are few. However it is worth mentioning two sensitive magnetometer applications which would be quite impossible with classical devices.

A major preoccupation of military men of the world's super-powers has been to keep watch on the other side's nuclear submarines, with their horrific loads of nuclear missiles, programmed to wipe out most of us.

It is almost impossible to detect underwater vessels as water is such a good absorber of microwaves, light, and sound which are traditionally used to locate targets. However underwater caches of superconducting magnetometers can detect small perturbations of the earth's magnetic field as a submarine arrives in the locality. They have to be connected to a surface buoy containing a transmitter which informs boffins in bunkers what is going past.

A more definite and much safer application is one which Oxford's Laboratory of Archeology works on and publishes freely. The silicaceous and clay-like materials in pottery are mildly paramagnetic. When they are fired in kilns the high temperature destroys the magnetism, and as they cool the permanent dipoles re-set themselves in the local magnetic field of the earth. When an archeologist uncovers an old kiln he can measure this magnetism in the bricks and thus find the direction of the earth's field when the kiln was last fired. The variation of the earth's field and angle of dip has been determined for several thousand years at some places. Thus it is possible to date kilns by accurate measurements. As large ceramic articles have to be kilned standing on their bases, accurate measurements of the dip angle can also date cups and statues, if their place of origin is known. With very sensitive magnetometers, this measurement can be done on a small, unobtrusive piece of ceramic removed from the base of the statue. It is a method considerably used by major museums.

Metrology. I mentioned earlier that one can determine one of our fundamental constants (velocity of light) with the aid of lasers. It turns out that Josephson junctions may be used for determining another fundamental constant. The relevant formula is eqn. (13.73). Measuring the voltage across the junction and the frequency of radiation, h may be determined. As a result the accepted value of Planck's constant changed recently from 6·62559 to 6·626196 10^{-34} J s.

Suspension systems and motors. The tendency to exclude magnetic fields may be used for suspension systems. Plate 2 shows a lead-coated hollow glass sphere supported by the magnetic flux trapped in two lead rings. This frictionless suspension system is made possible by the diamagnetic property of the superconducting sphere. If the sphere is pressed downwards it compresses the magnetic flux, thus amplifying the repelling force. Noting further that it is possible to impart high speed rotation to a suspended superconducting body, and that all the conductors in the motor are free of resistance, it is quite obvious that the ideal of a hundred-percent-efficient motor can be closely approximated.

Radiation detectors. The operation of these devices is based on the heat provided by the incident radiation. The superconductor is kept just above its critical temperature, where the resistance is a rapidly varying function of temperature. The change in resistance is then calibrated as a function of the incident radiation.

Heat valves. The thermal conductivity of some superconductors may increase by as much as two orders of magnitude when made normal by a magnetic field.

This phenomenon may be used in heat valves in laboratory refrigeration systems designed to obtain temperatures below 0·3 K.

Resonant cavities. This is a rather obvious application. If microwave cavities with very low electrical losses are required, they can be made of (or coated with) superconducting materials. The losses can be reduced by a factor of up to a hundred.

Examples

1. It follows from eqn. (1.15) that in the absence of an electric field the current density declines as

$$J = J_0 \exp(-t/\tau)$$

where τ is the relaxation time related to the conductivity by eqn. (1.10).

In an experiment the current flowing in a superconducting ring shows no decay after a year. If the accuracy of the measurement is 0·01 per cent calculate a lower limit for the relaxation time and conductivity (assume 10^{28} electrons/m^3). How many times larger is this conductivity than that of copper?

2. In the first phenomenological equations of superconductivity, proposed by F. and H. London in 1935, the current density was assumed to be proportional to the vector potential and div $\mathbf{A} = 0$ was chosen. Show that these assumptions lead to a differential equation in $\mathbf{A}$ of the form of eqn. (13.63).
3. The parameter λ defined in eqn. (13.60) may be regarded as the penetration depth for $\kappa \simeq 0$. A typical value for the measured penetration depth is 60 nm. To what value of ψ_0^2 does it correspond?
4. What is the frequency of the electromagnetic waves radiated by a Josephson junction having a voltage of 650 μV across its terminals?

Epilogue

Eigentlich weiss man nur, wenn man wenig weiss, mit dem Wissen wächst der Zweifel.

JOHANN WOLFGANG VON GOETHE

I hope these lectures have given you some idea how the *electrical properties* of materials come about and how they can be modified and exploited for useful ends. You must be better equipped now to understand the complexities of the physical world and appreciate the advances of the last twenty years. You are, I hope, also better equipped to question premisses, to examine hypotheses, and to pass judgment on things old and new. If you have some feeling of incompleteness, if you find your knowledge inadequate, your understanding hazy, don't be distressed; your lecturers share the same feelings.

The world has changed a lot since the first edition of this book. The quiet optimism reigning ten years ago is no longer the order of the day. Nowadays people tend to be either wildly optimistic, envisaging all the wealth our automatic factories will produce, or downright pessimistic, forecasting the end of civilization as we run out of energy and raw materials. The optimists take it for granted that the engineers will design the automatic factories for them and even the pessimists have some lingering hopes that the engineers will somehow come to the rescue. It is difficult indeed to see any alternative group of people who could affect the desired changes. I greatly admire physicists. Their discoveries lie at the basis of all our engineering feats but I don't think they can do much in the present situation. We don't need to pry any more into the secrets of Nature, we need to make them work for us. The current research of geneticists, microbiologists and biologists may well produce a new species of supermen but it is unlikely that we can wait for them. We cannot put much trust into politicians either. They will always (they have to) promise a better future but the power to carry out their promises is sadly missing. There is no escape. The responsibility is upon your shoulders. Some of you will, no doubt, opt for management but I hope many of you will employ your ingenuity in trying to find solutions to the burning engineering problems of the day. You are more likely to succeed if you aim high. And it is more fun too.

Appendix I

List of Frequently Used Symbols and Units

Quantity	Usual symbol	Unit	Abbreviation
Current	I	ampere	A
Voltage	U, V	volt	V
Charge	q	coulomb	C
Capacitance	C	farad	F
Inductance	L	henry	H
Energy	E	joule	J
(1 eV is the electron energy change caused by 1 volt acceleration ($=1{\cdot}60\times10^{-19}$J))		electron volt	eV
Resistance	R	ohm	Ω
Power	P	watt	W
Electric field	$\mathscr{E}$	volt/metre	V m^{-1}
Magnetic field	H	ampere/metre	A m^{-1}
Electric flux density	D	coulomb/(metre)^2	C m^{-2}
Magnetic flux density	B	weber/(metre)^2	Wb m^{-2}
Frequency	f	hertz	Hz
Wavelength	λ	metre	m

List of Physical Constants

Symbol	Quantity	Value
m	mass of electron	$9{\cdot}11\times10^{-31}$ kg
e	charge of electron	$1{\cdot}60\times10^{-19}$ C
e/m	charge/mass ratio of electron	$1{\cdot}76\times10^{11}$ C/kg
h	Planck's constant	$6{\cdot}62\times10^{-34}$ J s
$\hbar$	$h/2\pi$	$1{\cdot}05\times10^{-34}$ J s
N	Avogadro's number	$6{\cdot}02\times10^{26}$ molecules/kg molecule
k	Boltzmann's constant	$1{\cdot}38\times10^{-23}$ J/ K
		$8{\cdot}62\times10^{-5}$ eV/ K
kT	(at 293 K)	$0{\cdot}025$ eV
		$4{\cdot}05\times10^{-21}$ J
c	velocity of light	$3{\cdot}00\times10^{8}$ m/s
ϵ_0	permittivity of vacuum	$8{\cdot}85\times10^{-12}$ F/m
μ_0	permeability of vacuum	$4\pi/10^7$ H/m
hc/e	wavelength of photon having 1 eV energy	1240 nm
μ_{mB}	Bohr magneton	$9{\cdot}27\times10^{-24}$ A m^2

Appendix II

Variational Calculus. Derivation of Euler's Equation

THE problem of variational calculus is to find the form of the function $y(x)$ which makes the integral

$$\int_{x_1}^{x_2} F(x, y(x), y'(x))\, dx \tag{A.1}$$

an extremum (maximum or minimum), where F is a given function of x, y and y'.

Assume that $y(x)$ does give the required extremum, and construct a curve $y(x)+\eta(x)$ which is very near to $y(x)$ and satisfies the condition $\eta(x_1) = 0 = \eta(x_2)$. For the new curve our integral modifies to

$$\int_{x_1}^{x_2} F(x, y(x)+\eta(x), y'(x)+\eta'(x))\, dx. \tag{A.2}$$

Making now use of the fact that η and η' are small, we may expand the integral to the first order, leading to

$$\int_{x_1}^{x_2} \left\{ F(x, y(x), y'(x)) + \frac{\partial F}{\partial y}(x, y(x), y'(x),\eta(x) + \right.$$
$$\left. + \frac{\partial F}{\partial y'}(x, y(x), y'(x))\eta'(x) \right\} dx. \tag{A.3}$$

Integrating the third term by parts (and dropping the parentheses showing the dependence on the various variables) we get

$$\int \frac{\partial F}{\partial y'}\eta'\, dx = \left[\eta \frac{\partial F}{\partial y'}\right]_{x_1}^{x_2} - \int_{x_1}^{x_2} \eta \frac{\partial}{\partial x}\frac{\partial F}{\partial y}\, dx = -\int_{x_1}^{x_2} \eta \frac{\partial}{\partial x}\frac{\partial F}{\partial y'}\, dx \tag{A.4}$$

which substituted back into eqn. (A.3) gives

$$\int_{x_1}^{x_2} \left[F + \left(\frac{\partial F}{\partial y} - \frac{\partial}{\partial x}\frac{\partial F}{\partial y'}\right)\eta\right] dx. \tag{A.5}$$

We may argue now that $y(x)$ will make the integral an extremum if the integral remains unchanged for small variations of η. This will occur when the coefficient of η vanishes, that is, when

$$\frac{\partial F}{\partial y} - \frac{\partial}{\partial x}\frac{\partial F}{\partial y'} = 0. \tag{A.6}$$

The above equation is known as the Euler differential equation, the solution of which gives the required function $y(x)$.

When F depends on another function $z(x)$ as well, an entirely analogous derivation gives one more differential equation in the form

$$\frac{\partial F}{\partial z} - \frac{\partial}{\partial x}\frac{\partial F}{\partial z'} = 0. \tag{A.7}$$

These two differential equations, (A.6) and (A.7), are used in section 13.3 for obtaining the vector potential and the wave function which minimize the Gibbs free energy.

Appendix III

Suggestions for further reading

(a) *General texts*

R. P. Feynman *et al.*, *The Feynman lectures on physics* (Addison Wesley, 1965).
A. J. Dekker, *Solid state physics* (Macmillan, 1958).
C. Kittel, *Introduction to solid state physics* (Wiley, 1953).
B. I. Bleaney and B. Bleaney, *Electricity and Magnetism* (3rd edn. O.U.P., 1976).
H. M. Rosenberg, *The solid state* (2nd edn. O.U.P., 1978).
Materials, A Scientific American Book (W. H. Freeman, 1967).

(b) *Topics arising in specific chapters*

Chapter 4

W. Heitler, *Elementary wave mechanics* (O.U.P., 1956).

Chapter 5

A. H. Cottrell, *Theoretical structural metallurgy* (Arnold, 1948).

Chapter 7

J. M. Ziman, *Electrons in metals* (Taylor and Francis, 1962).

Chapter 8

J. P. McKelvey, *Solid state and semiconductor physics* (Harper and Row, 1966).

Chapter 9

A. S. Grove, *Physics and technology of semiconductor devices* (Wiley, 1967).
J. E. Carroll, *Physical models for semiconductor devices.* (Arnold, 1974).
D. A. Fraser, *The physics of semiconductor devices* (2nd edn. O.U.P., 1979).

Chapter 10

P. J. Harrop, *Dielectrics* (Butterworth, 1972).

Chapter 11

J. C. Anderson, *Magnetism and magnetic materials* (Chapman & Hall, 1968).

Chapter 12

I. P. Kaminow and A. E. Siegman, *Laser devices and applications* (I.E.E.E. press, 1973)
Amnon Yariv, *Introduction to optical electronics* (Holt, Rinehart, and Winston, 2nd edn. 1976).

Chapter 13

L. Solymar, *Superconductive tunnelling and applications* (Chapman and Hall, 1972)
E. Rhoderick and A. C. Rose-Innes, *Introduction to superconductivity* (Pergamon Press, 1969)

Answers to Examples

Chapter 1 (p. 25)

1. 10^{21} electrons m^{-3}; $2{\cdot}7.10^{-13}$ sec.
2. (i) $0{\cdot}015\ m_0$; (ii) $46{\cdot}8\ m^2\ V^{-1}\ sec^{-1}$; (iii) Sharp resonance; since $\omega\tau \simeq 15$.
3. $0{\cdot}14$ and $0{\cdot}014\ m^2\ V^{-1}\ sec^{-1}$
4. (i) $2{\cdot}5.10^{28}\ m^{-3}$, (ii) $5{\cdot}3.10^{-3}\ m^2\ V^{-1}\ sec^{-1}$; (iii) $0{\cdot}97\ m_0$; (iv) 3.10^{-14} sec; (v) $0{\cdot}99$.
5. $R = -\dfrac{N_e\mu_e^2 - N_h\mu_h^2}{e(N_e\mu_e + N_h\mu_h)^2}$; not necessarily if $\mu_e \gg \mu_h$
6. $[1-(1-\omega_p^2/\omega^2)^{\frac{1}{2}}]^2[1+(1-\omega_p^2/\omega^2)^{\frac{1}{2}}]^{-2}$; $4(1-\omega_p^2/\omega^2)^{\frac{1}{2}}[1+(1-\omega_p^2/\omega^2)^{\frac{1}{2}}]^{-2}$
7. Transmitted power $\simeq \exp[-2d/c\ (\omega_p{}^2 - \omega^2)^{\frac{1}{2}}]$; 2.10^{-5}; $\sim 10^{-53}$.

Chapter 2 (p. 40)

1. (i) $(6{\cdot}25/a^{\frac{1}{2}}).10^{-9}$ m; (ii) $7{\cdot}3.10^{-11}$ m; (iii) $4{\cdot}5.10^{-15}$ m.
2. (i) 5.10^{-12} m; (ii) $27{\cdot}3$ eV; (iii) lens aberrations, voltage stability.
3. Max at $\theta = \sin^{-1} 0{\cdot}42,\ 0{\cdot}83$.
5. $15{\cdot}6$ eV.

Chapter 3 (p. 59)

1. $\hbar^2/32\ ma^2$
2. 0; $\dfrac{a^3}{3}\left\{1-\dfrac{6}{(n\pi)^2}\right\}$; 0; $\hbar^2\left(\dfrac{n\pi}{2a}\right)^2$ where $n = 2, 4, 6$, etc.
3. 0; $\dfrac{a^2}{3}$; 0; p^2
5. Reflected current/incident current $= \left|\dfrac{k_1-k_2}{k_1+k_2}\right|^2$;

   transmitted current/incident current $= \dfrac{\mathrm{Re}(k_2)}{k_1}\left|\dfrac{2k_1}{k_1+k_2}\right|^2$
6. $0{\cdot}13I$; $6{\cdot}6.10^{-13}I$.
9. $E_n = (n+\frac{1}{2})\hbar\omega_0$.

Chapter 4 (p. 77)

1. $1{\cdot}22.10^{-7}$ m.
2. $4{\cdot}1.10^6$ m sec^{-1}.
3. 7.10^{-5} nm, Doppler and pressure broadening
4. $5{\cdot}3.10^{-16}$ sec.
5. $4\pi\hbar^2\varepsilon_0/me^2$.
6. -54 eV.
7. $\dfrac{-\hbar^2}{2m}[\nabla_1^2\psi+\nabla_2^2\psi+\nabla_3^2\psi]+\left[-\dfrac{3e^2}{r_1}-\dfrac{3e^2}{r_2}-\dfrac{3e^2}{r_3}+\dfrac{e^2}{r_{12}}+\dfrac{e^2}{r_{13}}+\dfrac{e^2}{r_{23}}\right]\dfrac{\psi}{4\pi\varepsilon_0} = E\psi$.

Chapter 5 (p. 97)

4. $0{\cdot}312$ nm.

Chapter 6 (p. 117)
1. 0·27; 1260 K.
2. 3·15 eV.
4. 40 mA; 1·29 A; about 1·9 times.
5. $1{\cdot}02.10^{-7}$ A; yes.
6. 7·3 eV; 392 J/kg K; about $\frac{1}{2}$%.

Chapter 7 (p. 146)
1. $1{\cdot}12\ m_0$.
6. electron.

Chapter 8 (p. 182)
1. 0·043 eV.
2. $\frac{3}{2}kT$.
3. 0·76 eV; $m_0/2$.
4. (i) 0·66 eV; (ii) $1{\cdot}9.10^{-6}$ m.
6. $\frac{1}{2}$.
7. $2{\cdot}0.10^{22}\ \mathrm{m}^{-3}$; $\frac{1}{2}$; $1{\cdot}0.10^{22}\ \mathrm{m}^{-3}$; $1{\cdot}4.10^{8}\ \mathrm{m}^{-3}$.
8. $kT \ln \dfrac{1-\alpha}{\alpha}$; $kT \ln \dfrac{\beta}{1-\beta}$; 197 K
10. $\dfrac{\partial N_{\mathrm{h}}}{\partial t} = \dfrac{N_{\mathrm{hn}}-N_{\mathrm{h}}}{\tau_{\mathrm{p}}} - \dfrac{1}{e}\nabla \cdot \mathbf{J}_{\mathrm{h}}$.
11. (i) 4; (iii) $0{\cdot}16\ m_0$, $0{\cdot}21\ m_0$, $0{\cdot}34\ m_0$, $0{\cdot}52\ m_0$; (iv) no; (v) $6{\cdot}5.10^{-11}$ sec, 5.10^{-11} sec; (vi) 0·29:1; (vii) deep level sparsely populated.

Chapter 9 (p. 246)
3. $X_{\mathrm{n}} = -(\epsilon_{\mathrm{s}}/\epsilon_1)d_1 + [(\epsilon_{\mathrm{s}}/\epsilon_1)^2 d_1^2 + 2\epsilon_{\mathrm{s}}\epsilon_0 U_0/eN_{\mathrm{D}}]^{\frac{1}{2}}$
4. $d = [4U_0\epsilon/eN_{\mathrm{D}} + d_0^2/3]^{\frac{1}{2}}$
5. 0·35 V; 0.77 V.
6. $1{\cdot}9.10^{19}\ \mathrm{m}^{-3}$; $8{\cdot}9.10^{16}\ \mathrm{m}^{-3}$.
7. 18 V.
8. $N_{\mathrm{h}} - N_{\mathrm{hn}} = N_{\mathrm{hn}}\{\exp(eU_1/kT)-1\}\exp\left\{\dfrac{-x}{(D_{\mathrm{h}}\tau_{\mathrm{p}})^{\frac{1}{2}}}\right\}$.
9. 0·94 mm.
10. $e\left(\dfrac{D_{\mathrm{h}}}{\tau_{\mathrm{p}}}\right)^{\frac{1}{2}} N_{\mathrm{hn}}\{\exp(eU_1/kT)-1\}\exp\left[\dfrac{-x}{(D_{\mathrm{h}}\tau_{\mathrm{p}})^{\frac{1}{2}}}\right]$.
11. $e\left\{\left(\dfrac{D_{\mathrm{e}}}{\tau_{\mathrm{n}}}\right)^{\frac{1}{2}} N_{\mathrm{ep}} + \left(\dfrac{D_{\mathrm{h}}}{\tau_{\mathrm{p}}}\right)^{\frac{1}{2}} N_{\mathrm{hn}}\right\}$ per unit area.

Chapter 10 (p. 275)
2. $1{\cdot}43.10^{-40}$ farad m^2.
3. $1{\cdot}23\ \mathrm{V\ m}^{-1}$.
4. $2{\cdot}07.10^{-14}$ sec; 1·03 eV.

Chapter 11 (p. 316)
1. 10^{-6}.
2. $1{\cdot}6.10^{-29}\ \mathrm{A\ m}^2$.
3. 3·8 K.
6. $2{\cdot}71.10^{9}$ Hz; $4{\cdot}3.10^{6}$ Hz.
8. $4{\cdot}5.10^{-2}$; 658 K; 0·44.

Chapter 12 (p. 316)

1. 890 nm.
2. $\Delta f = c_m/2L$, where c_m is the velocity of light in the laser medium.

Chapter 13 (p. 385)

1. $3{\cdot}1.10^{11}$ sec; $8{\cdot}8.10^{31}$ $\text{ohm}^{-1}\,\text{m}^{-1}$; $1{\cdot}5.10^{24}$ times.
3. $1{\cdot}8.10^{27}\,\text{m}^{-3}$.
4. $3{\cdot}15.10^{11}$ Hz.

Index

Abrikosov, 373
Acceptor levels, 155, 156, 182, 185
Acceptors, 154, 156, 157, 164, 185, 188
  ionization energy, 154
Acoustic
  amplifier, 267
  surface wave, 268, 269, 270
  wave, 266, 267, 268, 277, 289
Aigrain, 18
Alkali metals, 76
Allowed energy band, 123, 129, 131
Alnico, 292, 295, 307
Alumina, 263, 275, 322
Aluminium, 74, 77, 102, 140, 154, 155, 295, 306, 307, 356
Ampere's Law, 294, 361
Antiferromagnetic resonance, 309
Antiferromagnetism, 304–305
Antimony, 153, 154
Archeology, 384
Argon, 74, 275, 328
Argon laser, 328–329, 343
Arsenic, 153, 154
Avalanche
  breakdown, 219, 234, 332
  diode, 218–220, 263
Average
  value of operators, 56
  velocity, 3
Axial modes, 336–338, 349

Backward diode, 218, 219
Band-pass filter, 269, 270
Band theory, 119–146, 147, 156
Bardeen, 196, 352
Barium, 107
  ferrite, 295, 306, 307
  titanate, 265, 270
Base, 197, 198, 201, 202
BCS theory, 352, 374
Benzene, 258
Beryllium, 73, 140, 141
Bias
  forward, 190, 191, 192, 193, 194, 195, 214, 225, 233
  reverse, 190, 194, 195, 210, 211, 214, 223, 232, 233
Boltzmann
  distribution, 102
  function, 319
  statistics, 189, 301, 319
Bohr
  magneton, 298, 302, 316
  radius, 67, 69, 77
Bonds, 78–97, 303, 304
  covalent, 84–86, 119, 129, 148, 153, 155, 304, 305
  ionic, 82–83, 258
  metallic, 84, 304
  van der Waals, 86–87
Born, 43
Boron, 73, 154, 155, 299, 300
  bromide, 244
Bragg reflection, 125
Brattain, 196
Breakdown, 222, 248, 276
  avalanche, 219, 234, 332
  dielectric, 262–264
  Zener, 219, 222
Brewster angle, 327
Brobeck, 314
Built-in voltage, 187, 194, 196, 209, 224, 247, 248
Bulk elastic modulus, 80
Busch, 31

Cadmium sulphide, 181, 223, 225, 266, 267
Caesium, 76, 102, 107
Carbon, 73, 85, 258, 295, 299, 307
Carbon dioxide, 245, 329, 331, 332
Cavity dumping, 337
CCD, charge-coupled devices, 230–233
Charge neutrality, 152, 193
Chemical bond, 74, 78
Chemical laser, 332
Chlorine, 74, 79, 83, 85, 303
Chromium, 74, 299, 300, 302, 305, 322, 339, 340
CO_2 laser, 329

Cobalt, 295, 299, 300, 305, 307
Coercive field, 289, 293, 294
Collector, 197, 198, 199, 201, 202
Collision time, 3, 25, 26, 149, 165, 172, 183
Collisions, 3, 136, 160, 161, 175, 327, 328
Communications, 223, 274, 332, 334, 345
Conduction band, 144
Conductivity, electrical, 4
Contact
  ohmic, 209
  potential, 116
Continuity equation, 60, 183, 248, 276
Cooper, 312
  pair, 353, 375, 378
Copper, 76, 102, 117, 295, 299, 307, 380
Coulomb force, 79, 358
Coupled modes, 87
Covalent bond, 84–86, 119, 129, 148, 153, 155, 304, 305
Critical
  magnetic field, 355, 356, 362, 363, 372
  temperature, 351, 356, 357, 358, 368, 374
Crystal growth, 176–179
Crystallography, 1, 264, 287
Curie
  constant, 284, 316
  temperature, 284, 285, 316
Cyclotron
  frequency, 18, 20
  resonance, 19–22, 25, 133, 170, 172, 183, 310

Davisson, 27
de Broglie, 29
  wavelength, 29, 40, 47, 160, 354
Debye, 259
  equations, 257, 259–260
Deep depletion mode, 213
Degenerate semiconductors, 213
Delay lines, 233, 268
Density
  gradient, 165
  of states, 99–103, 106, 149, 150, 151, 250, 374, 375
Depletion region, 186, 187, 188, 194, 195, 204, 211, 222, 248
Deuterium, 345
Diamagnetism, 263, 304, 357
Diamond, 85, 86, 140, 145, 148, 258
Dielectric
  constant, 9, 19, 154, 155, 251, 254, 255, 256, 257, 258, 259, 265, 334, 374
  materials, 250–277
Diffusion, 242, 243, 244
  coefficient, 165
  current, 165, 199, 200, 246, 248
  equation, 165
  reactance, 199–200
Diode
  avalanche, 218–220
  backward, 217, 218
  Gunn, 235–240
  photo, 223–225
  tunnel, 213–217
  varactor, 220–223, 271
  Zener, 218–220
Dirac, 43
Direct-gap, 174, 175, 182, 225, 333
Dispersion equation, 10, 19, 24, 277
Donor level, 154, 157, 159, 225
Donors, 154, 156, 157, 159, 162, 163, 164, 173, 185, 188, 194
  ionization energy, 154
Double heterojunction laser, 334
Drift velocity, 3, 167, 168, 238, 267
Dye laser, 329, 338

Eddy currents, 289, 357
Effective
  mass, 22, 132–136, 146
    average, 146, 167
    semiconductor, 149, 152, 155, 157, 167, 182, 183, 196, 236
    measurement, 170–172
    negative, 135, 141, 235
    Table, 181
    tensor, 135, 146
  number of electrons, 136–138, 139, 140, 144
Einstein, 320, 321
  relationship, 247
Eigenvalues, 68
Electrical
  conductivity, 4
  noise, 4
Electron–hole pair, 163, 164, 175, 220, 224

Electron
  microscope, 31–32
  spin resonance, 308, 316
Electron volt, 66
Electrostriction, 270, 290
Emitter, 196, 197, 198, 200
Energy
  band
    allowed, 123, 129, 131
    forbidden, 123, 131
  gap, superconducting, 373–379
  surface, 165, 364–366
Entropy, 360
Epitaxial growth, 147, 179–181, 240, 243, 244
Etalon, 336, 337
Euler's equation, 389
Exchange interaction, 305, 306
Exciton, 132, 192
Exclusion principle, 70
Extrinsic semiconductor, 153–160, 162, 163, 173, 183, 196

Fabry-Perot etalon, 336
Fermi-Dirac distribution, 99–103, 318, 358
Fermi
  function, 117, 144, 145, 150, 151, 213, 250, 318
  level, 101
    extrinsic semiconductor, 158
    intrinsic semiconductor, 153
    superconductor, 375
Ferrites, 306, 307, 310–312
Ferroelectrics, 270–272
Ferromagnetic resonance, 309
Ferromagnetism, 305
FET (*see* field-effect transistors)
Feynman, 87, 89, 94
  model, 129–131, 139, 146, 161
Field-effect transistors, 226–230
Field-emission, 112–113
  microscope, 114
Floating zone purification, 179
Fluorine, 74, 332
Forbidden energy band, 123, 131
Forward bias, 190, 191, 192, 193, 194, 195, 214, 225, 233
Fourier transform, 38, 259
Free carrier absorption, 225
Free-electron theory, 98–118

Gabor, 347
Gallium, 154, 356
  arsenide, 166, 167, 181, 182, 225, 236, 240, 241, 258, 333, 334, 349
  phosphate, 225, 258
Garnets, 306, 314
Gas lasers, 327–329, 350
Gauge factor, 241
Germanium, 25, 119, 148, 258, 333
  breakdown voltage, 248
  covalent bond, 85
  crystal structure, 165
  cyclotron resonance, 172, 183
  diffusion constant, 248
  energy gap, 145
  impurities in, 147, 154
  indirect-gap, 175
  lifetime of holes, 248
  melting point, 179
  mobility, 248
  resistivity, 182
  Table, 181
Germer, 27
g-factor, 297, 298, 309
Giaever, 375
Gibbs
  free energy, 359, 362, 365, 366, 368, 369, 371
  function, 359, 360, 361, 367
Gilbert, 278, 284, 291
Ginzburg, 366
Gold, 209, 240
Goudsmit, 70
Graphite, 85, 167
Ground state, 67
Group velocity, 35, 133, 134, 146
Gunn, 236
  effect, 162, 235–240

Hall
  coefficient, 7, 26, 167, 170
  effect, 6–8, 140, 241, 315
Hamiltonian, 88
Hard
  magnetic materials, 294, 307
  superconductors, 380
Haynes-Shockley experiment, 168–170
Heat valve, 385
Heisenberg, 43, 305
Helicon, 19
Helimagnetism, 306

Helium, 114, 226, 327, 328, 332, 351
atom, 71
Helium-neon laser, 117, 327–329, 335
Hertz, 43
Hilsum, 236
Hockham, 273
Holes, 8, 131, 135, 140–141
heavy, 165, 172
light, 165, 172
Holography, 348–349
Homopolar bond, 84
Hooke's law, 82, 265
Hund's rule, 299
Hydrofluoric acid, 242
Hydrogen, 298, 299
atom, 37, 62–77, 154
molecular ion, 90, 94, 130
molecule, 90, 94–96
Hysteresis, 289, 294

Impurities, 153–160, 162
Indirect-gap, 166, 174, 175, 182, 225
Indium, 154, 155, 156, 182, 208
antimonide, 19, 25, 181, 241, 258
Infrared detectors, 225–226
Injection, 168, 192–194, 197, 225, 234, 248, 332, 345
Injection laser, 332–334, 345
Integrated circuits, 147, 179, 180, 242–246
Interdigital transducer, 269
Interference, 29
Inverted population, 323, 324, 327, 332, 333, 336, 337, 341, 349
Ionic bond, 82–83, 258
Ionized
acceptors, 154, 157
donors, 154, 157
Ionosphere, 17
Iron, 282, 284, 285, 287, 291, 295, 299, 300, 305, 306, 314, 316
ferrite, 306, 307
Isolators, 19, 310, 312–314
Isotope effect, 358
separation, 345

Josephson, 375
junction, 375, 382, 383, 386
tunnelling, 375, 378, 379, 382
Junctions
capacity, 194–196
Josephson, 375, 382, 383
laser, 332–334
metal–insulator–semiconductor, 209–213
metal–metal, 116–117
metal–semiconductor, 203–209
p–n, 184–195, 196, 199, 202, 210, 213, 214, 221, 222, 223, 224, 225, 226, 234, 235, 246, 247, 248, 332, 333, 334
tunnel, superconducting, 375, 376, 378

Kammerlingh Onnes, 351
Kao, 273
Kogelnik, 334, 335
Kronig–Penney model, 120–125, 128, 132, 139
Krypton, 76, 332

Landau–Ginzburg theory, 366–373
Landau levels, 310
Langevin function, 254, 282
Larmor frequency, 281
Laser, 115, 117, 256, 263, 318–350
applications, 343–349
chemical, 332
dye, 329, 338
fusion, 345
gas-dynamic, 331–332
gaseous discharge, 327–329, 350
injection, 332–334
machining, 345
modes, 334–338
parametric, 330–331
semiconductor, 332–334
solid-state, 325–327
welding, 345
Lattice ion, 121
Lead, 356
titanate, 265
Lenz's law, 281, 357
Li, 334, 335
Lifetime, 165, 167, 175
Limericks, 58, 59, 353
Liquid epitaxy, 181
Lithium, 73, 74, 76, 102, 140, 146, 299, 345
niobate, 271
Little, 352
Lodestone, 278
London, F., 386
London, H., 386

Loss tangent, 254, 273
LSI, large scale integration, 224

Madelung constant, 83
Magnesium, 7, 140, 141, 306
Magnetic
  amorphous material, 292
  anisotropy, 288, 290, 292
  bubbles, 314–316
  domains, 285–296
  materials, 278–317
  resonance, 307–310
Magnetostriction, 290
Majority carriers, 175, 182, 189
Manganese, 299, 300, 305, 306, 307
Marx, 43
Masers, 296, 318, 323, 338–341
Mask, 242, 243
Mass, effective, 22, 132–136, 146
Mass action, law of, 163
Matched filter, 270
Maxwell, 43
Maxwell's equations, 9, 43, 254, 276 373
Meissner effect, 356, 357
Mendeleev, 72
Mercury, 352, 356
  lamp, 326
Meson, 94
Metal–insulator–semiconductor junctions, 209–213
Metallic bond, 84
Metal–metal junctions, 116–117
Metal–semiconductor junctions, 203–209
Metrology, 344, 384
Mica, 248
Microelectronics, 147, 180, 242–246
Microwaves, 24, 171, 220, 306, 310, 329, 339, 342, 343, 344, 346, 349, 384, 385
Minority carrier, 175, 182, 183, 192, 194, 197, 223, 225
Mobility, 4, 170, 172, 181
  measurement of, 167–169
Molybdenum, 292
Momentum operator, 46
Moore's Law, 245
MOSFET, metal–oxide–silicon field-effect transistor, 229–230
MOST, metal–oxide–silicon transistor, 228–230, 248

Neel, 295
  temperature, 304, 305
Negative effective mass, 135, 141, 235
  resistance, 216, 220, 235–236, 240, 375, 378
  temperature, 318, 324, 341, 342
Neodymium, 263, 326, 327, 329
Neon, 74, 327, 328
Neutron, 94, 96, 309
Nickel, 299, 305, 306, 307
Niobium, 356, 380
Nitrogen, 73, 352
Noise temperature, 341, 342
n-type semiconductors, 156, 157, 159, 168, 180, 185, 187, 189, 191, 193, 206, 220
Nuclear forces, 94
  magnetic resonance, 309

Ogg, 353
Ohmic contact, 209
Ohm's Law, 4, 236
Operators
  average value, 56
  Hamiltonian, 88
  momentum, 46
Optical darwinism, 337
Optical fibres, 272–274, 296
Orthonormal wavefunction, 89
Oxygen, 73, 225, 258

Paramagnetic resonance, 307–308
Paramagnetism, 282, 284, 301–304
Parametric amplifier, 220–223
  oscillator, 330–331
Pauli, 43, 70
Pauli's principle, 70–71, 95, 100, 299, 300, 305
Penetration depth, 371
Periodic table, 72
Permalloy, 291, 292, 307
Permanent magnets, 293–296
Permeability, 9, 279
Phase diagram, 176–177
  transition, 362
  velocity, 35
Philosophical implications, 57–59
Photoconduction, 164–165, 274
Photo-detector, 223–224, 225
Photo-diode, 223–225
Photo-electric effect, 114–116
Photo-emission, 114, 164

Photo-engraving, 242–243
Photo-resist, 231, 232
Photo-voltage, 224
Piezoelectric constant, 265
Piezoelectricity, 264–270, 276
Piezomagnetism, 290
Pippard, 366
Planar technique, 179, 242
Planck, 29, 61, 320
Planck's constant, 29
Plasma diagnostics, 344
  frequency, 14, 15, 24
  heating, 344
  physics, 24
  waves, 22–24
Platinum, 107, 351
p–n junction, 184–196, 197, 200, 217, 219–220, 223–224, 233, 247, 248, 332–334
  capacity, 194–196, 199–200, 220–221
Poisson's equation, 23, 186, 237, 272
Poisson's ratio, 241
Polarizability, 252, 256–257, 259, 261, 262, 275
Polarization, 253–254, 275
Population inversion, 323, 324, 327, 332, 336, 337, 341, 349
Potassium, 74, 76, 102, 107
  chloride, 97, 277
  chromium alum, 302
Potential barrier, 48, 51, 110, 111, 112, 124, 206, 208, 213, 214, 219
  well, 51, 55, 61, 99, 120–122
Poynting vector, 26
Proton, 90, 91, 92, 93, 94
  spin resonance, 309, 316
p-type semiconductor, 156, 157, 159, 180, 203, 226, 228, 240
Pump, 222, 322, 323, 325, 326, 328, 331, 333

Q switching, 337
Quantum of energy, 29
  numbers, 68–70, 296–300, 327
Quartz, 181, 265, 266

Radar, 270, 343, 346
Rayleigh, 268
Recombination, 163, 183, 184, 192, 193, 194, 197, 212, 225, 249
Rectification, 189–191, 207, 215, 233–234
Rectifier equation, 191, 215, 248
Refractive index, 254, 255, 256, 258, 373
Relaxation time, 3
Remanent flux, 289
Resonator, 324–325, 326, 336, 337
Reverse bias, 190, 194, 197, 211, 214, 218, 219, 220
Ridley, 236
Rochelle salt, 265
Rubidium, 76, 102
Ruby, 322, 325, 326, 339, 340

Saturable absorber, 338
Scattering, 160–162
Schottky effect, 109–112
Schrieffer, 352
Schrödinger, 42, 43
Schrödinger's equation, 42, 43, 44–47, 87, 88, 89, 92, 92, 95, 103, 105, 120, 214
  time independent, 46, 47, 121
Seitz, 104
Semiconductors, 119, 147–183
  degenerate, 213
  devices, 184–249
  extrinsic, 153–160
  Fermi level, 151, 153, 156–160, 183, 189
  intrinsic, 148–152
  lamps, 223, 225
  lasers, 332–334
  measurements, 167–176
  mobility, 167–169
Shockley, 196, 226, 276
Silicon, 74, 119, 177, 180, 181, 182, 228, 258, 333
  covalent bond, 85, 148
  controlled rectifier, 233–234
  crystal structure, 148, 165
  dielectric constant, 155
  effective mass, 155, 181
  E–k curve, 165–167
  energy gap, 145, 166, 181
  epitaxial deposition, 180
  holes in, 166
  impurities in, 153, 154, 155
  indirect gap, 166
  in iron, 291
  melting point, 179

metallurgical phase diagram, 177
microelectronic circuits, 242–246
mobility, 181, 248
MOST, 228–230
strain gauge, 240–241
Table of properties, 181
Silicon dioxide, 228, 242, 243
Silver, 102, 299, 300
Single crystals, 176–178, 180, 181
Skin depth, 12
Sodium, 26, 74, 76, 102, 117, 303
Sodium chloride, 79, 82, 107, 304
Soft magnetic materials, 307
Sommerfeld, 96
Specific heat, 25, 103, 363, 364, 366
Spin, 70–71, 95, 196, 298, 299, 300, 304, 305, 306, 308, 316, 317, 339, 340, 349, 353
Split-off band, 165, 172
Spontaneous emission, 320, 341, 342
Stern-Gerlach experiment, 349
Stimulated emission, 320
Strain gauges, 240–241
Sulphur, 74, 291
Superconductors, 351–386
applications, 379–385
energy gap, 373–379
hard, 380
Josephson tunnelling, 375, 378, 379, 382
magnetometer, 383–384
memory elements, 380–384
surface energy, 364–366
switches, 380–384
type I, 372
type II, 372, 373, 380
vortex, 373, 380
Supermalloy, 292, 307
Surface
acoustic waves, 268
energy of superconductors, 364–366
states, 207–209
Susceptibility
dielectric, 251
magnetic, 279, 281, 282, 284, 296, 304, 316

Tamm states, 209
Tantalum, 107, 356
Tetrahedral bonds, 85, 86
Thermionic emission, 105–109
Thermal velocity, 2, 5
Thoria, 275
Thorium, 356
Thyristor, 234
Tin, 209, 356
Transistor, 147, 167, 179, 194, 196–203, 226, 244, 245
Transition
elements, 74, 304
region, 186, 187, 189, 192, 213 (see depletion region)
Transverse modes, 334–335
Tritium, 345
Tungsten, 117
Tunnel diode, 213–217
Tunnelling, 51, 92, 112, 113, 161, 213, 214, 215, 217
between superconductors, 374–379
Josephson, 375, 378, 379, 382
Type I superconductors, 372
Type II superconductors, 372, 373, 380

Uhlenbeck, 70
Uncertainty relationship, 37, 45, 56, 77, 297

Valence band, 144
Vanadium, 74, 356
Van der Waals bond, 86–87
Varactor diode, 220–223, 271
Viscous force, 6
Volume holography, 348–349

Water, 257, 275
Watkins, 236
Wavefunction, 44
Waveguide discontinuity, 52
Wave packet, 34, 48
Weiss, 283, 305
Weiss constant, 283, 316
Work function, 104–105, 107, 111, 116, 117, 204, 207, 208

Xenon, 325
Xerox process, 274
X-rays, 31, 245, 350

Yttrium, 306, 314
Yttrium-iron-garnet, 306

Zener breakdown, 218, 219, 220
Zener diode, 218–220
Ziman model, 125–128, 139
Zinc, 176, 177, 225, 306, 307, 356
Zone refining, 147, 178–179